Spatial Cultures

What is the relationship between how cities work and what cities mean? *Spatial Cultures: Towards a New Social Morphology of Cities Past and Present* announces an innovative research agenda for urban studies in which themes and methods from urban history, social theory and built environment research are brought into dialogue across disciplinary and chronological boundaries. The collection confronts the recurrent epistemological impasse that arises between research focussing on the description of material built environments and that which is concerned primarily with the people who inhabit, govern and write about cities past and present. A reluctance to engage substantively with this issue has been detrimental to scholarly efforts to understand the urban built environment as a meaningful agent of human social experience. Drawing on a wide range of historical and contemporary urban case studies, as well as a selection of theoretical and methodological reflections, the contributions to this volume seek to historically, geographically and architecturally contextualize diverse spatial practices including movement, encounter, play, procession and neighbourhood. The aim is to challenge their tacit treatment as universal categories in much writing on cities and to propose alternative research possibilities with implications as much for urban design thinking as for history and the social sciences.

Sam Griffiths is Lecturer in Spatial Cultures and a member of the Space Syntax Laboratory at the UCL Bartlett School of Architecture. His research interests include the spatial cultures of industrial cities, processional culture, high streets, the representation of spatial cultures in literary and historical writing, and interdisciplinary spatial-morphological theory and methods. He teaches a range of topics in spatial cultures on UCL's Arts and Sciences BASc and MSc 'Spatial Design: Architecture and Cities' programmes. He has published widely on a range of urban and suburban topics.

Alexander von Lünen is Senior Lecturer in Digital Humanities at the University of Huddersfield. He holds a degree in computer science and a doctorate in history, both from the Technische Universität Darmstadt, Germany. From 2007 to 2012 he was technical lead for the Great Britain Historical GIS and its website 'A Vision of Britain through Time'. He is co-editor, with Charles Travis, of *History and GIS: Epistemologies, Considerations and Reflections* (2013).

Design and the Built Environment

Series editor: Matthew Carmona

This series provides a means to disseminate substantive research in urban design and its allied fields. Contributions are welcomed which are the result of original empirical research, scholarly evaluation, reflection on the practice and the process of urban design, critical analysis of particular aspects of the built environment, or important conference proceedings. Volumes should be of international interest, although they may focus on the particular experience and practice in one country. They may reflect theory and practice from across one or more of the spatial scales over which urban design operates.

Total Landscape, Theme Parks, Public Space
Miodrag Mitrasinovic

The Design of Frontier Spaces
Control and Ambiguity
Carolyn Loeb and Andreas Luescher

Urban Design in the Arab World
Reconceptualizing Boundaries
Robert Saliba

Towns and Cities: Function in Form
Urban Structures, Economics and Society
Julian Hart

Urban Planning in Lusophone African Countries
Carlos Nunes Silva

Urban Planning in North Africa
Carlos Nunes Silva

Spatial Cultures
Towards a New Social Morphology of Cities Past and Present
Sam Griffiths and Alexander von Lünen

Spatial Cultures

Towards a New Social Morphology of Cities Past and Present

Edited by
Sam Griffiths and Alexander von Lünen

Routledge
Taylor & Francis Group
LONDON AND NEW YORK

First published 2016
by Routledge
2 Park Square, Milton Park, Abingdon, Oxon OX14 4RN

and by Routledge
711 Third Avenue, New York, NY 10017

First issued in paperback 2018

Routledge is an imprint of the Taylor & Francis Group, an informa business

British Library Cataloguing in Publication Data
A catalogue record for this book is available from the British Library

Library of Congress Cataloguing in Publication Data
Names: Griffiths,Sam,editor . | Lünen, Alexander von, 1971– editor.
Title: Spatial cultures : towards a new social morphology of cities past and present / edited by Sam GriffithsandAlexandervonLünen.
Description: Farnham, Surrey, UK, England; Burlington, VT : Ashgate, [2016] | Includes bibliographical references and index.
Identifie rs: LCCN 2015035565 | ISBN 9781472450296 (hardback : alk. paper) |
Subjects: LCSH: Cities and towns–Study and teaching. |
Cities and towns–History.
Classification:LCCHT109.S652016|DDC307.7609–dc23
LC record available at http://lccn.loc.gov/2015035565

ISBN 13: 978-1-138-60166-6 (pbk)
ISBN 13: 978-1-4724-5029-6 (hbk)

Typeset in Baskerville
By Out of House Publishing

For Lynne Hapgood

Contents

Figures

Tables

Contributors

Gareth Dean is Postdoctoral Research Associate at the University of York. He is the author of *Medieval York* (The History Press, 2008), and his PhD thesis (University of York, 2012) examined the social and spatial development of medieval York from c.600–1600AD. His research interests explore ways of using Geographical Information Systems and digital media to draw together published and unpublished excavation data with historical and cartographic sources to explore the social and physical development of medieval urban space.

Ava Fatah gen. Schieck is an architect, educator and researcher. She is a Reader (Associate Professor) on the MSc Adaptive Architecture and Computation at the Bartlett, UCL. Her research in the area of Architecture, Urban Interaction Design and Ubiquitous Computing has developed over the last 14 years through teaching and research positions she held at UCL. She is the Principal Investigator of the 'Screens in the Wild' project, which explores the potential of networked urban screens for communities and culture and offers a unique longitudinal 'in the wild' research and a living-lab environment since 2011. Ava is a chair and a member of the organising committee on the Media Architecture Biennale (2012, 2014 and 2016).

Dion Georgiou is a PhD student in the School of History at Queen Mary University of London, where he is currently completing a PhD on suburban charity carnivals in Edwardian London. Dion's broader research interest is in the social and cultural history of Britain from the late nineteenth century through to the present, focusing primarily on diverse popular cultural forms – including festive occasions, popular music, sport and film – and their geographies and temporalities, particularly with regard to processes of urban change. He has previously published articles in *The London Journal* and *Sport in History*.

Sam Griffiths is Lecturer in Spatial Cultures and a member of the Space Syntax Laboratory in the UCL Bartlett School of Architecture. His research interests include the spatial cultures of industrial cities, urban manufacturing, processional culture, high streets and spatial-morphological theory. Sam teaches Spatial Cultures on UCL's Arts and Sciences BASc and MSc *Spatial Design: Architecture and Cities* programmes. He has published on a range of urban and suburban topics.

David Jeevendrampillai is an anthropologist at UCL. His current research explores the relationship between transformations of the built environment and social life in suburbs. He has particular interest in practices of knowing and representing place including mapping, walking, parading and 'local' carnivals. He is interested in how contestations over representing the material environment intersect with different practices of knowing and showing such as when indigenous or 'local' claims are incommensurate with planning policy, academic research or systems analysis. Overall his interests include body–land relations, material culture, the politics of affect, geographies of knowledge, GIS, phenomenology and spatial thinking in anthropology.

Ole B. Jensen is Professor of Urban Theory at the Department of Architecture, Design and Media Technology, Aalborg University (Denmark). He holds a BA in Political Science, an MA in Sociology, a PhD in Planning and a Dr. Tech. in mobilities studies. He is co-founder and board member at the Center for Mobilities and Urban Studies (C-MUS) at Aalborg University. His main research interests are within Urban Mobilities and Urban Networked Technologies. He is the co-author of *Making European Space: Mobility, Power and Territorial Identity* (Routledge, 2004; with Tim Richardson), and author of *Staging Mobilities* (Routledge, 2013) and *Designing Mobilities* (Aalborg University Press, 2014).

Ditte Bendix Lanng is Research Assistant at the Department of Architecture, Design and Media Technology, Aalborg University (Denmark). She holds a MA in Urban Design, and recently submitted her PhD thesis on the topic of mobilities design. She has worked in architectural studios in Denmark, the Netherlands and China, as well as a municipal urban planner. Her main research interests are within urban design and mobilities – a combination of fields that she likes to explore through design speculations, ethnography, and as a collective enterprise with planning and architectural practice.

Ray Laurence is Professor of Roman History and Archaeology at the University of Kent. Ray taught Roman History and contributed to courses on the Classical Tradition at the University of Reading for over 13 years. In 2005, he moved to the University of Birmingham where he taught courses on Pompeii and Roman Italy; and contributed to the multi-disciplinary module on Cities. Since joining Kent in April 2010, Ray has been Head of Classical and Archaeological Studies and is currently Director of Research and Enterprise for the Faculty of Humanities. Ray has published widely in his areas of research expertise in Roman Social History and Roman Archaeology. Recent books include *Roman Archaeology for Historians* (Routledge, 2012) and the edited collection with David Newsome *Rome, Ostia and Pompeii: Movement and Space* (Oxford University Press, 2011).

Katy Layton-Jones received her undergraduate degree in fine art and art history from Goldsmiths, University of London. She went on to complete an MPhil and PhD in British history at the University of Cambridge. In 2006 she was appointed Research Associate on the ESRC-funded parks project at the University of Liverpool. In 2008, Katy moved to the Centre of Urban History, University of Leicester, where she is now a Visiting Fellow. Her monograph, *Beyond the Metropolis: The Image of Urban Britain 1780–1880* was published by Manchester University Press in 2015.

Lasse Suonperä Liebst is a Postdoctoral Researcher at the Department of Sociology, University of Copenhagen, where he also recently earned his PhD. His current research interests are in micro-sociology of street violence, and more broadly, the relationship between face-to-face interaction and urban space.

Alexander von Lünen is Senior Lecturer in Digital Humanities at the University of Huddersfield. He holds a degree in computer science and a doctorate in history, both from the Technische Universität Darmstadt, Germany. From 2007 to 2012 he was technical lead for the Great Britain Historical GIS and its website A Vision of Britain through Time. He is co-editor, with Charles Travis, of *History and GIS: Epistemologies, Considerations and Reflections* (Springer, 2013).

Wallis Motta holds a PhD in Anthropology (UCL), an MA in Material and Visual Culture (UCL) and a BSc in Media and Communications (Universidad Iberoamericana). She is an inter-disciplinary scholar and researcher who works with architects, artists, computer scientists, psychologists, sociologists, anthropologists, business strategists and media professionals in multi-stakeholder projects. She is currently a Research Officer at the London School of Economics and Political Science (LSE) in the Department of Media and Communications. Her work studies the intersection of place-making, multiculturalism, new media, social capital, entrepreneurship, socio-technical innovation and cultural change.

Katrina Navickas is Senior Lecturer in History at the University of Hertfordshire, UK. She researches the history of popular politics in England during the eighteenth and nineteenth centuries, with a particular interest in the geographies and landscapes of protest. Her next major publication is *Protest and the Politics of Space and Place, 1789–1848* (Manchester University Press, 2015), which follows her first monograph, *Loyalism and Radicalism in Lancashire, 1798–1815* (Oxford University Press, 2009). She has also published numerous journal articles on the Luddites, Swing rioters, Chartists and political material culture.

Vinicius M. Netto holds a PhD in Advanced Architectural Studies (The Bartlett School of Graduate Studies, University College London, UCL), and is currently Tenured Assistant Professor at Federal University Fluminense, in Brazil. Netto's work explores three fundamental dimensions in the social and material shaping of cities: (1) cities as systems of encounter: a new approach to urban segregation as segregated social networks operating in the city; (2) cities as systems of communication: a view of shared spaces as means to association; (3) cities as systems of material interaction: a systematic view of urban form as an effect of social interactivity.

Hannah Platts is a teaching fellow in the History department at Royal Holloway, University of London. Her PhD from the Classics and Ancient History Department of Bristol University examined the art, architecture and landscape of elite Roman villa residences of mainland Italy and she is currently developing this into a monograph. She has published on various aspects of Roman domestic space as well as on the reception of Roman villa dwellings in the architecture of Palladio. In addition to publishing chapters with Ashgate and Cambridge University Press on sensory perception and the domestic realm, she is also currently publishing articles on the ancient Roman domestic realm and the nature of Rome's elite, and the development of the elite Roman house in the city of Rome.

Anna Plyushteva is an urban geographer and doctoral candidate at University College London, UK. Her research focuses on mobility, transport infrastructure, change and the everyday life of cities. Her thesis discusses the expansion of the metro in Sofia, Bulgaria, and its role in reconfiguring habits and technologies of commuting. She has published on post-socialist urbanism, public transport socialities and public space.

Regner Ramos obtained his Bachelor and Master degrees at the University of Puerto Rico's School of Architecture. His PhD research conducted at The Bartlett School of Architecture at University College London focuses on the relation between embodiments, urban space and GPS mobile apps. He is also an Associate Lecturer at UAL – Central Saint Martins and London College of Communication, the Editor-in-Chief of *LOBBY* magazine and a writer for *Glass Magazine*.

Holger Schnädelbach works as Senior Research Fellow at the Mixed Reality Lab, School of Computer Science, University of Nottingham. He is an architect with more than 15 years of experience in Human-Computer-Interaction research focusing on the interface of information technology and architecture. Recently, he has focused on the nature of the embodied relationship between a building's inhabitants and their adaptive environment. He has held a Leverhulme Early Career Fellowship (2007–2009) and was PI on the EPSRC Creativity Greenhouse (EP/J006688/1) and Screens in the Wild projects (EP/I031413/1). He has recently been awarded a Nottingham Research Fellowship to investigate personal data in Adaptive Architecture.

Simon Sleight is Lecturer in Australian History at King's College London and Adjunct Research Associate at Monash University, Melbourne. His work explores the processes of 'making place', the histories of urban experience and the evolution of historical youth cultures. Recent publications include *Young People and the Shaping of Public Space in Melbourne, 1870–1914* (Ashgate, 2013), 'The Leary Larrikin: Street Style in Colonial Australia', *Cultural and Social History*, vol. 11, no. 2 (2014) and 'The City in the Child: Colin Ward, Urban Becoming and the Shift to Experience', in Catherine Burke and Ken Jones (eds), *Education, Childhood and Anarchism: Talking Colin Ward* (Routledge, 2014).

Hanna Stöger is Lecturer in Classical and Mediterranean Archaeology at the University of Leiden. Throughout her studies she developed a keen interest in the spatial dynamics of urban sites and the application of spatial analysis to Roman towns. She has published extensively on spatial aspects of Roman Ostia. Apart from Roman urbanism, her areas of interest include urban history, city development and today's methods of urban planning. She is currently conducting research into the spatial, social and cultural organisation of Roman neighbourhoods. The project seeks to create a dialogue between ancient and modern neighbourhoods by combining archaeological methods with contemporary urban studies.

Anna H. Walas is currently awaiting the award of her PhD at the University of Leicester. She specialises in the Roman Army and the social use of space within Roman military bases. She holds an MPhil from the University of Cambridge and a BA from the Jagiellonian University.

Frederik Weissenborn is a completing research student at the Bartlett School of Architecture, UCL. Inspired by Focillon's book, *The Life of Forms in Art,* his doctoral thesis explores the autopoietic formation and sociogenetic efficacy of urban structures, i.e. the way in which these are generated and subsequently shape social network formation and perpetuation.

Foreword

Richard Rodger

Stocktaking can be therapeutic. A calm, rational assessment of the state of play, whether in business, sport, personal targets or research demonstrates a certain degree of confidence to undertake a survey, assess the evidence, and formulate a course of action. Of course, it can be scary. It may reveal dearth or surplus in certain categories of stock. *Spatial Cultures* offers a stocktaking, an opportunity to reflect on how far a 'spatial turn' has permeated different disciplines, and the sub-title 'Towards a New Social Morphology of Cities Past and Present' indicates that there is a positive direction of travel and a degree of confidence amongst practitioners.

If proof was needed of the vigour of spatial culture then look no further than the range and reach of chapters in this collection. Contributors engage with at least one of a dozen facets of spatial relations. These include the material spaces of everyday life and how city spaces function to the varieties of the built environment and inhabited spaces; they also reflect on how physical arrangements in the city change over time, and how that affects social practices and meanings. Spatial form and social function are considered in temporal and performative terms. Routines and rhythms, inertia and unresponsiveness, each figure as themes on which contributors muse, as some also do in most dramatic terms arguing that the replacement of text by mapping is, or will be, the dominant cultural technique of the twenty-first century. As with any stocktaking, re-ordering is essential and several authors touch on how humanities and social science research can inform policy issues associated with architecture, planning and civic design. It is to be hoped that practitioners listen – or at least read and ponder.

The chapters are organized in time and space: the classical city; the long nineteenth-century (English) city; the contemporary city with its emphasis on urban mobility; and a final section on practice and policy, technology and social morphology. Some might argue for other temporalities, and if so, the point of the chapters would be lost on them. These themes, processes and structures are the drivers for the book; the chronological divisions are merely the stage set or laboratory for an investigation of the issues.

Space in an abstract sense and spaces as a set of relationships are amongst the most fundamental aspects of human existence, and so straddle many disciplines each of which develops its own methodology, theories and interpretive frameworks. We own or rent space in terms of accommodation; we 'need' space in terms of developing our own personalities; we share spaces in retail complexes, sports events and other leisure activities; we derive meaning and add to it in sacred spaces and ceremonial settings; we create space and attempt to create places through our new towns and housing estates in the hope that community and neighbourhood will follow. Social, cultural and emotional dimensions differ according to such a typology of space and as our individual approaches to it intersect. There are inter-generational aspects of the space, as reflected in the diurnal nature of city centre space as youth culture and pensioners' habits cause usages to diverge.

In considering these aspects of space and spatial cultures it is important to retain a sense of scale. Do the themes and processes work at all or only some scales of human interaction? If so, are they more intense or weak at the scale of the village or metropolis? Do villagers and suburbanites share the same spatial processes and relations, but do so at a different intensity or velocity? Stereotypes often have some basis in reality, so do New Yorkers, for all their lack of understanding of and familiarity with Mid-westerners, essentially experience the same set of spatial relations but to a different intensity? As some parts of cities become 'no-go' or transgressive according to some definitions does this not indicate a shifting locus within the urban rather than a new phenomenon? Slums and red-light districts shift, but the poor and prostitutes are always with us, and the underlying relationships of power are ongoing, if mutable.

Administrative scale, and rigidities associated with it, both facilitate yet impede the management of space and spatial cultures. Whereas in the past we might have shared Max Weber's view that local administration by unelected officials constituted an efficient and rational way to organize space and human activity, though dangerous if unfettered, nowadays it has connotations of inefficiency and has spawned political campaigns for greater accountability, a stronger audit culture, and fewer and weaker regulatory codes. The balance of public and private spheres of actions has been recalibrated over time, with powerful implications for different segments of society and the spaces they inhabit – the evolving social morphology of the sub-title.

In different ways, and with greater sophistication and complexity, *Spatial Cultures* tries to unravel these scale issues and systems approaches. Case studies always attract accusations of its exceptionalism yet have the merit of embedding empiricism in their conclusions, and this is contrasted with a generalizing approach that makes wide-ranging observations but overlooks the externalities, the unforeseen casualties, in such a blanket approach. It is this delicate balance between the strength and weaknesses of both approaches that remains an ongoing tension in social science research. *Spatial Cultures* provides a mix of the two approaches. A number of contributors are understandably excited by the potential of 'big' digital data for contemporary urban research. If this 'digital turn' is to be successful then a much greater emphasis on Open Data and Open Access issues is crucial. Too many public institutions pay lip service to the twin pillars of administrative transparency and data availability, and restrictive licensing (the absence of ODbL and SA or share alike licensing) means that, strictly, little of the data is re-usable to the public. It remains, in the United Kingdom at least, much easier to research nineteenth-century social structure and spatial relations than those of the twentieth century.

With twenty-one essays spanning two millennia and a multidisciplinary approach *Spatial Cultures* provides a timely signposting in a route *Towards a New Social Morphology of Cities Past and Present*. The energetic confidence of the contributors makes for varied and engaging reading, and a welcome, enjoyable and useful stocktaking.

Preface

Sam Griffiths and Alexander von Lünen

Spatial Cultures is intended to facilitate a process of transdisciplinary framing for increasing scholarly interest across the humanities and social sciences in the ways in which the materiality of the urban built environment is agential in the production of social life and social meaning. Early versions of many of the chapters in this volume were presented in a workshop at UCL in 2013.[1] Speakers were asked to present research that would articulate, from their own disciplinary perspectives, what often appears to be a missing conceptual link between the description of urban-material infrastructures and accounts of the practice and experience of everyday social life. This request was intended to make theoretical and methodological demands of those disciplines (including human geography, history, sociology and anthropology) that tend to view the built environment either as derivative of socio-economic processes or as a purely representational domain, and those (such as architecture, urban design, urban geography, archaeology and urban GIS[2]) whose specialized knowledge of urban form does not necessarily explain or even seek to explain the question of what the material arrangements of a society have to do with quotidian social practice.

To adapt Dennis' (2008, 3) pithy summary of contrasting disciplinary perspectives on the city the project of spatial cultures has to do with conceptualizing the relation between how built environments *work* as complex material-informational systems and how they come to *mean* as inhabited spaces in particular social-historical contexts. The epistemological difficulty is that there is no basis to map one neatly onto the other – indeed to do so is to close down what Tschumi (1996) calls the 'disjunction' between material and conceptual spaces through which the agency of the former emerges. Westin (2014, 178–85) has described as the problem of 'inference' the traditional epistemological shortcomings of spatial science in attempting to translate from material form to social process. At the same time she argues that without attempting to explain the effect of society's material organization on social processes then social interaction – the very stuff of society – seems to emerge implausibly 'out of nothing or in some kind of vacuum' (2014, 187). Rather than treating 'doing' and 'meaning' in two separate knowledge domains then, the challenge of spatial cultures is to give precise articulation to their disjunction.

Ramadier (2004, 433–4) argues that transdisciplinarity seeks to 'confront' multiple disciplinary perspectives and 'articulate' their contradictions rather than resolve them into a single model (in interdisciplinary mode) or juxtapose their differences (in multidisciplinary mode). 'Articulation' in this sense involves an active acknowledgement of the multidimensional complexity of reality in order to reconstruct the object of study such that contrasting and sometimes contradictory disciplinary realities are preserved rather than simplified: it aspires to a knowledge whole, in other words, that is greater than the sum of its parts. According to Ramadier trandisciplinarity can flourish in knowledge domains such as urban studies where there is already a tradition of interdisciplinarity and multidisciplinary engagement. While these knowledge domains reflect the existence of genuine areas of overlap between ever-more specialized academic disciplines with regard to the research object, they are also sites where the paradoxes and incommensurabilities between different disciplines are likely to become firmly entrenched (see Chapter 17 in this volume). It is not surprising that Ramadier has developed his argument in the context of urban studies for there are few such pervasive themes of contemporary scholarship that so thoroughly encapsulate the necessity (some would say the necessary evil) of an interdisciplinary nexus as that of *the city* – and, it follows, few objects of research which are so characterized by difference and contradiction in conceptualization and methodological definition. At the most elementary level empirical research into cities embraces the whole spectrum of quantitative and qualitative approaches. In normative terms the city is represented as a source of wealth and poverty, community and alienation, creativity and exploitation, cultural diversity and homogeneity, environmental sustainability and despoliation, and health and sickness to name just a few of its many contrasting images. Each of these 'realities' has its champions and disciplinary home.

The notion of 'spatial culture' itself be might thought to contain a contradiction in conceptual terms – embracing the opposing epistemological claims of the material (spatio-temporal) and representational (cultural-semantic) domains. Yet as a transdisciplinary proposition spatial culture seeks to articulate the multidimensional complexity of urban realities that are all too often reductively couched as 'space versus place'. This contrived debate tends to pitch 'space-first' disciplines (systems, objects, movements, mappings and synchronic phases) against 'place-first' disciplines (bodies, subjects, identities, texts and the temporalities of memory). The stubborn persistence of the 'space–place' binary becomes intellectually suspect when it becomes an excuse for scholars *not* to engage with each other's research epistemologies or, even more seriously, with their own conceptual apparatus. In the social sciences and humanities researchers seeking to decode the *meanings* of space tend to find the hermeneutic concepts they need in the phenomenological discourse of *place*, in Lefebvre's account of the *social* 'production' of space or Bakhtin's literary chronotope, since to access formal modes of description for material arrangements, for example through GIS or space syntax techniques, seems largely a question of methodological heuristics rather than hermeneutics, raising Westin's problem of inference (though see Lünen and Travis 2013; Griffiths 2012; Psarra 2013). A challenge for research in spatial cultures then, is to access modes of discursive description for complex patterned phenomena such as networks, densities and time-space distributions that are in themselves 'non-representational' (Thrift 2008) and to show, through precise empirical research, how these inflect the actions through which the social world of meaning is 'assembled' (after Latour 2005) or made and 'remade' (after Law 2007), such that the agency of the material becomes evident and more open with regard to questions of 'meaning' than has hitherto been the case. This transdisciplinary aspiration justifies the subtitle of the book, *towards a new social morphology of cities*, recalling Durkheim's foundational idea of 'social morphology' in order to revisit the conceptualization of what he regarded as the relationship between the form and content of society (Andrews 1993; see also Chapter 18 in this volume).

The occasional appearance of the term 'spatial cultures' in the scholarly literature tends to be in the field of architectural studies (for example Reeh *et al.* 2012; also McCreery 2005) reflecting its intellectual genealogy in this disciplinary context.[3] The complex epistemological challenge it presents in requiring the articulation of both *space* and *culture* recommends it as an appropriate framing for productive engagement between built environment research, and work in the humanities and social sciences (see Chapter 20 in this volume).[4] The development of spatial cultures as a theoretical proposition owes much to early work in the field of space syntax. Peponis (1985) deploys the term in his account of the effects of architectural layout on the social and organizational dynamics of factories as workplaces. The first exposition of 'spatial culture' in relation to cities is provided by Hillier (1989, 6) who argues that:

> human societies order their spatial milieu in order to construct a *spatial culture*, that is, a distinctive way of ordering space so as to produce not *actual* social relations (the essential error of modernist architectural determinism) but the *principles for ordering social relations*.

Hillier draws attention not to architectural form per se (what he calls the modernist error) but to the rules and regulations, codes and conventions that generate spatial form and govern its usage. He sees these rules as internal not external to material form such that space appears not as a derivative 'function' of a society but as 'intrinsic', a 'necessary part of the social morphology' (1989). For Hillier the key is the way in which space is deployed socially 'sometimes to generate and sometimes to restrict the *field of encounter* of human beings and their symbols' (1989). It is this that allows space to avoid definition entirely in terms of socio-economic structure (Hillier *et al.* 1976, 179–80). The full implications of this powerful insight for a broad theory of spatial cultures have not, however, been systematically developed by Hillier for whom patterns of movement and encounter remain largely an epiphenomena of spatial morphology. The de facto separation of the material (built environment) and embodied (human movement) dimensions in much space syntax research serves to elide rather than to articulate the disjunction of space and culture by reducing the multidimensionality of social practice to a function of morphological structure, as this is defined by the observer-analyst, and consequently depriving it of temporality (Griffiths 2011). Peponis (2001, xvii) makes greater allowance than Hillier for the contested nature of social practice in the production of spatial culture, which he conceptualizes as the regulation of 'interface' between fields of 'co-presence and co-awareness as by-products of movement':

> Spatial culture can be defined in terms of how intense or dissipated such spatially sustained domains and interfaces are, how they become invested, claimed, qualified, contested or challenged by social practices and discourses, and above all, what characteristic forms they assume and what underlying purposes they serve.

Peponis' emphasis on social practice is welcome in this context but the analytic separation of spatial morphology (interface) and practice (doing, meaning) still pervades his conceptualization. We would go further in claiming the production of spatial cultures to be a fundamentally performative and temporal process. Performative, to be precise, of *form* in which Hillier's notion of a syntactical morphic language of space becomes infused with material *and* semantic information; and temporal because the materiality of lived space is continually being negotiated, made and remade in the light of this information (Netto 2015; Chapter 7 in this volume). This is the conceptual domain of Lefebvre's (1991, 38) 'spatial practice'. As Forty (2000, 256) notes, Lefebvre's theory of (social) space confronts the modernist separation of physical space and the perceptual space of the body. Yet developing this critique requires further articulation of the material, as well as the perceptual, nature of embodied social experience than Lefebvre himself acknowledges.

Spatial cultures understood in this performative sense draws attention to the need for a fuller articulation of the collective dimension – what Hillier (2003) refers to as the *allocentric* qualities and Koch (2012, 10) the 'allocentric subjectivity' of sociospatial experience – that is the necessary construction of *social* experience from a point of view that extends beyond that of the embodied individual. This allocentric perspective makes it possible to conceive of collective practices realizing material forms without implying a structuralist equivalence between the positioning of the social subject and the analytical perspective of the observer, and also of semantic 'forms' that emerge socially from particular material conditions – the 'material semiotics' defined by Law (2007). To deploy an analogy: in the theatre the idea of 'performance' implies the primacy of the playwright's text in relation to which the stage set is a background that will be dismantled following every production. By contrast, in spatial cultures, the manufacture of the set by the actors themselves (though not entirely in circumstances of their own making) is an integral part of the performance. Here the discursive imperative of performance is permeated with the ongoing (non-discursive) business of constructing and negotiating the material world in which the semantic 'action' takes place.

In a somewhat broader context the study of spatial cultures belongs in a tradition of urban studies with a research focus on the uses of urban space. This tradition goes back at least to the ethnographic work of the Chicago School (Deegan 2001) and includes classic ethnographic studies by the likes of Whyte (1943) and Wilmott and Young (1962), as well as more architecturally focused studies on streets (Anderson 1978) and public spaces (Whyte 1980; Gehl 1987). More recently Hebbert (2005) has shown how the urban streetscape is a vital source of quotidian collective memory, and Stevens (2007) has developed the notion of the informal, playful 'ludic' city (Chapters 2 and 9 in this volume both address the theme of 'play'). The relationship of the physical street with the digitally enabled environment was formulated as a question by Mitchell (1995) and forms a current strand of research in spatial cultures (see particularly Chapters 11 and 15).

One of the formative contributions to research in spatial cultures is Jane Jacobs' famous study of the built form characteristics and community vitality in Greenwich Village: *The Death and Life of Great American Cities* (Jacobs 1961). Jacobs' study encapsulates a remarkable interdisciplinary 'moment' at the point when she realized how the conceptual basis of her largely intuitive-ethnographic study could be formally described in terms of the cybernetic theory of Warren Weaver. Weaver (1948) distinguishes between simple (cause and effect) problems, problems of disorganized complexity (random) and problems of *organized complexity*. In the last chapter of *Death and Life* Jacobs identifies cities as problems of organized complexity; that is, problems of emergent, self-organizing systems. Hers was an original and profound epistemological connection to which much contemporary urban science can be traced. Yet Jacobs' (1993, 65) influential notion of the 'sidewalk ballet' as a 'complex order' is much more than a theoretical justification for urban-systems modelling because her primary interest was in what made places *work* and *mean* for people. The metaphor of dance can be found in the work of the human geographer Pred's theory of the 'choreography of existence' (Pred 1977) and the architectural phenomenologist Seamon's notion of 'place ballet' (Seamon 1979) – a dance in which, one is led to think that the built environment itself must, to some extent, play the role of choreographer. Seamon's rich environmental phenomenology of movement, encounter and rest is essential to accounts of spatial culture. There is also something of this notion of choreography in Lefebvre's (2004) theory of 'rhythmanalysis' in which the embodied rhythms of urban life are said to be culturally formative. It is also present in de Certeau's (1984, 117) imbrication of walking and text and his notion of 'space as practised place' – meaning in this sense as produced by the everyday act of walking.

Recent scholarly interest in mobilities (Urry 2007) has brought an opportunity for thinking afresh about space as a category of social analysis, an issue that has not yet been satisfactorily addressed by developments in the 'spatial turn' in the humanities (Kingston 2010). While this label usefully brackets much valuable research exploring spatial themes across a range of disciplinary domains it has remained largely unproductive from a theoretical and methodological standpoint. One reason is that the dominance of Henri Lefebvre's (1991) notion of the social *production* of space has tended to focus research on the social *construction* of space in terms that are overwhelmingly normative and representational. An effect has been that the epistemology of inhabited space as a material field of meaningful social practice has remained rather restricted. As Jerram (2013) has convincingly argued the irony of Lefebvre's position is that the social space that he theorizes remains largely dematerialized. For Hillier (2008, 223) Lefebvrian space presents an 'abstracted' domain that prioritizes the spatialialization of socio-economic structures over the agency of material arrangements themselves. While he views these as a primary agent of social experience they are viewed as largely derivative in the Lefebvrian tradition. By shifting the analytical focus from space as product to space as infrastructure, from meaning as language to meaning as practice and from embodied subjects to embodied subject–object relations unfolding in time and space, the 'mobilities turn' promises much by focusing research on the fundamental problematic of the elementary material elements of intelligible social practice.

Spatial cultures do not only refer to patterns of human movement. Urry notes how the mobilities of people, objects and information enabled by complex infrastructural networks calls into question any categorical distinction between space as system and place as meaning. It is also raises the question of what (Hayles 1999) has labelled as the 'posthuman' condition, as the embodied subject becomes increasingly mediated by technological systems – though as Urry notes 'we have never been entirely "human" ' in the sense that human activities have always been mediated in relation to the material world (Urry 2007, 45). This emphasis on transport and communications infrastructure as being formative of social practice presents a powerful critique of social theory that insists on separate epistemological domains for the study of (material) object and (human) subject. It also invokes the temporality and contingency of such practices and calls into question organicist notions of social interaction as the primary integrative function of a projected social totality (for examples in Giddens' (1984) theory of structuration) where this is not materially mediated. Corfield (2007) has argued an historical perspective on 'space' requires reflection on the particular 'shape' or temporal framing it takes in historical time. Insisting on such a temporal framing suggests how strongly spatial concepts such as distance, proximity and scale are properly regarded less as universal constants than as historically particular assemblages of practice.

Castells' (1996) account of the rise of 'network society' raises the question of the transformation and, to an extent, de-materialization of social practice through the collapsing of distance into what he refers to as the 'space of flows' – a proposition that strongly informs Urry's work on mobilities. Here we might put a different methodological inflection on this question by following Jameson's (1984, 92) discussion of Lynch's (1960) famous work on urban imageability. Jameson asks what sort of mapping exercise is needed to 'grasp our positioning' both as individual and collective subjects in a globalizing world order. How, in other words, to become sufficiently orientated to how things work in order that we might begin to understand what they mean? (An issue touched on by Chapters 11 and 12 in this volume.) One such instance of 'mapping' becoming realized as a cultural technique are location-based services (LBS) which, since the introduction of GPS-enabled smart phones are now ubiquitous. Real-time services such as LBS have the benefit of not having to deal with change over time, as they are solely concerned with the present. For most scholars, the mapping over time poses a serious challenge, both in terms of availability and consistency of data, but even more so in term of methods. How exactly a meaningful mapping of historical data can be achieved is still a topic under discussion and highlights some general issues with the space–place binary (see Chapter 16 in this volume). It seems that greater epistemological sophistication is required to make headway with *mapping* social processes. Another response is to map how the material world itself is realized socially, exploding the architectural object and revealing it as a meshwork of negotiated pathways, a method adopted by Yaneva (2012; see also Chapter 19 on Adaptive Architectures).

To summarize the preceding discussion, the chapters presented in this book touch on at least one of the following themes which also serve as a useful checklist of the research agenda for spatial cultures.

1. To articulate the inhabited material spaces of everyday life as a fundamental sociological problematic in articulating the collective dimension of human experience, including its digital dimension.
2. To articulate through precise empirical research the relationship between how city-space 'works' and what city-space 'means'. This entails extending the conceptualization of infrastructural systems to encompass the intrinsic spatiality of everyday social life and the conceptualization of quotidian space in relation to other infrastructural systems.
3. To foreground the distinctive temporalities of urban spatial cultures, noting that to conceptualize inhabited space exclusively in terms of its 'inertia', 'fixity', or 'unresponsiveness' is to misunderstand its temporality and to universalize historically constructed notions of proximity, distance and scale.
4. To articulate through precise empirical research an appropriate discourse for describing the temporal patterning of built-environment phenomena and to deploy this critically in decoding the time-space configuration of inhabited space, its symbolic regimes and cultural productions.

5. To articulate through precise empirical research the intrinsic ontological ambiguity of space-in-time as a source of disjunction and hermeneutical possibility, and to examine how the temporality of material arrangements inflects on patterns of continuity and change in urban social practices.
6. To propose that spatial cultures are implicated in the production, perpetuation and dissolution of semantic domains as (social) networks that extend across time and space, and to note that such domains are strongly coded with temporal and spatial descriptions.
7. To critically examine the binary relationship of (spatial) form and (social) function in temporal and performative terms, noting that the idea of performance implies a collective action of making and remaking the world such that semiotic elements of performance are infused with material descriptions and vice versa.
8. To articulate material-informational and sensorial arrangements as an extra-somatic or *allocentric* domain centred *between* bodies rather than on the individual body, such that subjectivity arises in relation to experience of the material world rather than egocentrically, in isolation from it.
9. To explore themes such as routines of work and play, production and consumption, formality and informality of movement, rest and encounter, quotidian and ritualized performance and the patterning of co-presence and absence in historical space, as activities which demand particular articulations of the material environment in order to explore the relation between doing and meaning.
10. To note that arrangements of material phenomena do not neatly map onto the semantic representation of social meanings in a linear fashion and to consider what alternative 'mappings' might be appropriate to articulate the disjunction between the material and the meaningful.
11. To contribute to developing mapping as a hermeneutical method. It is important to overcome the distinction between mapping (space) and writing (place) to incorporate the writing of maps and the spacing of writing. Harley's (1989) proposition that 'text' is a more accurate metaphor for maps than viewing them as representations of nature is indicative of this trend, that 'mapping' is about to replace 'writing' as the dominant cultural technique in the twenty-first century.
12. To open a transdisciplinary research space for urban studies in which perspectives and methods from the humanities and social sciences can inform and be informed by those originating in architecture and urban design research.

The book is divided into four parts. The first three are organized on a chronological basis while the fourth examines a range of interdisciplinary research contexts in spatial cultures. In the first part of the book, the contributors look into the spatial cultures of the ancient and medieval worlds, also addressing the usage of 'space' as trope by historians. Chapter 1, by historian Ray Laurence, deals with mobility in ancient Rome. Laurence addresses a lacunae in recent mobility studies: ancient Rome as the first metropolis in Europe and its 'development of new forms of mobility'. More than anywhere else at the time, Rome became the urban experience *par excellence*, letting visitors and its citizens alike walk paved streets detached from nature. In Chapter 2 archaeologist Hanna Stöger considers a typical Roman citizen: the *homo ludens*. As Stöger points out, urban culture encompasses spatial manifestations of play in a variety of ways, but nowhere more apparent in the Roman world. She examines the urban experience of Romans through the prism of 'the playability of urban space' to identify the contingency of public places. In Chapter 3 historian Anna Walas follows this with an examination of the spatial organization of Roman and modern military bases. Her study exemplifies community-building enshrined by control and dependency, and how the infrastructure of a base reflects contemporary notions of congregation and segregation. Historian Hannah Platts in Chapter 4 highlights the importance of employing multi-sensory analysis of ancient Roman domestic space. She argues that sensory responses to the home have largely been restricted to examining sightlines within and outside the house as well as ways of movement and access within it. Platts proposes the use of 'smellscapes' and 'soundscapes' in concert with methodologies that investigate vision and movement within elite Roman houses to enable a more nuanced study of the owner's aims of personal display within his domestic realm. In Chapter 5 historian Gareth Dean then points to the questions of what constitutes a neighbourhood, and what a neighbourhood meant to medieval city residents. Contemporary studies of cities, he argues, have shown the development and uses of urban space, social identity and the definition of neighbourhoods as highly complex. He proposes that by building on approaches from social sciences, archaeology, history and urban geography urban space within the medieval city can be investigated to attempt identification of the spatial and social framework within which social groups like neighbourhoods could form and their association with wider social networks of residents across the city.

The second part of the book brings forward debates about the spatial culture of the 'long nineteenth century' in the context of the urbanization associated with the upheavals of industrialization. Historian Katrina Navickas discusses in Chapter 6 some more general issues (modern) historians might experience with the use of 'space' as category of investigation. The debate about the 'spatial turn' in history is still strong, and Navickas details some of these historiographical challenges under discussion. She puts forward the use of 'space' for historians in a case study of political protest and graffiti in the early nineteenth century. Urbanist and historian Sam Griffiths expands on this in Chapter 7 by looking at processional routes as an aspect of the 'processional culture' of nineteenth-century Sheffield. He draws on the theory and methods of space syntax to argue that the development of this culture should not be seen in isolation from the quotidian spatial culture of the city as a whole. In Chapter 8 historian Dion Georgiou focuses on carnivals held in suburban London during the late Victorian and Edwardian era – specifically on the procession routes compiled for these events and how they functioned as place-making procedures. He considers how carnival organizers mapped them against a backdrop of rapid suburbanization, and how processionists and spectators might have themselves engaged with procession routes on carnival day itself. In Chapter 9 historian Simon Sleight considers 'youthscapes', i.e. at how young people appropriate urban space in their lives. Youngsters, he argues, do not just live in cities and use public places, they mould them. As Sleight points out most scholars have regarded the city as a 'container' of youth. In Chapter 10 historian Katy Layton-Jones deals with the British urban parks. As Layton-Jones outlines, the urbanization of nineteenth-century Britain led to a longing for supposedly 'wild' nature where the woes of the modern city were at a distance and modern parks still serve as some kind of 'wildscapes'. Layton-Jones draws on the romantic notion of green spaces as a point of departure to discuss contemporary debates about the future of parks.

The third part of the book shifts the focus from historic to contemporary spatial cultures – with an emphasis on digitally-enabled and urban mobilities. Architect Vinicius Netto in Chapter 11 investigates the relationship of urban spaces and digital networks, their modes of coexistence and their roles in the emergence of social practice. He examines the conditions for integration of social and material reality and offers descriptions of the ways in which social action circulates between the urban and the digital. Netto argues that as practice unfolds in diverse networks, space is becoming increasingly removed from its original role as a material setting of social reproduction to become one among many possible environments. In a similar vein sociologist Frederik Weissenborn argues in Chapter 12 that the widespread introduction of digital technologies in the social sphere means that the social sciences are currently experiencing a moment of significant fluctuation. In this context, social science researchers – including those working with the socio-spatial problematic – are faced with the challenge of reassessing existing methodologies and theories in the light of new findings. Weissenborn aims to explore the phenomenon of rhythm – arguably one of the central tenets in the socio-spatial problematic – within the context of urban digitization and digital urbanization. In Chapter 13 anthropologist Wallis Motta and architect Ava Fatah gen. Schieck engage with a practical consequence of the two preceding chapters: how to study urban anthropology in a digital age. They take a *mediated spatial cultures* approach to examine how interactive urban screens are understood and utilized in two neighbourhoods in London. Qualitative data was collected over a year of ethnographic fieldwork and six workshops. Their study then focuses on how residents construct a sense of belonging through both material culture and mainstream digital media. In Chapter 14 geographer Anna Plyushteva is also interested in contemporary urban life and its quotidian routines. Adopting a mobilities perspective her research engages with commuters in Sofia, Bulgaria, to explore the changing spatial cultures of urban commuting, particularly on public transport such as underground railways. Commuting, she points out, is to a good extent 'learned', a form of spatial knowledge that can form the basis of a distinctive community identity between mobile workers. Architect Regner Ramos juxtaposes virtual with physical worlds in Chapter 15. He looks at the location-based service Grindr, an online dating service on smartphones specifically aimed at homo- and bi-sexual men. Here, more than probably anywhere else, the dichotomy of digital urban spaces becomes visible. Grindr links its users in virtual space for physical encounters. Using these virtual-yet-physical spaces as a point of departure, Ramos uses the 'cyborg' principle to discuss technology-mediated realities and its implications for gender identities.

'Practices' and the digital are also at the heart of the chapters in Part IV but here the focus is on articulating issues in spatial cultures arising from a wider range of interdisciplinary and multidisciplinary engagement, including the fields of architectural and urban design. Computer scientist and historian Alexander von Lünen takes a critical look in Chapter 16 at what is usually referred to as 'Historical Geo-Informations Systems' (GIS). GIS have not only transformed the academic discipline of geography in the last decades; more recently, variants of GIS such as Google Maps have challenged the very concept of space and its experience. GIS have been applied to historical scholarship for some time now, but, as Lünen argues, under a rather limited and unimaginative precept. By discussing the history of mathematics in conjunction with philosophy of technology, he points out that the main reason for these deficiencies is not a gap between geography and history, but between the digital and history. In Chapter 17 anthropologist David Jeevendrampillai explores a particular moment of interdisciplinary 'incommensurability' on a collaborative research project when it proved impossible to identify cartographic categories that would serve both the ethnographic aims of the community-based researcher while also being consistent with the implementation of a Public-Participatory GIS community-mapping social enterprise. Jeevendrampillai argues that such moments of incommensurability should not be read as 'failures', but rather that such 'failures' can be regarded as productive in producing contrasting multidisciplinary interpretations of the research object. In Chapter 18 sociologist Lasse Suonperä Liebst examines the significance of 'space' for sociology. The social morphological research programme outlined by the founding father of French sociology, Émile Durkheim, recognizes space as a key element in the structuring of the social world. Yet, as Liebst outlines, this early Durkheimian sociology of space is often neglected in the older as well as more recent sociological literature. His chapter examines the historical background of this neglected status of social morphology and considers which steps need to be taken in order to allow social morphology to further fulfil its space- sociological potential.

Architect Holger Schnädelbach in Chapter 19 introduces the reader to 'Adaptive Architecture'. As he discusses, the introduction of sensing, computation and actuation technologies into the built environment pervades buildings from the mundane to the experimental. Beyond simply being adaptable, architecture is becoming adaptive to its environment and to its inhabitants through these technologies. His chapter focuses on the relationship of human movements to environmental 'movements' in the growing field of Adaptive Architecture. In Chapter 20 sociologist Ole B. Jensen and urban designer Ditte Bendix Lanng investigate how mobilities studies are relevant for urban design. They argue that on the heels of the 'mobilities turn' there should be further development towards a 'material and design oriented turn': 'mobilities design'. The cross-pollination of mobilities studies and design is twofold in their view: for mobilities studies to have a legitimate claim to have a grasp of contemporary mobilities; and for urban design to cater for a highly dynamic constituency and to conceptualize the 'mobile' as an immanent design issue.

The contributions of chapters from historians, archaeologists, geographers, anthropologists, sociologists and architects reflect the transdisciplinary agenda of spatial cultures as a knowledge domain in urban studies. Spatial cultures are presented from a range of historical and geographical contexts from ancient Rome to contemporary Sofia, from London to Melbourne. A smaller number of theoretical and methodologically focused investigations examine how diverse thinkers including Durkheim, Heidegger, Hillier, Lefebvre, Harvey, Latour and Leroi-Gourhan to this area of study. An approach to spatial cultures that is comparative in disciplinary, as well as chronological, geographical and theoretical terms provides greater scope to *articulate*, in Ramadier's sense of the term, both points of contact and points of contradiction as these emerge from research in particular disciplinary niches without, however, displacing their distinctive disciplinary contributions.

Notes

1 *Spatial Cultures – Past, Present, Future: a one-day workshop exploring the relationship of urban and architectural built form to movement, rest, encounter and social interaction past and present,* hosted by UCL Bartlett Faculty of the Built Environment, 19 September 2013.
2 Geographical Information Systems – see Chapter 16 in this volume.
3 Middlesex University currently runs an MA in Spatial Cultures in its Arts and Design School: www.mdx.ac.uk/courses/postgraduate/spatial-cultures, accessed 14 April 2015.
4 The scholarly journal *Space and Culture* (Sage) published since 1997 occupies similar disciplinary terrain. Accessed on 14 April 2015 from http://sac.sagepub.com/.

Acknowledgements

This book has its origins in the *Spatial Cultures* workshop held at UCL in 2013. This event would not have been possible without the support of Professor Alan Penn, Dean of the UCL Bartlett Faculty of the Built Environment. I am also grateful to Professor Matthew Carmona for his support of the *Spatial Cultures* project and to the students and staff of the Space Syntax Laboratory at UCL for their encouragement. The help of Frederik Weissenborn and Lasse Liebst (then research students) in organizing the workshop, both practically and academically, was indispensable. I would also like to thank everyone who participated in the workshop, particularly those scholars who contributed papers, many of which are included here. I would also like to thank my co-editor Alexander von Lünen for his hard labour and expertise in the editorial process. Valerie Rose, then at Ashgate, was supportive and understanding of the editors early on in the process, for which we are very grateful. Thank you also to my wife Rachel for keeping the domestic show on the road and for my children Benji and Abby for making sure I sometimes think about things other than *Spatial Cultures.*

Sam Griffiths

Part I

Spatial cultures in the ancient and medieval worlds

1 Ancient Rome

Mobility in Europe's first metropolis

Ray Laurence

Introduction

Ancient Rome has been largely omitted from recent work on mobility in geography and the social sciences (e.g. Urry 2007). Rome as a growing metropolis from 300 BCE became a location for the development of new forms of mobility that developed in response to the size and continuing growth of the city. Urban journeys of several miles could be taken without seeing the countryside – so different from the urban–rural experience found for example in Pompeii, where within twenty minutes you could cross the city. Rome is one of the first documented cities that was detached from nature. Walking on Rome's paved streets shifted the urban experience towards 'clean' streets and the avoidance of dirt associated with nature. This chapter seeks to engage the study of mobility in the city of Rome with some recent developments in the study of space and movement.[1]

Rome and conceptions of space

Mobility is a human practice that is culturally specific and, when we focus on ancient Rome we are examining mobility in a large preindustrial metropolis. Understanding mobility in its historical context involves engaging with theories of society and space, using the theory to recognise how Rome produced new concepts of space, new technologies of mobility, new urban spaces, a new language of space, new critical genres of urban discourse, new ideas on the nature of walking, new systems of traffic flow, and a gradation of space (from an *angiportus* or alleyway to a *locus celeberrimus* or famous place). Thus, David Harvey's (2006) definition of space as a keyword provides a basis for considering how ancient Rome might be conceived within the systemic mode of thought at the heart of the Lefebvrian spatial turn: 'material space', that is experienced representations of space, such as maps, and spaces of representation, that refer to the psychological or *mentalité* of space. Lefebvre's triad can be used in conjunction with Harvey's own tripartite scheme: absolute space, relative space and relational space to create a conceptual grid that has implications for understanding our sources on the city of Rome (cf. Table 1.1; see also Laurence 1997).

We would place our textual evidence for movement in the category of relative space. These are all textual creations with complex literary and linguistic structures to them, that have a linkage to the material through the deployment of strategies of subjective realism- causing readers to feel that the fictional text might correspond to a lived experience. The material evidence from Rome has major limitations when

Table 1.1 'Space' as a keyword

	Material space (experienced space)	*Representations of space (conceptualised space)*	*Spaces of representation (lived space)*
Absolute space	Walls, bridges, doors, streets, buildings, cities, mountains, etc.	Cadastral maps, landscape description	Feelings of contentment at home, empowerment
Relative space (time)	Circulation and flows of water, people, information, etc.	Antonine Itineraries Exchange value = value in motion	Anxiety of congestion, exhilaration of time-space compression
Relational space (time)	Social relations, rental, odours, pollution	Surrealism = 2nd/3rd style wall painting	Visions, memories, dreams, stories, etc.

Source: Harvey (2006)

Table 1.2 Principal foci of the spatial turn

Dimension of socio-spatial relations	*Principle of structuration*	*Observable patterns*
Territory	Borders, boundaries, parcels, enclosure	Inside/outside (e.g. pomerium)
Place	Proximity, areal differentiation, area	Core to periphery (forum to *angiportus*)
Scale	Hierarchy	Dominant to marginal (elite to *plebs sordida*)
Networks	Interconnectivity	Nodal points and their connections

Source: author

compared to Pompeii, a site that has become a laboratory for the study of urban space over the last two decades (e.g. Poehler 2006). Moving across the category of relative space in Table 1.1, evidence for the exchange value of movement can be located in the measurement of distance by milestones and within a mentality of space-time that relates journey time to distance (for example, in Martial's *Epigrams* or in the Antonine Itineraries, cf. Laurence 2011). The dependency on evidence from literary texts for movement, of, say, itinerant traders, causes our sources to be categorised in the Lefebvrian sense as either representations of space or spaces of representation. Actual practice eludes us, however much historians and archaeologists may attempt to reconstruct 'daily life' from these texts. Thus, we can only ever expect to have a history of the *conception* of mobility, as opposed to a history of mobility as a lived experience. We can say though, that the conception of mobility was spatially varied across the city of Rome, according to whether you viewed the city from a litter or sedan chair carried on the shoulders of others, or made your own way through crowded streets.

The relationship between space and society has become a major area of debate in social theory as each reformulation of the spatial turn attempts to add another dimension. The original focus on territory was followed by a greater interest in place which was then supplanted by new foci; first on scale and then on networks (see Table 1.2). Each refocusing of the spatial turn privileged one aspect of space–society

Table 1.3 Integrating approaches to space and society

Structuring principle	**Field of operation**			
	Territory	*Place*	*Scale*	*Networks*
Territory	Past and present Emerging frontiers, borders, boundaries	Places in territory	Multilevel government	Interstate system
Place	Core-periphery, borderlands, empires	Locales, cities, localities, globalities	Differentiation of empowerment	Local/urban governance partnerships
Scale	Division of political power within the state	Scale as area rather than level	Vertical ontology of nested or tangled hierarchies	Parallel power networks
Networks	Origin-edge Ripple effects	Global city networks, polynucleated cities	Flat ontology with multiple entry points	Networks of networks, spaces of flows

Source: after Jessop *et al.* (2008)

Table 1.4 Application of the intersection of territory, place, scale and networks to ancient Rome

Structuring principle				
	Territory	*Place*	*Scale*	*Networks*
Territory	Relational content to the past – static continuities	Relational content in the present	City prefect Neighbourhood Magistrate	Metropolis to empire (Urbis–Orbis relationship)
Place	Edge of the city, extension of the city	Constitution of place through monumentality	Alleyway to famous place	Locale to emperor
Scale	Fragmentation of Rome into regions and neighbourhoods	Competing famous places Decentring of place	Neighbourhood to forum	Paved streets Regions
Networks	Relationship between neighbourhood and emperor	Creates many competing famous places	Communication forum to crossroad	Movement of knowledge and rumour

Source: author

relations while causing the others to be neglected. Theoretical developments emanating from the core disciplines of the spatial turn (principally geography) gradually found their way into archaeological thinking, which has witnessed equivalent shifts in emphasis from territory, to place, to scale and networks.

Recently, Jessop *et al.* (2008) have developed a toolkit to cross-reference all four foci, so that the distinctive contribution of each approach is embedded within the context of the other three approaches (see Table 1.3). This integrated approach is of use not just in the study of contemporary cities, but also in the exploration of historical forms of urbanism. It is a system that can be readily adapted to the study of ancient Rome and the work of historians and archaeologists and integrated with the new toolkit of Jessop *et al.* (see Table 1.4).

Territory and space: four regions and, then, fourteen regions

The size of Rome meant it was an entity difficult to measure or conceptualise except via breaking it down into constituent territories (Haselberger 2007, 18–23, 224–237). There appears to have been a register or even a census of the people from the time of Julius Caesar based on the *vici* (neighbourhoods) and information derived from the owners of the blocks of apartments (*insulae*). The area within the republican walled circuit of Rome had been divided into four regions by Servius Tullius: the Subura (NB nothing to do with suburbs), the Esquiline, the Hills (Collina) and the Palatine. Yet, there was a hierarchy of tribes, Palatine being the highest, the Hills (Collina) the second, with the Subura and the Esquiline being distinctly inferior (Taylor 1952–4, 229). In 7 BCE this system was replaced by fourteen regions, including the fourteenth across the Tiber (Favro 1996, 135–140). These new regions were disconnected from the voting units and would seem to have been a system for enumerating the city, but also were a means to allocate resources to the city – for example, a cohort of *vigiles* (firefighters) was assigned to every two regions of the city (Nicolet 1991, 194–198). Each person was located by the census in a region and within a neighbourhood of that region. The territorial relationship, for the inhabitant, was mediated through neighbourhood magistrates and was focused on the altar(s) of the local gods known as the *Lares Augusti*. The territorial division of Rome created knowledge of the city, as much as administrative function, and it was with the measurement of the population of the city and its division into neighbourhoods that gave the emperor the power to 'control' Rome.

Place-territory interfaces – seven hills and the expanding city

There are, however, other competing territorial divisions in Rome. A conception of the city as comprised of seven hills (Aventine, Caelian, Capitoline, Esquiline, Palatine, Quirinal and Viminal) that were a real and imagined territorial definition of Rome focused on those hills bounded by the Servian Wall (Figure 1.1).

The hills as raised ground create a sense of territory, but also can be named and thus become places in the city. Whilst at the same time, the phrase 'seven hills' can substitute as a metaphor to signify the whole city. There is a link between hills, walls and territory in the definition of Rome that can also be found in the establishment of *Roma Quadrata* defined by a wall around the Palatine Hill. Interestingly, the city extension under Servius Tullius, re-told by Livy, involved a growth in population and inhabited area that was to be defined by a wall and/or sacred boundary – the *pomerium* (Vout 2012, 77).

Intriguingly, these accounts from the end of the first century BCE imagine an expanding Rome from earlier times in which the city defied definition or required definition as comprised of fourteen regions – an increase not just from the four regions of the republic, but also a doubling of *seven* hills defining the city to *fourteen*

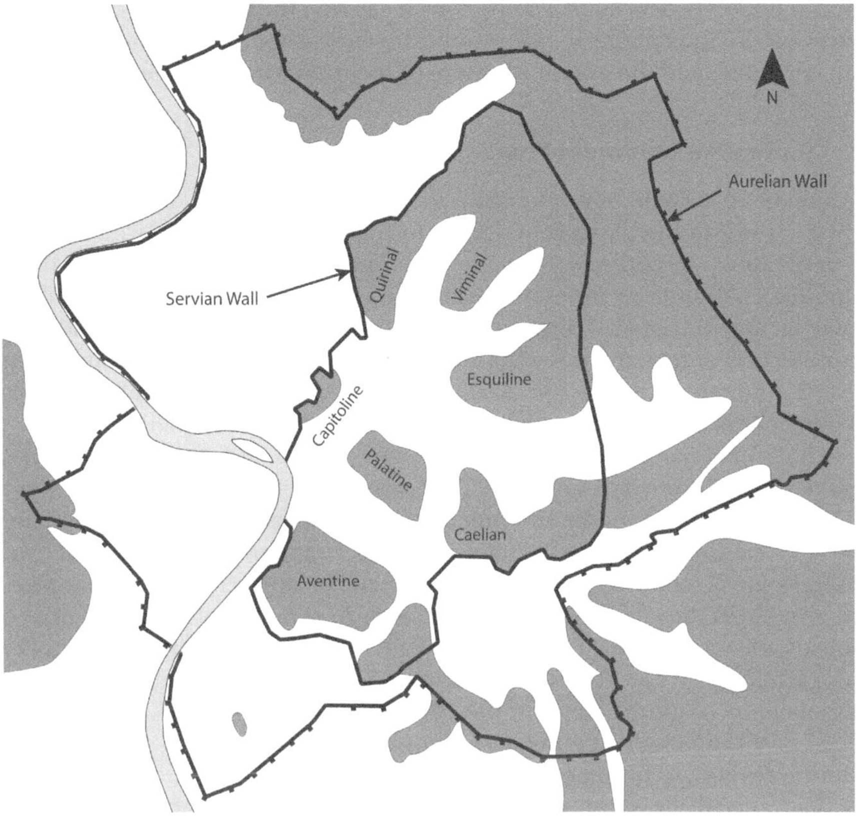

Figure 1.1 The seven hills of Rome and the Servian and Aurelian Walls
Source: Lloyd Bosworth

regions. The principle of division is by seven: a magic number associated with seven celestial bodies known to the ancients. Just as the seven hills could be used to denote the whole of Rome, so could the fourteen regions. More importantly, there is a sense in which the regions that were at the edge of the built-up area of Rome could expand and become bigger, whereas the seven hills developed a canonical or fixed territorial short-hand for the city of Rome.[2] The city as an expanding space was in many ways legally undefinable, causing the jurists to define the territory of the city by the term *continentia* (see Frézouls 1987 for discussion of definition of Rome's territory) or as an agglomeration.

The great fire of Rome in 64 CE is an example of how the city's territory included both the enumeration of space by the hills and by the regions. Tacitus measures the destruction with reference to the fourteen regions: four remained intact; three were levelled to the ground and in the other seven nothing survived but half-burned relics of houses.[3] The linkage between regions and fire prevention in the distribution of the seven cohorts of fire brigade (*vigiles*) causes Tacitus' linkage of regions to destruction by fire to invert the order of the city into disorder under Nero (Sablayrolles 1996, 245–289). The spread of the fire though was written with reference to the hills: it begins in the Circus abutting the Palatine and the Caelian and is stopped at the foot of the Esquiline six days later. The regions measured the territory of the city (and by implication the inhabitants in their homes), whereas the hills provided a definition of the beginning and the end of the story of the fire with reference to place.

Place: movement to monuments

The absence of a single obvious centre to the city Rome from the time of Augustus (27 BCE – 14 CE) caused places in the city to compete for attention with one another. The geographer Strabo saw how monumental buildings created centres in the Campus Martius and the three forums of the city and caused the rest of the city to become an appendage or of little consequence.[4] At the same time, the older narrow crooked streets of the city were regarded as the opposite to the wide straight streets that were developed after the fire of 64 CE. In Rome we can find a dialectical relationship between the use of, or movement to, places in Rome and their celebrity – these are each defined by the term 'a famous place' or *locus celeberrimus* (Newsome 2011, 20–26 developed from Trifilò 2008). This definition of place in Rome depends less on the actions of the emperor and much more on the use of space by the inhabitants of the city – causing Pierre Gros (2005) to see a *locus celeberrimus* as a *lieux du consensus*. Place in Rome is founded on use and movement to a place, creating that place as a point or node in the city. What is difficult for us to conceive of is how often and how far the inhabitants of Rome travelled to use these famous places in their city. Perhaps it is not actual usage that constitutes a *locus celeberrimus*, but knowledge of its availability and memory of using a place in the city in the past.

The constitution of place through 'movement to' causes the issue of boundaries (and territory) to become less important, since it was the movement of people itself that created the sense of *locus celeberrimus* – and such movement included a journey through streets that led to the destination. The definition of place through spatial boundaries varied from the enclosed space of the Forum of Augustus to the space defined by structures built around or within its confines that defined the *Forum Romanum*, and onto the less defined spaces such as the Sacra Via (Sacred Way) with its associated monuments. Interestingly, today we see these places as plans or in reconstructions that create empty spaces defined by monuments. However, it was the people moving, stopping, chatting within these spaces that constituted them as places of fame, as much as the monuments and the images of those who did not move – the statues that needed occasionally to be cleared out to create greater space for movement (Stewart 2003, 136–140).

Scale: time and movement

Rome was the first city in which urban journeys could be made over a number of miles, rather than a 15–20 minute stroll across the city. Travel through its streets created a particular form of urban living not found in other cities. In Rome time spent traversing city space could be measured as a burden and resulted in places within the city being seen as distant from one another. Unlike in other cities the speed of travel in Rome was disrupted by crowded streets narrowed by the encroachment of shops. This meant the metropolitan journey time was slower than expected – producing a different urban speed than found elsewhere in Italy (Laurence 2011). The combination of scale, distance and slow travel speeds to be found in Rome must have restricted the desire to travel from place to place in the city and the increased the potential for locating individuals in neighbourhoods. As a result, we may describe Rome as a city with a greater friction of distance than the smaller cities found across Italy, such as Pompeii.

This conception of Rome as a city with a higher friction of distance is important for the historical understanding of Rome as a specific form of urbanism. It causes us to recognise a need for the elite to be near the forum, but not too near in order to ensure that they are seen to cross at least part of the city followed by their clients and supporters. It also helps to explain the location of the emperor's bodyguards in the praetorian camp on the edge of the city. The camp is located out of the sight of most of the population, whereas the earlier billeting of soldiers across the city was more prominent, yet the inhabitants of Rome would have been aware of the presence of the Praetorians and their camp (Busch 2011, 31–72). The concentration of soldiers at the edge of the city caused them to move across the city to be deployed in their role as the emperor's guards or for policing within the city itself. The appointment of Claudius as emperor involved the removal of the future emperor from the palace to the praetorian camp and a movement across the city of the *locus* of power (Levick 1990, 21–40). The separation of military power to appoint an emperor from the location of the emperor in the palace clarifies the role of urban space in creating a sense by which the emperor in the palace was first amongst equals; whereas Claudius in the praetorian camp was an emperor backed by the military might of these soldiers – even if in some ways a hostage of the soldiers. Movement to and from different locations in Rome could restructure how the emperor's power was seen by those watching his gestures, listening to his words, and subject to his power.

Networks: streets and their surfaces

Scholars have used many techniques to elucidate the development of Rome, from simple demographic estimation of population growth through to recording the emergence of public buildings and the development of the public city. However, there is an alternative history of the city that can be written to chart the development of paved surfaces beginning with the creation of the forum as a level public space in the city. The paving of the forum in limestone can be seen as the first creation of a space for walking, in which walking was not disrupted by pot holes, puddles, mud and other filth. The extension of surfaces of this type created a network of paved streets that extended across the city and into the countryside. This process was stimulated not by the simple extension of streets into the countryside but by a change in Rome's approach to space encapsulated by the construction of the Via Appia from Rome to Capua in 312 BCE as a public space of over 100 miles (see Laurence 2001, 2013c). Five years later, in 307 BCE, roads were built into the public land of the city (*ager publicus*). The Via Appia was paved initially from the Porta Capena to the Temple of Mars in 295 BCE and then to Bovillae in 292 BCE. Individual streets in Rome were paved over the course of the third century, for example to allow vehicles to gain access to the Aventine Hill (via the *clivus Publicus* – so called because its paving in 238 BCE was publicly funded). The process was somewhat slow with more or less a century passing before all the streets of the city were paved in a standard hard stone and the roads extending into the public land were upgraded to compacted gravel in 174 BCE. The

Figure 1.2 Excavated street (Via Biberatica – Trajan's Markets) – this street appears to have been pedestrianised with game-boards found in the middle of the paved surface

Source: Lloyd Bosworth

effect was to cause movement in the city to be of a quite different nature to that found on the roads into the countryside, but equivalent to that on the Via Appia. The result by the middle of the second century BCE was a city that had paved streets and was connected to its rural hinterland by roads of compacted gravel, with longer-distance highways (Via Appia and Via Flaminia, etc.) forming corridors of movement from the city (as discussed by Witcher 2005).

The importance of the surface of Rome's streets is highlighted in the bronze tablet known as the *Tabula Heracleensis* that is one of the few surviving town charters from the first century BCE and provides detailed information on the upkeep of public space (Crawford 1996, 355–391). The charter refers to the repair of urban streets and of roads up to a mile from the city as a duty of annually elected magistrates. These magistrates could require those residents or owners of buildings adjacent to the street to maintain the roadway and above all to ensure that standing water did not occur and the roadway could be used for movement. For streets that were in front of public buildings, there was a requirement to let a contract for their maintenance. In addition, magistrates were responsible for cleaning the streets in the city and duumviri were responsible for cleaning the roads up to a mile from the city. The powers and responsibilities of the magistrates extended not just to the streets themselves, but to another milieu for walking: the porticoes. The emphasis on ease of movement is clear in that the porticoes were not to be blocked off and heavy wagons (*plaustra*) were not permitted to utilise the streets of the built up area of Rome between dawn and the tenth hour of the day (Kaiser 2011). These laws were drafted to ensure that mobility in the city was maintained; the secure surface on which feet trod was a feature of the city of Rome. We encounter in literary sources throughout the period of the Empire that good emperors repaired the streets of Rome that had become neglected (Robinson 1992, 59–82).

Rome – territory-place-scale-network

The streets of Rome formed the most extensive urban network of smooth surfaces in the whole world as understood in antiquity. Pliny the Elder opens his description of Rome in the *Natural History* (3.66–67) not with an account of the public monuments of the city but with an account of the extent of the city: 13 miles and 200 yards in circumference, divided into 14 regions, with 265 *compita Larum* (crossroad shrines), with roads leading to its gates measured at 20 miles and 765 paces and a total length of all roads leading through the neighbourhoods coming to 60 miles (Figure 1.3).

Rome, for Pliny, was a city of road surfaces that was more extensive and of greater magnitude than any other city in the world for this reason.[5] It was the roads that defined the city and only as an afterthought does Pliny add to this two-dimensional

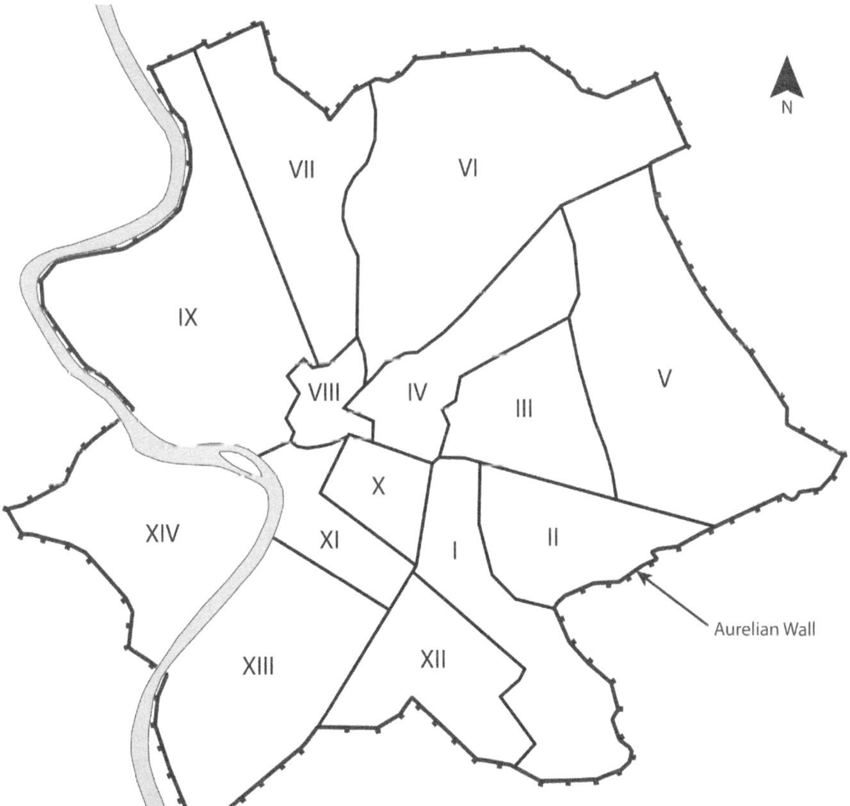

Figure 1.3 The fourteen regions of Rome
Source: Lloyd Bosworth

Figure 1.4 The network of paved streets in Rome
Source: Lloyd Bosworth

aspect the third dimension of the greater height of Rome's buildings. Similarly, Strabo in the *Geography* (5.3.7–8) notes the regulation of the height of buildings to no higher than seventy Roman feet and also that the talent of the Romans was in building roads, aqueducts and sewers – all items associated with movement that were either surfaces or sub-surface features. It was the networks of roads, sewers and aqueducts that provided these authors with a means to demonstrate that Rome was a city with a different scale from other cities in the ancient Mediterranean. The streets were punctuated by networked places such as the crossroad shrines that were places of worship in the neighbourhoods the roads traversed.

So far, the discussion has separated the four structuring principles of space and it is now time to set out how these principles intersect with each field of operation (Table 1.4). Territory obviously has a present-past aspect, revealed in the actions of creating territorial space. In terms of place the *vici* develop a monumental aspect as foci for religion and administration. The intersection between territory and scale can be seen by the nesting of people within a neighbourhood and a neighbourhood within a region. The network radiating from a neighbourhood was symbolically represented as a relationship with the emperor and the worship of the *Lares Augusti* at the crossroad shrines. A famous place (*locus celeberrimus*) was defined territorially as a core element of the city, but was a place in its own right – which in turn was scaled with reference to other places right down to the alleyway (*angiportus*) or perhaps even to the brothels and bars of the city. Inevitably, a famous place became a *locale* for the expression of power and was subsequently enhanced through the building of monuments that included the writing of the elite or the emperor onto that space. The issue of scale revolves around the role of time that fragments and separates the places of and territories of power (senate, praetorian camp, palace, and people in neighbourhoods), but unified around the figures of the emperor and the city prefect. The intersection between scale and place causes the city to be defined by 265 *compita Larum* (crossroad shrines) and a series of competing famous places that cause us to view Rome as a tangled series of spatial hierarchies that were underpinned by a variety of power networks leading from the local to the elite with their property (wealth) held outside city and often at a distance from the city. The expanding street network created new nodes (even becoming famous places), not least of these were the crossroads of the city that were formally constituted as sacred places. However, we should also allow for the possibilities of the development of new nodes through usage or informal spaces that could become formal places associated with elements of monumentality (Trifilò 2013). Across the city of Rome networks of information flowed and conceptualised as rumours spread not to the major houses from where it was disseminated to clients, but, to the crossroads (Laurence 1994). This overview of the intersection of territory, place, scale and networks in ancient Rome points to the polyvalency of all socio-spatial activities within Rome that, inevitably, were underpinned by movement.

Rome – the moving city

The exploration of Rome via the overlapping themes of territory, place, scale and networks alerts us to the many meanings that space can acquire or have projected onto it – multiplying its meanings through cross-referencing each of the four themes with each other. This allows a greater analytical depth to be maintained and also suggests how the meaning of space can alter. A city the size of Rome refuses to be defined or for the meaning of its space to be fixed. As a result, a sense of place could be subject to movement and change – especially in texts. More importantly, the scale of Rome – like the new scale of empire – needed to have a network of all-weather surfaces to facilitate the creation of physical networks of movement, both in the city and outside, that allowed for travel to and from a person/person's home. Movement, in this context, should not be seen as entirely focussed on the forums at the centre, but also include movements through space to other places set at a distance from the centre. It was the scale of Rome that created a different form of urbanism, in which the city can become differentiated with movement defining the nature of segmentation. The building of new monuments, perhaps, *responds* to movement and, certainly, redefines the places of movement (for example building the Colosseum in the middle of the city – *in media urbe*), but is not perhaps a *cause* of the reconfiguration of movement within the city. Thus, when Rome burnt down and was rebuilt, parts of the city were altered, for example with wider or straighter streets, but many of the places of the city remained in the same locations and continued to have movement to them – and to sustain relationships to the earlier buildings on the same sites. Thus, place may be adjusted through the addition of new monuments or the alteration of the physical fabric of the streets in Rome, but the changes to the network of streets or network of competing famous places remains hidden and difficult to recover. However, it is perhaps the changes to these networks of people and places, as well as the changes in scale associated with the growth of Rome that may be the most interesting phenomenon in Rome.

Notes

1 The chapter builds on earlier work and is particularly influenced by the work of David Newsome. I was privileged to supervise his PhD thesis and I owe him many debts of gratitude – not least his enthusiasm for addressing difficult subjects and resolving loose ends (see Newsome 2011, 4–9). This chapter is intended for readers outside of Classics and Ancient History; with this readership in mind I have included for the most part references to accessible modern studies of antiquity. For those working in Classics and Ancient History wishing to access full references to ancient texts, etc., I refer them to Laurence (2015).

2 Note that there were 425 hectares within the Servian Walls compared to 1,400 hectares within the Aurelian Walls – the latter a reduction of the urban area of the city.

3 The narrative can be found in Tacitus *Annals* 15.38–40, compare Suetonius' *Life of Nero* 38 and Dio Cassius *History of Rome* 62.16–18. The latter creates a narrative without reference to hills or regions, but notes that two-thirds of the city was destroyed, including the Palatine and the amphitheatre of Taurus (no other specific places are mentioned). Suet.*Ner.*38 provides no topographical indicators. Compare the linkage between regions and hills in Varro *On the Latin Language* 5.45–54.

4 For text see Strabo *The Geography* 5.3.7.

5 *Fateaturque nullius Urbis magnitudinem in toto orbe potuisse ei comparari.*

2 The ancient city and Huizinga's *Homo Ludens*

Hanna Stöger

Introduction

Play is a vital but often overlooked aspect of how urban space is experienced by people. Encompassing a wide variety of activities, play may be entertaining and delightful, but also impulsive and even dangerous. Today's urban landscapes gave rise to 'urbanathlons' and the action-packed 'parkour' movement.[1] Both make use of the urban environment through playful engagement with public space and architecture: walls, stairs, fences and street furniture serve as obstacles providing playing grounds and physical challenges for a new generation of 'urban athletes'. The idea of play allows us to take a fresh look at urban landscapes; moreover, play provides the motive for critically reviewing established assumptions about a well-functioning city in past and present times. Furthermore, it challenges the normative notion of 'good' city form which is traditionally linked to efficient way-finding, economic vitality, safety, and necessary amenities and activities (Stevens 2007, 198). Quite the opposite, the 'playability' of urban space reveals the contingency and perhaps also the fragility of public and semi-public spaces.

This chapter draws on two publications which at first glance might not have much in common with Roman cities: the first is Quentin Stevens' *The Ludic City* (2007), which provides a detailed depiction of *play* within the context of today's urban public space. Almost eighty years earlier Huizinga began to develop his profound ideas on play, later collected in his seminal work: *Homo Ludens: A Study of the Play Element in Culture* (1955).[2]

This chapter aims to emphasise the significance of playful activities within the cities of the past and present and seeks to underline the importance of Huizinga's work for archaeology. The playability of Roman cities is central to this study and is explored by means of a comparison between studies from modern urban contexts and the urban environments of Roman cities. To begin with, Huizinga's characteristics of play are introduced to create the wider philosophical framework within which he defines this concept. Next, organised play in Roman cities and the venues for official performances will be briefly presented and contrasted to the textual and iconographic evidence for informal, or spontaneous, playful behaviour. The main part of the chapter focuses on a comparison between the empirical observations from today's cities and the potential for playful activities in Roman cities. The spatial aspects of play, as defined by Huizinga, take priority and guide the search for urban playgrounds in the ancient city. Roman Ostia serves as a case study (see Figure 2.1).

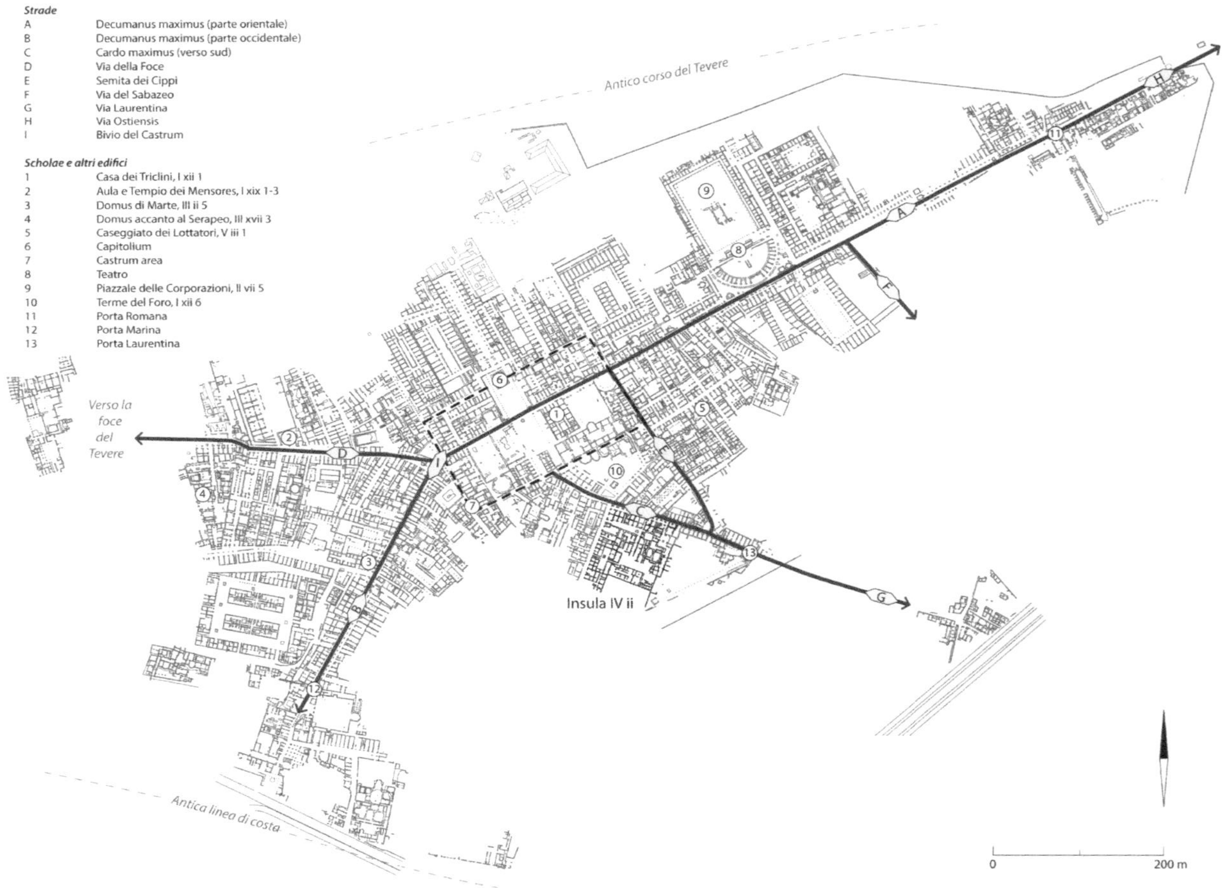

Figure 2.1 Site plan of Ostia, including the excavated areas
Source: Stöger (2011)

By examining the city's spaces from an archaeological and a space syntax perspective, the potentially available urban playgrounds of the Roman *Homo Ludens* can be identified.[3]

Huizinga's influence

While Huizinga's work has enjoyed considerable influence on studies in cultural history (for example Jardine 2010), it appears difficult to apply his philosophical thoughts to archaeology. The notion of play has been addressed only sporadically in archaeological research, some examples being the Mesoamerican ball games (for example Scarborough and Wilcox 1991) and play and games in Greek and Roman societies (e.g. Christesen and Kyle 2014). However, following trends in the humanities, archaeology is beginning to acknowledge the cultural relevance of play beyond its purely biological function (see Pellegrini 2009).

Huizinga argues that play is one of the fundamental facts of human life, placing it at the root of poetry, music, philosophy – even jurisprudence and war. More than discussing play, Huizinga contemplates human nature and understands the play-factor as the central drive in human life. He argued (1955, 173):

> The spirit of playful competition is, as a social impulse, older than culture itself and pervades all life like a veritable ferment. Ritual grew up in sacred play; poetry was born in play and nourished on play; music and dancing were pure play ... We have to conclude, therefore, that civilization is, in its earliest phases, played ... it arises in and as play and never leaves it.

Huizinga (1955, foreword) contends that play is the primary formative element in human culture and claims that *Homo Ludens* deserves a place in our nomenclature, next to *Homo Sapiens* and *Homo Faber*.

The characteristics of play according to Huizinga

Huizinga's (1955, 7–12) definitions are helpful to gain a deeper understanding of the complex nature of play. Hence, before we start to explore the playability of the Roman city it seems worthwhile to look at Huizinga's characteristics of play. He argues that play is, first and foremost, a voluntary activity, thus play under orders is no longer play. Huizinga asserts that play distinguishes itself from the course of the natural process by this quality of freedom alone. He adds that for the adult and responsible human being play is a function which he could equally well do without. Hence play is superfluous and it is never imposed by physical necessity or moral duty. It is never a task and it is done at leisure. Only when play becomes a recognised cultural function, as a rite or a ceremony, is it bound up with notions of obligation and duty (Huizinga 1955, 7–8). The second characteristic of play is closely linked to its voluntary status; Huizinga emphasises that play is not ordinary or real life, but rather a stepping out of real life into a temporary sphere of activity. He reminds us that any game can at any time completely run away with the players, whereby the contrast between play and seriousness is always a fluid one. The third main characteristic is its limitedness and secludedness. Play according to Huizinga is distinct from ordinary life both as to locality and duration; it is carried out within certain limits of time and place, and contains its own course and meaning. Huizinga writes (1955, 9–10):

> Play begins and then at a certain moment it is 'over'. It plays itself to an end. While it is in progress all is movement, change, alternation, succession, association, separation. Once played, it endures a new-found creation of the mind, a treasure to be retained by the memory. It can be repeated at any time, whether it is children's play, or a game of chess. In its faculty of repetition lies one of the most essential qualities of play. It is transmitted, and it becomes tradition.

Huizinga (1955, 10–11) underlines that next to its temporal limitation, play has a spatial limitation. He explains that all play happens within a playground marked off beforehand either physically, symbolically, deliberately or naturally.

No gain is made from play

Huizinga's explanations make it very clear that play has no productive output and has no instrumental side to it (Stevens 2007, 30–31).[4] These very characteristics have also been identified by a number of other authors including Lefebvre (1991), Benjamin (2000), and above all Caillois (1961, 5), who goes so far as to consider play an occasion of pure waste of time, energy, ingenuity, skill and money. For the argument presented here, the spatial aspects or the spatial limitations of play take priority, and will be helpful in identifying potential informal playgrounds within the ancient city.

Custom-built structures: the organised side of play

Most Roman cities featured areas specifically dedicated to performance and entertainment; these activities required a number of custom-built structures to accommodate spectators who attended circus-games, theatrical performances and gladiatorial combat (Coleman 2000, 220). This was the organised side of play, and hence would not meet the formal criteria of voluntary play as defined by Huizinga. However, it is the only form of play which leaves a clear mark in the archaeological record of Roman cities. Typical buildings associated with these spectacles were theatres, amphitheatres, circuses and stadia. By comparison, less formal play and street-entertainment have left behind few traces. Smaller groups of spectators could cluster around the performer on street-corners or gather in the courtyard of taverns, without requiring specific premises (Coleman 2000, 220).

The development of custom-built structures to accommodate spectators was a gradual process that reflected the ambitions of wealthy and powerful sponsors (Coleman 2000, 245). Coleman adds that the main building types and venues associated with public entertainment came into use at distinct stages during the development of the city of Rome. Entertainment was monumentalised by the last generation of the Republic; this marks a shift in the attitudes of the ruling class towards the provision of public entertainment. Consequently performances and their venues became a status marker for the sponsors and a regular expectation of the beneficiaries (Coleman 2000). A concentration of entertainment buildings was found in the Campus Martius (see Figure 2.2).

At the end of the Republic the Campus Martius was still available for large-scale public building projects, including structures for mass entertainment. In Rome's port-town Ostia, the theatre built by Agrippa signifies the organised side of play. It was large enough to accommodate a representative section of the city's population.

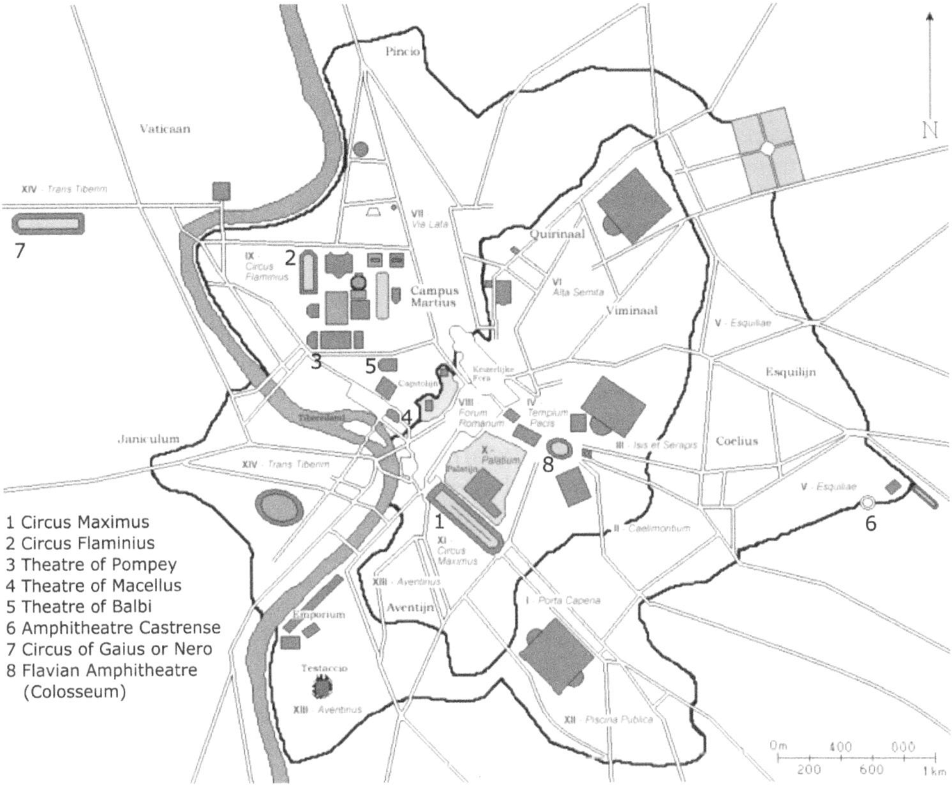

Figure 2.2 Entertainment buildings in imperial Rome
Source: Wikimedia Commons

Several large-scale imperial entertainment buildings became Rome's greatest surviving landmarks; in contrast, only few iconographic sources and written texts bear witness to non-organised, informal play. To give a few examples, the carved game boards found on the pavements of the *Forum Romanum* excellently illustrate the informal side of play which took place in Rome (Trifilò 2011, 312–313). Above all, the specific location of these game boards recalls the spatial significance of 'the marked off playground' as stated by Huizinga (1955, 10). A further reference to informal play is found in a passage from Marcus Minucius Felix's *Octavius.*[5] The text describes the leisurely walk of a group of friends along the shoreline of Ostia, the port city of Rome. On their way back the friends stop and watch a group of boys playing ducks and drakes using flat shells, a description of play in the most genuine sense. The simple play of letting flat stones or shells bounce off from the water's surface had found its way into this early Christian text, and in its simplicity it manages to describe one of the most fundamental playful encounters, involving nature and individual skill. The text takes into account both the players (the boys) and the observers (the group of friends) who enjoyed watching the play, and in this way the passage even hints at varying levels of play, including those of active and passive participation.

Urban playgrounds

Turning our attention now to the Roman city we should begin to investigate the whereabouts of the Roman *Homo Ludens* and his presumed playgrounds. Ostia, Rome's port-town, is one of the few Roman cities where the past built environment and the street network are well preserved; moreover the city has been excavated to an extent which allows us to explore Ostia's playability. Stevens' study of modern cities reminds us that the continuous system of public spaces and the diversity and density of activity patterns bring together different people in today's urban spaces. Altogether these factors increase the likelihood of social encounters and create opportunities for play (Stevens 2007, 39). However, unlike Stevens who based his study on direct observation of playful activities in today's cities, a study of play in the Roman city can only rely on the built and non-built environment and its scarce evidence or, more likely, its potential for playful behaviour. From the studies conducted in today's cities it can be established that play is stimulated by diversity (Stevens 2007, 97), and hence playful activities thrive in areas where there are choices between types of urban spaces. Stevens' case studies identified a number of urban environments which are conducive to play; these are most notably paths, intersections, boundaries, thresholds and props or street furniture. These specific urban environments will be reviewed and their role in the ancient city will be explored in the following sections.

Paths and intersections

As far as streets or urban paths are concerned it should be noted that a street has two main characteristics: it is, at one and the same time, both place and path (Moughtin 1991, 62). Hence urban street networks not only offer fixed physical conditions, but also stimulate playful behaviour through the dynamic element of human movement encountered along the streets. People are brought into close proximity with each other where paths intersect. At intersections people are exposed to the highest density of other people and the greatest range of sensory phenomena and opportunities for action (Stevens 2007, 99). Intersections punctuate urban routes; they are the point of connection and decision for the human in motion (Lynch 1960, 98). The need to change direction normally causes people to slow down. Directional changes prompt pedestrians to heighten their attention and take in sensory information from the surrounding buildings and spaces; moreover pedestrians intensify their awareness of things which are not in line with their movement. Therefore intersections not only broaden the field of vision, but also create occasions for new opportunities. This increases the chance for engaging with attractions encountered at crossing points. Intersections are thus sites where people can be distracted from instrumental purpose (Stevens 2007, 99), and, in Huizinga's sense this creates time and space apart, where play is possible.

Boundaries

Boundaries or edges have been identified as another urban environment where playful activities can flourish (Stevens 2007, 114). Boundaries define what is visible and accessible, and by presenting a physical boundary they provide structure to social relations. Physical boundaries within spaces are often explored and even tested through play. Solid edges (e.g. walls and fences) constitute physical barriers to move against, and they generate a tension which can provide physical challenges often experienced through the body as vertigo.[6]

Roman Ostia had natural boundaries marked by the ancient shoreline and the banks of the Tiber. In addition, stretches of the city walls, as well as architectural enclosures of various zones constituted man-made boundaries. Both environments must have been areas which inspired people to playful behaviour. The group of boys, who played at the ancient shoreline of Ostia, demonstrate a playful activity, not only occurring along a natural boundary, but also interacting with the boundary itself (see above). Next to solid physical boundaries we also need to think of soft boundaries created by edges. Stevens explains that the edge of public space is a popular place for people to position themselves during their free time. Edge positions offer protection, while allowing controlled exposure to outside stimuli (Stevens 2007, 115). Ostia boasts a number of public places, mainly squares and forums; there people could situate themselves at the edges, often colonnaded passages, and could watch from a secure position or engage in playful activities to a degree they felt comfortable with.

Thresholds

Thresholds represent yet another type of urban environment which has been identified as being favourable to playful activities (Stevens 2007, 152). According to Huizinga's (1955, 8) definitions of play, thresholds provide a 'space apart', being physically distinguished both from inside and outside. Thresholds are designed and scaled to negotiate and mediate people's passage. Many architectural elements function as thresholds: doorways, porticoes, colonnades, steps, landings, terraces and porches are created to serve functional purposes, but they also offer a great invitation for informal use and give rise to a large variety of play. At thresholds stimuli from the internal space can spill over to the outside and vice versa. For example, music performed inside a building can prompt people on the outside to dance and sing along. In contrast, less clearly defined thresholds can provide seamless transitions between inside and outside, or between public and private spaces. One should keep in mind that there are also symbolic thresholds between organised play like theatre performances and games and the unplanned reactions to it, or the simulation of it. Hence spontaneous, active forms of play often arise when visitors leave the venue of organised performances. Quite understandably, after a long period of mental and bodily passivity, people are keen to release unexpressed, stored energy. Stevens observed that various games of tag, running, jumping on each other and similar kinds of exuberance often occur in places outside of cinemas or theatres (Stevens 2007, 161). From modern festival and football culture we know of large-scale activities happening outside of football stadiums, before and after the matches. A growing phenomenon are the so-called tailgate parties outside of sport venues, where thousands of people gather before the actual game, wearing the team colours, having barbeques and drinks all from the back of pick-up trucks.

Ostia's built environment offered many architectural elements which served as thresholds. Most of Ostia's streets were lined with porticoes while a number of public and private buildings could be accessed from stairs leading from the street level to the building. Potentially, all these architectural thresholds could have provided

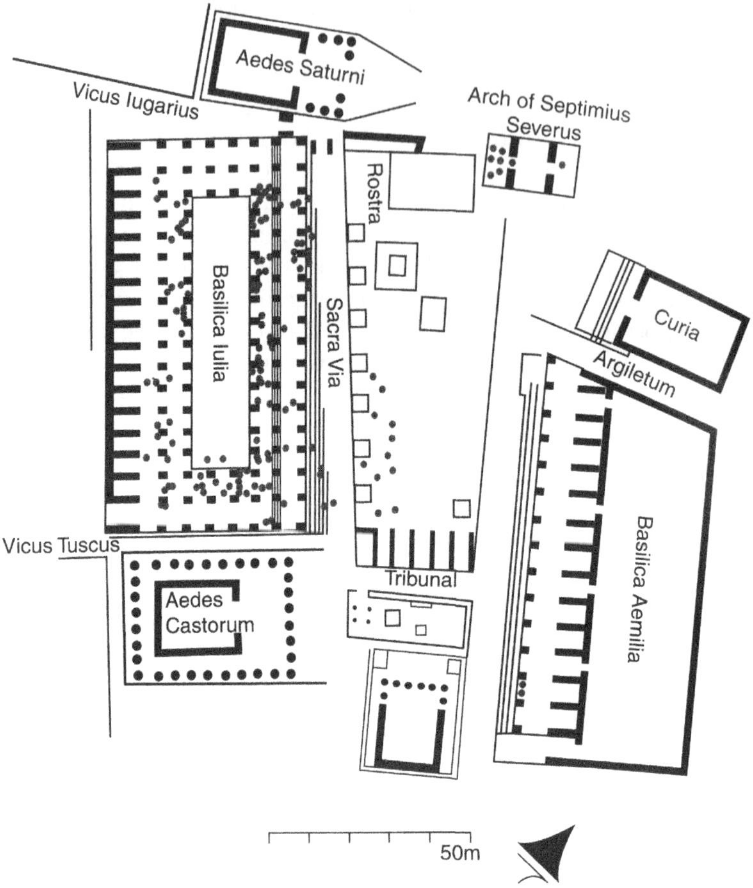

Figure 2.3 Incised game boards recorded in the area of the *Forum Romanum* (after Thédenat 1923)

Source: Trifilò (2011, 316, fig. 13.2)

areas which invited playful behaviour. The carved game boards from Rome's *Forum Romanum* fit well into this category of space (see Figure 2.3).

They are found concentrated in areas which marked the boundaries between busy passages and the more quiet areas of the *forum* in Rome (Trifilò 2011, 328). While the games themselves frame an escape into a world with rigid rules of encounter, the locations where the carved boards were found suggest that the players did not want to escape entirely, but also wanted to be exposed to people and events outside the game board (cf. Stevens 2007, 119). Ostia's porticoes and stairs could have invited play and informal entertainment in ways comparable to the public buildings of Rome. Likewise, the areas surrounding Ostia's theatre were well equipped to accommodate unscripted and spontaneous playful activities. Traces of board games have been found on the threshold slabs of the so-called *Foro della statua eroica*,[7] on the steps of Ostia's Capitolium (see Figure 2.4) and in one of the heated rooms of the Forum baths.

Figure 2.4 Game board incised on the steps leading up to Ostia's Capitolium
Source: author

The semi-public space of Ostia's baths are excellent venues for informal play. The location of the incised game board at the Forum bath is telling. The board is found right in the centre of the room, assuming a prominent position, allowing the players to take centre stage and leaving enough room for an audience (see Figure 2.5).

Moreover, the board captures the maximum of visibility from three neighbouring rooms which are connected through wide door openings.

Props or street furniture

As a final element, props should also be discussed since they successfully promote playful behaviour in today's cities. Under this category fall public art works, play equipment and street furniture (see Figure 2.6), all of which are fixed objects placed within public space (Stevens 2007, 178).

In the case of Roman cities one might think of honorific statues, and statues dedicated to imperial worship; also fountains, exedra and niches with benches. Together, these elements form an urban micro-geography within the body's reach; hence engagements with these elements of urban form are often tactile (cf. Stevens 2007, 179). Although generally not designed with play in mind, street furniture and

Figure 2.5 Incised game board in the centre of a heated room in Ostia's Forum Baths
Source: author

Figure 2.6 Street furniture in Barcelona, back to back with 'Iron Man'
Source: author

fountains often encourage playful activities and even provoke responses which might involve physical risks.

Fountains hold a specific position among the urban props; in today's cities they regularly fall under planned play equipment. Moving fountains in particular have been enjoying great popularity and can be found in many cities throughout the world. These fountains are 'on stage' and pedestrians often gather around them, either to passively observe the dancing water jets, or to step out onto this stage and begin spontaneous interactions with the moving fountain and with other people (Stevens 2007, 187). Fountains also played a significant role in Roman cities (Schmölder-Veit 2009, 92–98; Ricciardi and Scrinari 1996, Vol. II, 194–237). Fountains are present in a large variety of forms and functions (see Figure 2.7), ranging from water basins and fountain houses which supplied neighbourhoods with drinking water to large-scale urban

Figure 2.7 Fontano con Lurcerna. Located on the western decumanus in Ostia, the fountain offered a focal point for visitors coming in from the southern gate, the Porta Marina

Source: author

prestige objects like the impressive Antonine fountain from Sagalassos (Waelkens and Poblome 2011, 108–112).

Street intersections and props in Roman Ostia

To gain insights into the playability of Ostia's public spaces this section concentrates on street crossings and their immediate surroundings. The city's intersections are locations which concentrate crowds of people with different needs and interests (cf. Stevens 2007, 100). In today's cities, intersections have been identified as conducive to play since they provide occasions for people to step out of their purpose-driven itineraries and be sidetracked to engage in playful activities. Pedestrian behaviour observed in today's cities, when used with caution, can provide some indications about similar situations in past urban contexts. When approaching an ancient city a valuable starting point for research is a survey of the physical makeup of urban spaces with their related props and street furniture. However, in order to reach beyond an intuitive understanding, and to add a layer of objectivity to the research, the study presented here takes into account the results of a previous space syntax analysis of Ostia's street system to help explore the playability of street crossings and their settings (Stöger 2011, 220–226).

One of the tools for spatial analysis provided by space syntax is a visibility graph analysis (VGA). VGA allows insights into the visual structure of the streetscape, drawing on the relationship between visibility and movement (cf. Hillier and Vaughan 2007). The positive correlation between visual integration and observed movement is one of the important findings from syntactic studies of building layouts and urban spaces, and has been confirmed by empirical studies. VGA provides clear indications of where the city's most visually connected spaces were located; these are also the spaces where high concentrations of movement typically occur. With specific regard to play, visibility is of particular interest since most play seeks an audience and occurs or even thrives in the presence of other people.

Starting from the results of the VGA, a number of interesting observations can be made through a reverse mode of archaeological investigation. The analysis identified several locations within Ostia's streets and public spaces with a very high degree of visual integration (see Figure 2.8).

Visual integration is normally displayed like a heat chart along a colour range from red for the most integrated spaces to blue for the least integrated spaces. The greyscale display does not sufficiently indicate the places which are visually well integrated. The places identified from left to right: bivio del castrum, centre of the forum, the junction between the decumanus and the Via del Pomerio, and the area in front of the theatre (VGA Stöger 2011). Most notably these are the intersections along the eastern *decumanus*, the areas concentrated in front of the theatre, in the centre of the *forum*, as well as the so-called '*bivio del castrum*', referring to the intersection between the *decumanus* and the Via del Pomerio. The '*bivio del castrum*' is also the starting point for the Via della Foce and the western *decumanus*, hence a place where five streets meet.

When comparing the results obtained from VGA against the site map of Ostia and the architectural remains which are still extant, it can be observed that most of the intersections which have been identified as areas of high visual integration have also been marked by props or street furniture, such as fountains, arches or *compita*

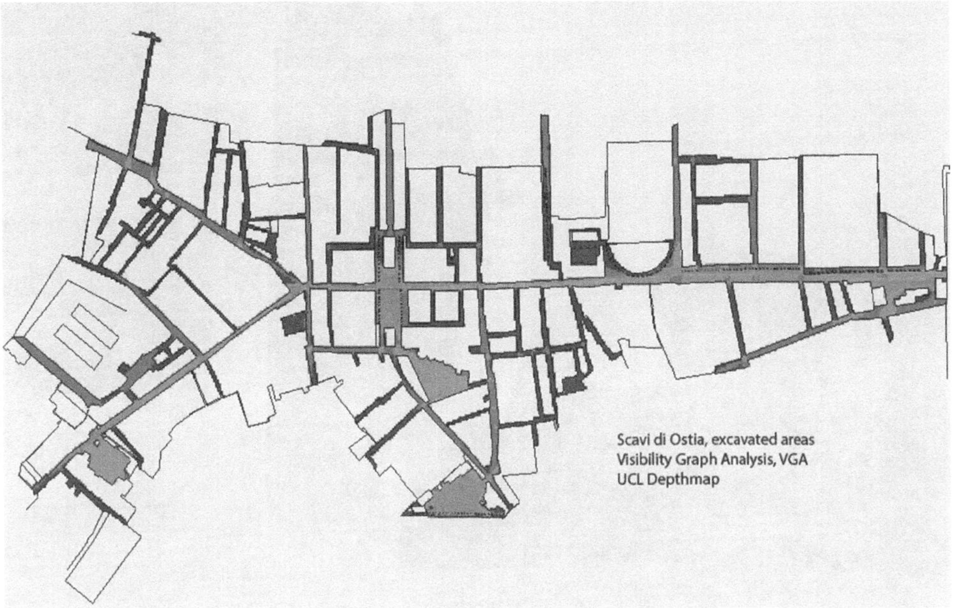

Figure 2.8 Visibility graph analysis of Ostia's street network
Source: author

(crossroad shrines). Since they were constructed at different periods of time these architectural elements do not conform to a coherent programme of urban embellishment, but seem to act as visual markers suggesting an intuitive understanding of the spatial particularities encountered at street crossings. In the specific setting of Ostia where buildings reached up to four or even five storeys, leaving most streets relatively dark and narrow, one should add that intersections offered a sky-lit space of relative openness and brightness and served to highlight whatever occurred there (cf. Stevens 2007, 101).

Playgrounds in front of the theatre

As indicated by VGA (see Figure 2.8), visibility converges in front of the theatre, where the Via delle Corporazione meets the *decumanus maximus.* This fact seems to have been exploited architecturally through the construction of a *nymphaeum* (fountain) placed there (see Figure 2.9), and a similar one on the other side of the theatre.

The *nymphaea* mark the location of greatest visual integration and give the area social cohesion. Water features, including water basins and fountains, are important architectural elements in ancient and contemporary cities. Next to their functional use of dispensing fresh water they provided a stimulating environment which invited passers-by to take a break and seek distraction from their daily chores. The walled basins of the *nymphaea* were kept conveniently low to encourage people to sit down, offering a great opportunity to watch spectators coming out of the theatre or visitors passing-by along the *decumanus.* The open setting in front of Ostia's theatre seems to

Figure 2.9 *Nymphaeum* located along the *decumanus*, west of the theatre of Ostia, the walled basin was kept low to encourage passers-by to sit down

Source: author

combine the key characteristics of an environment which allowed informal, playful activities to flourish.

The centre of the forum

Further along the *decumanus* the next area with a high integration of visibility is the central area of the *forum*. VGA shows that visibility is intensified only along the portion of the *decumanus* where it intersects the *forum* in an east–west direction. Interestingly enough, we again find an architectural marker close to the area of intense visibility. The circular structure forming the basin of a fountain, located in the centre of the *forum*, was placed to the south of the area with highest visibility, so as not to impact the flow of traffic along the *decumanus*. While assuring that traffic could pass the *forum* unhindered along the *decumanus*, the *forum* itself with its adjacent public and religious buildings brought many citizens to the centre to carry out their daily duties. The fountain seems to mark a spot where diverse activity patterns converged; in addition it added a nodal point to an otherwise open and architecturally undefined forum space. From today's cities we know that intersecting paths and varied activity patterns are favourable to playful behaviour. The fountain in the centre of the *forum* seems an ideal place for playful activities, attracting active spontaneous performers and passive observers alike. In addition, the porticoes along the *forum*'s edge would have allowed people to watch activities which took place at the fountain from a comfortable distance and a protected space.

The bivio del castrum

The next focus of concentrated visibility is the '*bivio del castrum*', an important crossroad, located west of the city centre. Throughout Ostia's urban development this junction must have been one of the busiest, making it one of the most sensitive urban areas. It remained a crucial spot throughout the entire imperial period, gradually assuming a more monumental character (Gering 2004, 349–353). Once again we find architectural markers, first and foremost a *compitum*, a crossroad shrine. The *compitum* held its position since Republican times, providing religious protection and assuming the role of a nodal point around which streets converged that came from several directions. Later it was accompanied by a monumental *nymphaeum* dated to the end of the second century AD. We can observe that the location of the *compitum* coincides with the area where visibility converges (see VGA graph of Figure 2.8); while the *nymphaeum* fills the interstices between the Via della Foce and the intersection. The *nymphaeum* offered comfortable seating on the low walls forming the basin. From there spectators could watch the spectacle of the busy street crossing and whatever spontaneous informal play arose from the density of encounters provided by the overlapping activity patterns.

All spaces identified by VGA as being the visually most integrated locations within Ostia's urban area possess a number of shared spatial properties instrumental for creating an environment that stimulates playful behaviour. As stated before, Ostia's busy street crossings were identified by VGA; in addition, these junctions were embellished or accompanied by street furniture, most notably fountains. The fact that we find *nymphaea* located next to the street crossings suggests that the city responded to the needs and wants of pedestrians. Fountains and other forms of street furniture offer sensory *stimuli* at the points of intersections which punctuated their daily itineraries. We should remember that the attention of people is heightened at crossroads and pedestrians can more easily be sidetracked to engage with distractions which are not directly in line with their intended purpose or destination. The degree of diversity in an urban environment adds to its playability.

'Strolling, gaming and pleasure'

Having identified the city's most likely spots for inviting playful behaviour, one should also take a look at Roman informal games which were played in public,[8] and although informal were able to attract a good crowd. Spontaneous stand-up comedies or cheerful mocking and boisterous imitation would leave no archaeological traces but were an aspect of the urban environment described so vividly in Martial's epigrams or Juvenal's satires. In addition, we have evidence for board games and simple coin tossing or various games derived from 'odd and even' (*par impars*) using nuts, pebbles or knucklebones often thrown into small oval or pear-shaped pottery vases (Pavolini 1980, 1010–1011). Carcopino's (1962, 271–276) study on daily life in Ancient Rome incorporates 'strolling, gaming and pleasure' into the day's routines. He tells us that the Roman version of chess (*latrunculi*) never lost its appeal with the public; it absorbed both those who played and the observers who stood around commenting on their moves.[9] Again we find elements of active and passive participation, performers and their audience; ultimately both elements imply urban playgrounds which were visible and accessible to the public, hence pointing to the type of spaces which have been identified in our case study of Roman Ostia.

Concluding thoughts and future research

This chapter cannot do more than outline a tentative approach, opening up the debate about the Roman *Homo Ludens*. The study would profit from a more exhaustive look into Roman popular culture, as this field of research has received only limited scholarly attention in recent years (see Toner 2009). Nevertheless this brief excursion into the playability of the Roman city has allowed us to gain a more nuanced understanding of how public space might have been used and enjoyed by the urban population. The traces of game boards and the numerous opportunities for playful activities offered by public (and semi-public) spaces add a new dimension to our often overtly instrumental approach to the Roman city. In the case of Ostia, the city's commercial image has always dominated its research and perception. The traditional view of the thriving port city busy with its traffic of goods and people has largely obscured the playful appeal of urban space which exits alongside its purely functional use. By tracing the Roman *Homo Ludens* one encounters spaces which encourage informal interaction and invite passers-by to pause and rest, and to engage in spontaneous encounter and play. Finally, the *Homo Ludens* topic should also be approached in much broader terms, to examine how ancient cities became intertwined with the business of leisure and began to market themselves as centres of proto-tourism, heritage and culture.

Notes

A first version of this chapter was presented during an Honours Class workshop on 'The City: Past Present and Future' at the University of Leiden in March 2011; I wish to thank the participants for their helpful comments and specifically thank John Bintliff, Natascha Sojc, Janet DeLaine, Gertrudis Offenberg and Alexander Verpoorte for encouraging me to develop these ideas further, and Christoph Pieper for advising me on the ancient sources.

1 'Urbanathlons' are endurance races that incorporate city landmarks and urban obstacle courses set on the streets of the three largest cities in the United States (Chicago, New York and San Francisco); similar urban runs are taking place in Europe and Asia. Originally developed in France, 'Parkour' is a holistic sport activity which uses movement that developed from military obstacle course training. Parkour athletes use only the human body and the surroundings for propulsion, with a focus on maintaining as much momentum as possible while still remaining safe. Parkour is mainly performed in urban environments using buildings and street furniture as obstacles.

2 First published in German in 1938 and translated into English in 1955.

3 For an introduction to space syntax see Bafna (2003), for theory and method see, above all, Hillier and Hanson (1984), and also Stöger (2011).

4 This statement needs to be qualified since we have to take into account that activities which involve physical action will lead to regular release of endorphins or 'pleasure hormones' which are an incentive in their own right.

5 Minucius Felix, *Octavius*, III (5); in Meiggs (1973, 490–491) [...*ubi subductae naviculae substratis roboribus a terrena labe suspensae quiescebant, pueros videmus certatim gestientes testarum in mare iaculationibus ludere*...]. See Meiggs (1973) for an English translation.

6 The parkour movement exemplifies a type of physical activity which thrives on the sensation of vertigo. Since it is played out in today's urban environment it is difficult to find parallels from past cities. One could think of people being chased through the crowded streets of Rome (for play or for serious motives), and, by taking advantage of architectural obstacles, they are able to outplay and outrun their pursuers.

7 Personal communication from Luke Lavan, Ostia Project, University of Kent, 2011.

8 Gambling, betting and other forms of illicit play are not considered in this chapter, but would also warrant a spatial assessment to investigate their particular spatial requirements within the urban environment.

9 *Laus Pisonis* 190–208, refers to a verse in praise dedicated to a certain Piso, probably a consul with Nero in AD 57. Piso was known for excelling in playing *latrunculi* and was able to attract a large audience (Carcopino 1962, 275). The author of the *Laus Pisonis* is unknown.

3 'Spatial culture' of an institution

Preliminary thoughts on Roman military bases

Anna H. Walas

The greatest difficulty was met in taking the Praetorian Camp, which the bravest soldiers defended as their last hope ... Soldiers' special pride is in their castra, that is their country, that is their Penates.

(Tacitus, *Histories* 3.84)

Introduction

In his account of the Praetorian Guard defending their camp in the final days of Emperor Vittelius' rule in the eternal city, Tacitus conveyed a sense of special attachment of the soldiers to their *castra*; the *penates* in Roman culture were the divinities watching over a family, sometimes denoting one's place of origin (Morwood 2000, 135). This affectionate reference to structures traceable in archaeological record has long inspired scholars' interpretations of the relationship between the Roman army and the spaces it created and occupied. In the last decade several scholars have theorised the Roman military environments from the standpoints of identity and *habitus* (Gardner 2007 and Collins 2012), community and routine (Haynes 2013) and that of their unifying aspects (see Driessen 2005). This chapter takes the case study of a Roman auxiliary base, places it in the context of a modern military complex and adopts the concept of a *spatial culture* defined as 'the associations of urban form and social behaviour' (Hillier 1996, 263) to discuss how the community of the soldiers fits with the urban form created by the institution of the army. The standpoint taken in this chapter draws on James' (2011, 172) understanding of the military bases as 'social pressure cookers', with control over soldiers being a factor in their functioning and suggests a two-way relationship of both control and dependency.

In classical scholarship the Roman army is understood to be both a community and an institution (James 2011, 23). The theme of the army as a community has a long track of outstanding research (Goldsworthy and Haynes, 1999). Military service and a strong sense of social purpose derived from it profoundly shaped the identities of both those on the rolls of the units and the wider extended community of servants, slaves, traders, women and children associated with the soldiers, who were present in and around the military bases (Mattingly 2006, 176). These military societies consisted of an array of interest groups based around religious and ethnic identities, military rank and occupational specialisation among others (James 1999). With both geographical and social displacement at the heart of their shared experience, Roman military communities created subcultures of their own. The bases were the physical signifiers of the community's military identity. Unlike in the west of the

Roman Empire, the troops in the east were sometimes stationed in military bases situated inside cities. This was thought to encourage softness among soldiers, testified by instances of soldiers from the western empire not respecting greatly their peers in the east (Wheeler 1996).

1960s Alconbury – a comparison with a modern military base

Military bases are spatial manifestations of one of the most studied institutions in history. To begin the discussion of how a community of the soldiers was fitted within the institutional spaces of the Roman military base, I turn first to a comparison with a modern military base. Examining the affordances and limitations of a plan of a twentieth-century military base is helpful in illustrating the contrasting spatial culture of the Roman military.

For this purpose I have used the accommodation area of the 1960s United States Airforce (USAF) and Royal Airforce (RAF) military base at Alconbury (Figure 3.1) as an example with an easily available contemporary plan significantly different from a Roman military base. As an American base located in the UK, Alconbury is an enclave of a professional minority with a distinctive national culture in a foreign territory. Founded in the 1930s, the base was strategically important during the Second World War and the Cold War. It remains in use, controlled by the USAF (Fletcher and Maxwell 1989, 99).

The triangular arrangement of the runways defined the layout of the base, with the accommodation site situated to the east of the runways. The accommodation

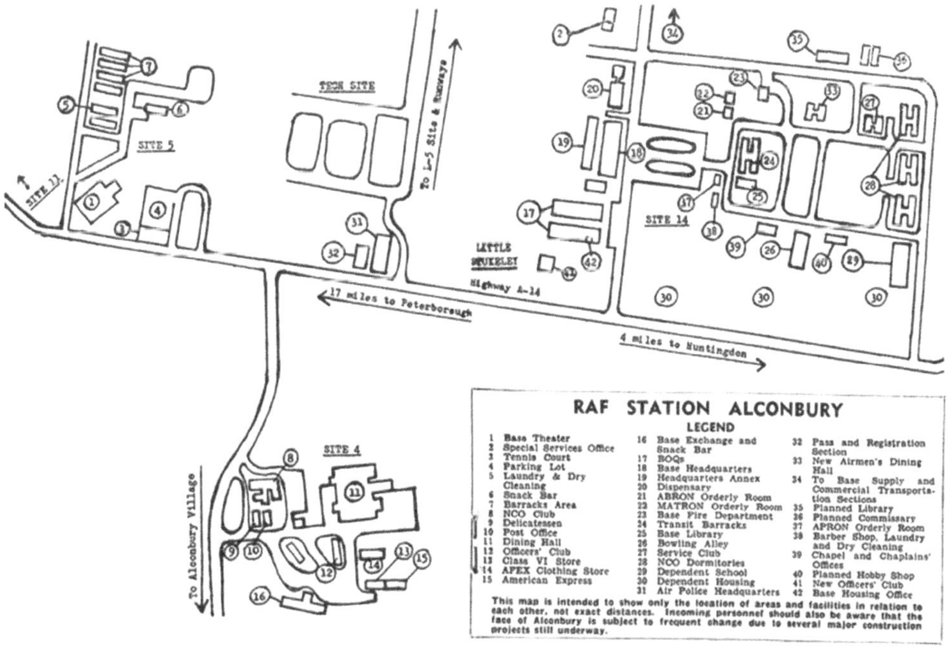

Figure 3.1 RAF Alconbury accommodation area
Source: RAF Alconbury Public Affairs Office

quadrant depicted in Figure 3.1 is irregular and spacious with four distinctive clusters of facilities and open areas in between the buildings. The 1960s Headquarters building (18) was located at the end of a central road running across the main site, with the Bachelor Officer Quarters (17) adjacent to it and the non-commissioned officers' (28) accommodation at the opposite end of the axial road. An abundant provision of facilities for the soldiers is noticeable, including facilities for organised social gatherings; Officers' Club (12), New Officers' Club (41) and Service Club (27). There was also a separate commercial site with diners and shops (site 4) to the south and an array of other buildings scattered on the site; a base theatre (1), a bowling alley (26) and an Airmen's Club (NCOs Club on the map – Figure 3.1, location 8). In the 1960s the Airman's Club functioned as a nightclub known as the 'Aquarius Club', open to the local civilian population with night bus service over the weekends providing transportation link with the nearby towns (Green, pers. comm.).

There were also ample sports facilities: a large sports ground (to the north of the area in the map), a tennis court (3) and later five baseball grounds. The nearby barracks, while maintaining a regular arrangement, were fairly dispersed. The roads between the barracks were, until recently, named after American states, enabling easier navigation around the garrison for recently arrived soldiers through the use of familiar cultural references. The architecture of 1960s Alconbury was unmistakably military with its metal barracks and the oval roofs of aircraft hangars (Figure 3.2). In terms of the routine off-duty activities of the soldiers, however, it also encouraged social life and gatherings, enabled contact with the civilian population and allowed for free movement

Figure 3.2 RAF Alconbury Airmen's Club
Source: B. Greene

within the accommodation site. Nevertheless, while relaxed, social life was still structured by the institution and in that respect resembled more a campus than a civilian town (Green, pers. comm.).

During the Cold War the UK was not a frontier zone, but modern NATO bases on the front line, such as Camp Bastion in Afghanistan, sought to preserve the veener of normal life through the provision of social facilities in a similar manner. At its peak between 2010 and 2012 the 'military metropolis' of Camp Bastion (Syal 2014, 4) provided its resident garrison with cafes, sports bars, a mall, a Pizza Hut and a provisional football pitch made from pads for helicopter landing (Hopkins 2012, 7). The base was situated in desert terrain that was difficult to approach without being seen from many kilometers away. Even so, the camp was subject to overt and intensive surveillance facilitated by modern technology, consisting of ground sensors, infrared cameras, thermal cameras, motion detectors, guard dog patrols and 9m fences with razor wire. The objective of the surveillance was the safety of the garrison. The outstanding record for safety in comparison with other NATO camps had an impact on news correspondents and soldiers alike – the base reportedly felt like the 'safest place' (Rayment 2012, 12). While the levels and technologies of security were very different between the contemporary military base in Afghanistan and that of 1960s Alconbury, in both cases once inside the base awareness of surveillance regimes was absent from the soldiers' experience (Green and Panteli, pers. comm.). The theme of social control is familiar in the sociology of twentieth-century military bases (Anderton 2002, 192), including airbases (Lake 2002, 175). The spaces of Camp Bastion and Alconbury did not, however, manifest unequal power relations between the soldiers and the state in the way Roman military bases did. Instead they encouraged social behaviour reminiscent of freedom of association in civilian life as evidenced by the provision of facilities by the institution of the army.

Elginhaugh – a Flavian military base

The analysis now moves onto the plan of Elginhaugh, a first-century CE Roman cavalry auxiliary base in Scotland (Figure 3.3). One immediately notices how Alconbury seems much more dispersed in comparison with the more rigid plan of the Roman military base; a lack of large open spaces is also clearly evident. The dispersed layout of modern military airbases developed in response to the need to accommodate appropriate fuel facilities, safety and approach zones, as well as the danger of attack from the air (Vick 2015), but the lack of open spaces is culturally specific to the Roman military. Roman auxiliary bases tend to be very cramped. Excluding instances when the natural contour of the hill precluded the site from being larger, the bases indicate a decided agoraphobia (James, pers. comm. and James 2011, 172), usually not allowing for empty plots inside of the perimeter. For a cavalry base, whose main axis lay atop Dere Street – a major Roman military supply route leading into Scotland – Elginhaugh's plan is rather dense. While the widest two roads – the *via principalis* and *via praetoria* – provide sufficient room for two-way traffic, the plan does not provide sufficient space for easy circulation of carts or horses in all of its parts. It would have been extremely difficult to drive a cart around the perimeter of the base as in some places the street (*via sagularis*) narrowed to less than 1.5 metres (Hanson 2007, 173). Roman horses were smaller than modern ones, thus requiring only 1.2 metres in width (Hodgson and Bidwell 2004, 133). To this figure one can

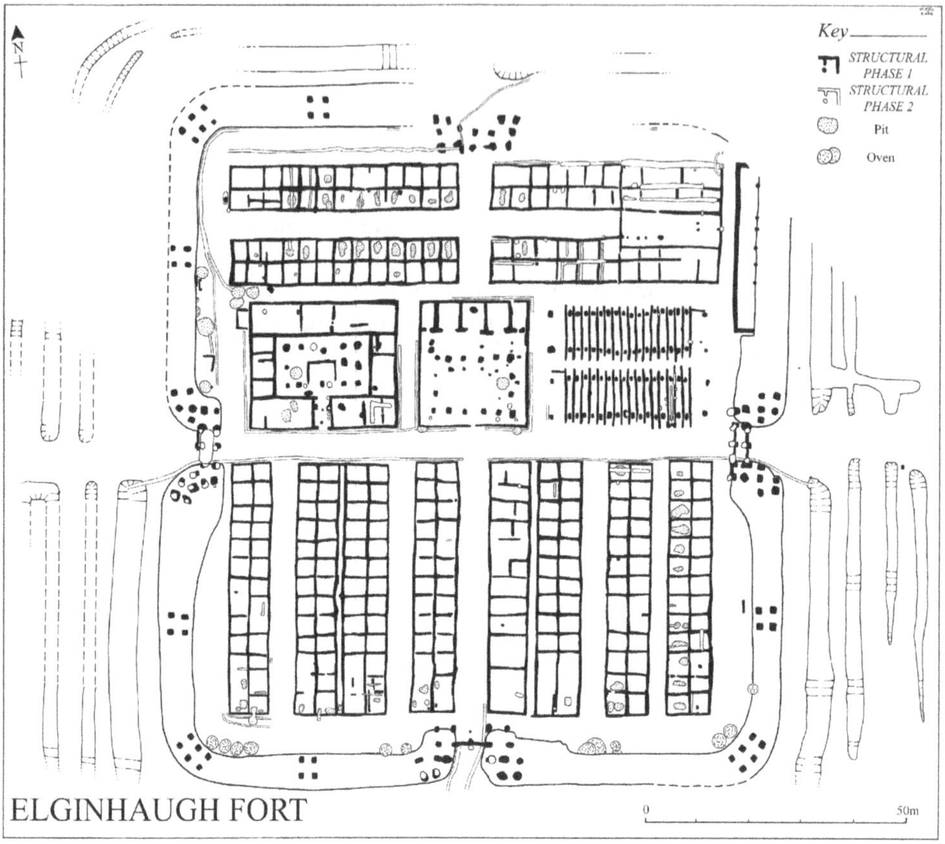

Figure 3.3 Flavian fort at Elginhaugh
Source: Hanson (2007, fig.12.3)

add another metre needed for a person to be able to lead a horse between buildings. The gap between the southern gate and the barrack in front of it was only two metres, limiting the possibility of two horses being drawn simultaneously. The *via quintana*, located behind the central range of rectangular buildings, due to the pits and sheds located there, was most likely a pedestrian zone. Across the plan, of what by Roman standards is a spacious base, one can easily identify a number of bottlenecks.

The other important difference in comparison with modern military bases is the paucity of facilities for social gatherings of soldiers within the perimeter. At Elginhaugh we find a bathhouse located outside of the main base on the slope of the hill. The practical benefit of this location was an easy supply of water to the building. The social functions of bathhouses are underlined by the finds of gaming pieces such as at Vindolanda, Northumberland (Birley 2010, 220). Auxiliary bathhouses are usually situated outside the perimeter and only sometimes in the base (for examples see Johnson 1983, 183). There is limited evidence for the existence of *tabernae* (bars) within the perimeter of Roman bases. One exception is found at a much larger legionary fortress at Vindonissa, where a letter addressed to a female innkeeper operating within the perimeter of the base was discovered (Speidel 1996, 180). This is, however, a rather unique and also a very early example. Instead, known archaeological examples of inns identified by large amounts of pottery and a high incidence of coins are usually located outside the perimeter, such as at Vindolanda, Building XXIX (Birley 2010, 139) and at Housesteads, Building VIII (Birley *et al.* 1933, 88). In comparison with a modern example at Alconbury far fewer social venues have been identified, especially inside the base.

Space, community and institution

In the Roman military there existed formal associations of soldiers, known as *collegia*. These were permitted among veterans for the purpose of a burial fund by the first century CE. But only since the reign of Septimus Severus (third century CE) would participation in the *collegia* begin to be opened up to soldiers on active service, with the associations now defending the professional interests of their members (Le Bohec 1994, 192). This is also the period when the associations began manifesting themselves in the spatial fabric of Roman military bases in the form of rooms specifically designed for their meetings – the *scholae*. The length of time it took for the Roman military to open up to *collegia* is has to do with their potential to threaten military authority (Haynes 2013, 322). The obedience and loyalty of the armies and of their officers was instrumental to the emperor's ability to maintain control over the empire (Lendon 1997: 237). The location of *scholae* closely overlapped with institutional spaces, consisting of small rooms with benches usually placed within the headquarters building. The location of the rooms facilitated control and supervision over the associations and prevented unsupervised, and therefore potentially subversive, networking (James 2011, 171). Not much could remain secret within the four walls situated at the administrative heart of the unit. Similar surveillance over those higher up in the hierarchy seems to feature in the spatial culture of the imperial court in Rome. Tacitus provides an image of a group of senators in fear of the possibility of being overseen or overheard and hiding with their ears stuck to the holes and cracks in the walls (Tacitus and Moore 1950, 115). Frederick (2003, 201) elucidated how some of Dominitian's building projects facilitated the invading of privacy of the elite of Rome through surveillance in order to stop threats to the emperor.

There were few spaces within the perimeter of auxiliary bases where the whole military unit could gather. As a consequence such gatherings were uncommon, except during official celebrations in the headquarters building under the religious auspices of the commander as priest. The calendar at Dura Europos provides a number of possible dates when this may have happened (Welles *et al.* 1959, 206–208). It is likely that on holidays, after the official celebrations in the *principia*'s courtyard took place (Fellman 1958, 83) the whole of the unit, or that part within the perimeter, would return to their usual duties (Gilliam 1954, 187). In other circumstances the whole unit would most likely have gathered outside of the walls of the base for manoeuvres in the open field (Vegetius and Milner 1996, 26) or potentially in an exercise ground (*campus*). There is evidence for the existence of such an exercise ground in the east of Dura-Europos in Syria (Speidel 1984, 168) and at Lambaesis in Northern Africa, two kilometres away from the base itself. Having accounted for the men detached to outpost duties, the 200m square space is estimated to have been sufficient to hold all of the legionaries remaining in the base (Speidel 2006, 3–4).

There are no proven examples of such structures in the auxiliary context in Britain (Waldock 1998, 101). Even recent work at Maryport provides no substantial evidence that a parade ground ever existed there (Haynes 2013: 207). In Britain several types of buildings have been identified as drill halls in the past. Among those for which the interpretation still holds is Birdoswald, which would have been sufficient in size only for small mounted groups to practice (Wilmott 1997, 98). *Basilicae* attached to the *principia* seem to have most likely been built for practical reasons (Schönberger 1969, 169), with the best understood example at Wallsend accepted as too small to be used for exercise (Hodgson 2003, 182). Some of the armour was kept in the general store (*armamentaria*) situated in the main headquarters building (*principia*) (Bishop and Coulsdon 2006: 263), where equipment belonging to different men would be stored – as ownership inscriptions indicate (MacMullen 1960, 23). The soldiers, however, would most likely have worn their weapons in camp, their signature being their military *tunica* and belt to carry the sword (James 1999: 19–21). Military bases were thus places where an armed community resided.

This evidence is consistent with James' argument that officers were concerned to prevent large, unauthorized assemblies of soldiers with the lack of large spaces for exercise inside the base being indicative of this tendency (James 2011, 171). In fact, James suggests that with the contradictory aim of keeping the soldiers ready for war, but also obediently waiting for the next mission, and with the progressive lengthening of periods in which the soldiers would occupy the base, the aim of the Roman base was increasingly to facilitate surveillance and control over soldiers, making them akin to 'wolf cages' (James 2011, 169). The soldiers' wartime service involved violence in battle with a large source of their motivation being a quest for glory and wealth (Sabin 2000). Soldiers could also be unruly, with many reported incidents of Roman soldiers' acts of extortion, unjustified acquisitions of goods and beatings performed on civilians (Pollard 2006: 224). It was no surprise that, dangerous at the command of the emperor, they could also turn against their superiors when unhappy with their situation or manipulated by officers (James 2011, 167). The acts mostly feared by officers were those of desertion, sedition and mutiny (Phang 2008). In the first century CE mutiny was more of a real concern than it is in modern armies in developed, Western countries. Troops in mutiny are often portrayed in classical sources as being in a frenzy or in a state of a temporary loss of sanity (Frontinus and Bennett 1950, 65–66). Psychologically speaking, anonymous crowds diffuse responsibility and the fear of being caught committing illicit acts (Brown 2000, 12), with anger known to spread more quickly through bigger crowds (Kelly 2001). Personal clashes between soldiers could result in killings; judging by the context of some fragments these could take place within the camp or its environs (Tacitus and Moore 1950, 67). After the revolts of 69 CE there were no more double legionary bases, containing up to 10,000 men. A similar concern for security may have been a reason why the Claudian auxiliary base at Valkenburg I, from the period when auxiliary non-citizen units were still a novelty, has not been proven to contain granaries (Groenman van Waateringe 1986, 156). In fact, Manning (1975, 115) observes that early auxiliary bases were usually not able to store one year's supply of grain, which became the standard only later. In case of a mutiny, the limited amount of supplies dispersed in small storage rooms at the end of barrack blocks meant the troops would soon run out of grain. In a civilian context the Romans were wary of potential problems caused by discontent of large crowds, too. In Jerusalem, during bigger Jewish religious holidays, extra troops were stationed along the colonnades of the temple watching for signs of popular discontent (Josephus and Williamson 1981, 305).

Theoretical interpretation

In a contemporary context an institution that makes similar efforts to limit unsupervised networking of its occupants is that of a prison (Steadman 2014, 292; Foucault 1995, 282–285). The plan of a Roman fort does in some respects resemble that of a penitentiary. There is a similar concern for spatial proximity of different components within the complex and a cellular layout expressive of a desire for control. There is also the absence of large open spaces and social facilities inside the base, and a dominating presence of the perimeter wall. The similarity seems to be explained by the more supervisory role of Roman military bases when compared to their modern counterparts. In general there is an underlying level of similarity in the way spatial complexes run by institutions, whether they are hospitals, schools, military bases or prisons, manage their occupants (Steadman 2014, 294; Foucault 1995). Foucault's *Discipline and Punish* (1995) offers the most penentrating theorisation of the functioning of an institution, but also portrays it at its most extreme, as a 'total' institution. The panoptical prison was a rather idealistic social experiment of an emerging industrial society (Semple 1993, 152) that was difficult to realise in architectural practice (Steadman 2014, 294). In studies of the Roman army the concept of a 'total institution' has been used to theorise a particular view of relations between the soldiers and local civilians as being characterised by social control (Shaw 1983, Pollard 1996). While inapplicable in its entirety, however, (because it evokes a rather simplistic image of the army as a fighting 'machine') certain elements of institutional control elucidated by Foucault can be identified in the practices of the Roman army.

These include a form of administrative control over the soldiers' whereabouts, both inside and outside the camp. This control did not take the form of drills in the training field (Phang 2008, 8), but instead was embedded in the army's bureaucracy. A document sketching conditions reminiscent of the panoptical arrangement described by Foucault (1995, 190–200) is the duty roster of Legio III Cyreneica (Fink 1971, 112–116). Its efficient tabular format allows for the individual treatment of each soldier designated for work on the particular day, who is assigned either a location or the name of a supervisor. The Vindolanda's *renuntio*, an inspection of men and their equipment taken each day in the morning (Bowman and Thomas 1994), necessitated the soldiers to be present and ready for inspection at a particular place and at a particular time each day. The administration also kept records of men's errands outside the base and of those on leave, for which often a formal permission had to be sought (Phang 2008, 209–213). The latter was an attempt to prevent desertion (Phang 2008, 204). Many duties taken up by *milites caligati* (ordinary soldiers) while in the base would have been relatively stationary either within the perimeter of the base or in the vicinity of the base. At Vindolanda the most common category was production work in the *fabrica* (Bowman and Thomas 1994, 99), at Bu Njem at the bathhouse or by the ovens (Marichal 1992, 92), in Cyreneica at the bathhouse or doing the upkeep of the base or manual labour in or near the base (Fink 1971, 113). To this can be added an awareness of being observed that was present in the disciplinary culture of the Roman military with mutual surveillance among the soldiers and discipline induced self-surveillance playing a part (James 2011, 171; Lendon 1997, 249).

There are, however, several important differences between the Roman military base and the prison in terms of the relationship between the subjects and the institution. While there was no place for formal buildings for social gatherings in the institutional framework of the auxiliary units, archaeology provides rich evidence for social activities taking place in the open and in an *ad hoc* manner. A good example is gaming, perhaps while on duty, as evidenced through gaming pieces at the ramparts of the fort at Vindolanda (Birley 2010, 222). Elsewhere cooking utensils of Legio XX show their shared ownership either within a unit of approximately 80 men (*a centuria*) (Frere and Tomlin 1992, 96; 1995, 15, 78) or a *contubernium* (the smallest eight men squad – *ibid.*, 77), indicating that food preparation (perhaps with the companionship of servants or families) would often take place within the subunits. This underlines that 'spatial culture' as the association of spaces and social behaviour may not always be readable through the plan alone. It also suggests that the institution of the Roman military was no different from modern institutions, such as factories (Peponis 1985, 386), in the way in which spatial proximity and institutional divides framed much of the social experience of the people working within the institution.

The second difference is that prisoners do not approve of the ideology of the prison, thus increasing the need for the institution to legitimize and assert its power over the imprisoned (Foucault 1995, 225). Since some were volunteers Roman soldiers were proud to be part of their unit. They actively internalised values proposed by the army. In some circumstances the soldiers identified more strongly with the institution of the army than the communities from which they were originally drawn (Shaw 1983, 143). Similar to how the functioning of a prison depends on a silent agreement between the imprisoned and the institution (Foucault 1995, 223), the soldiers of the Roman army participated in a game of social 'give and take', subordination and reciprocity with the state. The soldiers' superiors were maintaining a status quo through manual labour, fear, corporal punishment, reward and flattery (Phang 2008). The institution and the soldiers both needed each other. On the one hand the institution would become at times lax in its implementation of discipline. Marichal (1992, 88) quotes a letter found in Oxyrhynchus from Theonas to his mother in which Theonas tells his mother not to worry because in fact he is not really sick. Marichal sees the letter as a case of a soldier pretending to be sicker than he actually was to avoid work and the army's lax discipline allows him to get away with this deceit. If Marichal's interpretation of the letter is correct it presents an image of the hospital as a space of resistance to routines of the army. Parallel examples of individuals and groups creating subnarratives within the military system have been noted in the context of women in anti-aircraft batteries working in the Auxiliary Territorial Service. Through their everyday practices these women made efforts to break down social divides, according to which they were not seen as fully functional members of military society (Anderton 2002, 193-4).

On the other hand, the control of soldiers at some level would likely render itself invisible and be internalised as a feature of military life. Centurions' housing is an example of socially aimed design with a strategy of control at its heart. James notes that the houses were always placed at the far ends of the barrack blocks forming a circle around the perimeter of the base. James suggested that one of the reasons for this was very likely to prevent desertion (James 2011, 171). Justinian's code indeed stipulates specific punishment for soldiers attempting to desert the base over its ramparts or jump the fosse (Justinian 1998, 410). The centurions' houses were also strategically connected with the headquarters building. One would need to walk in between two facing barracks to get to the headquarters building with control over the barracks exerted at all times as a by-product of this movement. As the centurions could reasonably be expected to be seen in and around their houses, they could casually retain visual control over the area without their presence necessarily being overtly oppressive. Since their houses were dispersed, the centurions could not maintain their coherence as a group without controlling their subordinates in the barracks, which was organisationally their main function. There were no visibility lines between any two pairs of barracks. There was also an inherent element of surprise for the soldiers present inbetween two barrack blocks as to whom and when they could expect to see someone walk down the *Via Principalis*, which at its various points offered unobstructed views of the spaces between facing barracks.

Circitores (inspectors of sentries) provided a more formalised means of control of the behaviour of soldiers implemented through the movement of patrolmen around the base checking on the guardsmen stationed at various sentry points (Vegetius and Millner 1996, 79; Justinian 1998, 410; Josephus and Williamson 1981, 176). Spatial dispersion and mobility of supervisors in modern factories are shown to make the control more difficult to avoid for the employees (Peponis 1985, 368). The disciplining role of *circitores*' movement is visible in a situation portrayed by Sallust (2002, 88). A situation of chaos within the camp is depicted, which the new commander brings back to order. Much of the chaos has to do with people being in places where they should not be, or at an hour when they should not be there, and leaving posts when they are wanted. As a remedy for the lack of discipline, the new commander created more sentry posts and placed them at shorter distances, with the commander with his legates performing the role of the *circitores* themselves (Sallust 2002, 82). This increased the likelihood of being spotted and ultimately punished if inappropriate behaviour was detected. Implementation of control through *circitores* is confirmed in the context of *hiberna* (winter quarters) in papyrological evidence at Dura-Europos (Welles et al. 1959, 274, 281).

Movements of people performing their duties and the discipline-induced movement such as the *circitores*' rounds filled the base with predictable movement and assigned rhythm to places and tasks. The supposed effects of abandoning the practice as a means of punishment are illustrated in Tacitus:

> He (Valens) forbade the centurions to visit the sentinels, and discontinued the trumpet calls by which the troops are summoned to their usual military duties. Thereupon all stood paralysed, and gazed at each other in amazement, panic-stricken by the very fact that there was no one to direct them.
>
> (Tacitus and Moore 1950, 112)

This illustrates the institutional dependency of soldiers on the systems which were primarily implemented to control their behaviour. While it is possible that Tacitus exaggerated, the construction of the depicted situation is informative. Military psychology observes similar side effects of obedience (Eriksen 2010). Armies socialise soldiers to obey orders and do not foster individual agency or decision-making (Kelman and Hamilton 1989, 104–106). Soldiers function as part of a larger organism, outside of which the men in Tacitus' account were left not knowing what to do.

The role of sound in this passage is also important. Studies on the relationship of sound and the social organisation of space indicate that sound is both a product of cultural processes and a cultural process in its own right (Rodgers 2006, 49; Wollscheid 1999, 6). In prisons the cyclical, repetitive nature of the soundscape is expressive of domination (Paglen 2006, 56). In the case of the Roman army it is more likely that the soundscape of official signalling within the camp was both constitutive of, and constituted by, the daily routines of the base. The rhythm engrained military life in soldiers alongside a sense of security and familiarity as the passage illustrates.

Conclusions

In conclusion, an image has been presented of Roman military bases as a particular case of institutionalised space which posed context specific limitations to the generation of social behaviour. The comparison with a modern military counterpart made it possible to define the Roman military as a controlling and limiting institution, which did not play an active part in fostering the community spirit among its occupants in the way the army authorities did in the 1960s at Alconbury. Arising from a history of unrest and mutiny among troops in the first century CE the plan of the base facilitated control over gatherings of people to prevent networking that might have led to action against the military authorities or the emperor. The obsession with control may seem striking when compared with the idea of the Roman military as a community, that would suggest a more consensual image of the relationship between the institution and the soldiers. Military bases, however, were places where both the institution of the army and the individuals forming the military community functioned side-by-side. This social dynamic comprised a three-way process involving negotiation, coercion and sub-narratives created opportunistically by individuals within the institutional system. The element of control also explains why the community visible in a wide range of archaeological and textual material does not feature in the spatial layout of the military base. The effect of the strong power relationship exerted on military personnel seems likely to have led to the internalisation of the unequal power relationship and the development of an institutional dependency of the soldiers on the socio-spatial systems implemented to control them.

4 Approaching a lived experience of ancient domestic space

Hannah Platts

> Imagine what a variety of noises reverberates about my ears! I have lodgings right over a bathing establishment. So picture to yourself the assortment of sounds, which are strong enough to make me hate my very powers of hearing! When your strenuous gentleman, for example, is exercising himself by flourishing leaden weights; when he is working hard, or else pretends to be working hard, I can hear him grunt; and whenever he releases his imprisoned breath, I can hear him panting in wheezy and high-pitched tones ... Besides all those whose voices, if nothing else, are good, imagine the hair-plucker with his penetrating, shrill voice, – for purposes of advertisement, – continually giving it vent and never holding his tongue except when he is plucking the armpits and making his victim yell instead. Then the cakeseller with his varied cries, the sausageman, the confectioner, and all the vendors of food hawking their wares, each with his own distinctive intonation.[1]

Introduction

The analysis of spatial organisation encompassed under the umbrella term 'space syntax' devised by Hillier and Hanson (1984) has encouraged an increased emphasis on space and spatial configuration across the social sciences and humanities. Hillier and Hanson's study helped raise questions regarding the function of spatial patterns in society by focusing on the role of movement within and between spaces in buildings from different cultures. Through an examination of wide-ranging settlement types and individual buildings, Hillier and Hanson investigated how and why diverse societies utilise differing spatial forms.[2]

More recently this research into relationships between society, space and place has moved towards understanding the spatial organisation of housing (Hanson 2003). Thus through analysis of a house's room configuration, size and decoration, attempts have been made to understand residences as being more than merely a series of spaces grouped together which people inhabit. Moreover, just as archaeology has proved a fruitful discipline for spatial syntax studies, the developing employment of spatial syntax for interpreting the domestic realm has also been popular within this discipline.[3] This has posed questions about how dwellings from different cultures have been organised and what we can learn from these patterns with regard to how residents and visitors relate to each other in these spaces. Houses are thus imbued with the ability to 'express cultural and lifestyle preferences' (Hanson 2003, 1). This chapter argues that such investigations are dominated by the senses of sight and movement within the understanding of spatial patterns. As such they present us with a simplistic impression of the domestic realm that fails to reflect the complex physical experiences to be had within the home.

Archaeological interest in space syntax is considered in this chapter by focusing on the domestic realm of ancient Pompeii to explore alternative ways of interpreting elite domestic space.[4] Rather than emphasising the visual and kinaesthetic within the Roman home, I suggest also a consideration of non-sight reliant sensory responses. Thus I am concerned with how multiple sensory experiences of the Roman home described in literature, such as the colourful descriptions from Seneca above on the plethora of sounds he hears when at home 'which are strong enough to make me hate my very powers of hearing', might be interpreted from archaeological remains.[5] By studying human physical experience of ancient residences, this chapter suggests that perceiving the domestic realm through a lens of multi-sensory experience was key to how visitors encountered and were guided through the myriad of spaces, social identities and hierarchies within ancient Roman domestic space. Moreover, interrogation of the way in which owners attempted to control the complete physical experience of their residence both externally (i.e. from the street) and internally (within the residence's walls) helps develop a deeper understanding of the construction of elite Roman identities through sensory display and manipulation.

The primacy of sight and vision in the interpretation of the past

At the outset it is necessary to consider the dominant role that vision has played within our approaches to evidence from the past. The continued dominance of sight within the Western sensorium becomes clear when one considers attempts to promote the other senses through scholarly interpretations of human expression, experience and material culture. These, too, have struggled to dislodge the perceived pre-eminence of sight.

A champion of the need to re-engage scholars with studying bodily experience and the array of human senses through which the world is perceived was the French philosopher Merleau-Ponty (1962, 219–220). At the heart of his approach to interpreting the physical, visual, literary and dramatic experiences of human life in the world, was an emphasis placed upon the ability to 'read' various types of evidence. This essentially proposed a logocentric approach that considered words and verbal communication as the superior mode of expression, which in turn enabled uniform reading of all forms of evidence. Thus the study of music, dance, religious ritual or the visual and dramatic arts employed a relatively standardised approach similar to that used to study the written word (Howes 2005, 1).

The consequence of Merleau-Ponty's logocentric methodology is that whilst it aims to develop engagement and interpretation of the varied sensory aspects of human corporeal experience, prominence is inherently placed upon vision through the process of reading evidence. The other sensory responses – touch, sound, smell and taste – fade into insignificance. Merleau-Ponty, then, attempted to rehabilitate the multitude of human senses; but it is only from criticism of his methodology together with developments in exploring the wide-range of corporeal experience from scholars such as Serres (2008), that academic disciplines are now developing more detailed considerations of the *haptic, olfactory, gustatory* and *auditory*, together with the *ocular*. It is this recent move to consider all the senses, not merely that of sight, that is utilised in this chapter.

Vision and movement in spatial syntax and domestic space coding

The sensory revolution and its attempt to restore the whole human sensory realm to studies in the humanities remains incomplete, since domination of the senses by sight and movement continues. This is well demonstrated in attempts to reconstruct imagined journeys through the city of Rome, which have generally focused on changing views as individuals moved through the city's streets (cf. Betts 2011, 118–119). Attempts to take advantage of increasing interest in sensory approaches within archaeology have started to encourage new questions and frameworks for studying the tangible remains of ancient societies. However, Hillier and Hanson's justified access-maps, which render visually the accessibility of spaces, remain popular within archaeology as a means of representing space.[6] Through this method of mapping building location or room distribution, networks of movement and vision continue to be emphasised, since these access-maps are merely another way of representing a space and its subdivisions in a two-dimensional form, rather like a floor or archaeological plan.[7] Once again it is the sense of sight that dominates.

A significant development in the space syntax analysis of dwellings comes from Hanson, who suggests that the manner in which houses are variously organised presents a method of communicating social and cultural information about inhabitants to each other and to guests. She has termed the way in which houses are structured and decorated as being domestic 'space-codes', emphasising that, 'The manifest variety of ordinary people's lifestyles seemed to point away from behavioural universals and basic human needs and towards a view that, if space had a purpose, this was to encode and transmit cultural information' (Hanson 2003, 109). What is particularly interesting is the examination not just of access networks between rooms within a residence but also considerations of architecture and décor. For Hanson, it is through the social coding of domestic space, together with its architecture, décor and the relationship of the house's interior to its exterior that social relations experienced within the house could be identified.[8]

The theory that domestic space codes developed to articulate the spatial layout and hierarchies of a residence to inhabitants and visitors is not restricted to the study of residences from the modern historical period. This aspect of the interpretation of domestic space has been applied to ancient Roman housing. Andrew Wallace-Hadrill's investigations of Pompeii and Herculaneum have presented a detailed insight into Roman housing, spatial organisation and its role in the articulation of social identity. As part of his examinations of Pompeii and Herculaneum, Wallace-Hadrill (1988, 77) discussed the development of space codes and the role they played in the formulation of the Roman domestic realm. Once again architectural form, decoration and artefacts play an important role in the interpretation of these spatial codes.

Yet just as Hillier and Hanson's process of spatial mapping encouraged focus on vision and movement, investigations into decoding domestic space also rely on sight and movement-through-space and the role these play in reading different facets of the domestic realm and its layout. For Hanson and Wallace-Hadrill, approaches to interpreting the domestic realm continue to highlight the importance of vision and the permeability of spaces within dwellings. As Hanson (2003, 54) explains, 'In moving around in buildings, people orientate themselves by reference to what they can see and where they can go'. Similarly for Wallace-Hadrill emphasis is placed upon the ability of individuals to see into the heart of the house from the street as well as the plotting of visual axes throughout residences.[9] Floor plans, axial views through buildings, the size and shape of objects are emphasised in archaeological reports, but the effects of smells and sounds on individuals and their response to their surroundings are rarely considered, even when, as Seneca shows us with his references to the sounds of men in the baths exercising and wheezing and the cries of street vendors outside in the extract above, literary descriptions of buildings are fully immersed in the body's complete sensory experience.[10]

Once again sight dominates scholarly approaches to interpreting spaces from the past or present. Herein lies a methodological problem, for though the sense of vision undoubtedly plays a crucial role in how people move around and interpret their environs, space itself is not only experienced via sight, but engages various sensory responses. These in turn can affect the way in which individuals perceive and experience the spaces they inhabit, and can render as simplistic visually dominated approaches to understanding space and spatial organisation.

Visibility and permeability into (and out of) the Roman home

Critical to Hanson's (2003, 115–125) investigation into domestic space codes was the relationship between a house and the street on which it is located. Thus she examines the presence (or lack) of net curtains or other objects that conceal views into the dwelling from the pavement and the manner in which entry into a residence is (or is not) controlled. For Hanson (2003, 132) variations between these aspects from domestic examples emphasise fundamental differences that can be found within a community's social relations.

Interpenetration between urban and domestic realms in the Roman world has equally proved key to understanding social relations and the role of domestic space in elite identity. An important aspect of the relationship between urban and domestic spheres in the Roman world has revolved around the public nature of the Roman house and the role it played in the construction of social identity.[11] At a basic level, an emphasis on the visibility of both elite urban and country residences can be observed within ancient literature. Thus we hear about the desire of Rome's elite to dwell in visually prominent locations demonstrated by Cicero's description of his house on the Palatine being 'in sight of almost the whole city'.[12]

The visible nature of Roman elite housing was not limited to prominent location. Roman houses themselves had a crucial openness as is evidenced in Vitruvius' architectural treatise where he discusses the layout of houses.[13] An often-cited part of his text is the description of housing for different social classes, which emphasised not only the types of dwellings suitable for rich and poor but which also differentiates between public and private areas of the Roman home:

> we must go on to consider how, in private buildings, the rooms belonging to the family, and how those which are shared with visitors should be planned. For into the private rooms no one can come uninvited, such as the bedrooms, dining rooms, baths and other apartments which have similar purposes. The common rooms are those into which, though uninvited, persons of the people can come by right, such as vestibules, courtyards, peristyles and other apartments of similar uses.[14]

For Vitruvius the Roman home was divided into areas that were accessible to all, whether or not invited, and parts that were not. One of the crucial 'public spaces' outlined by Vitruvius was the *atrium*, which was generally located in the front half of the house and was the first, most accessible room within the homes of Rome's elite. Here an owner presented, and visitors had their first contact with, the public image he wished to parade.[15] In this space were located the wax portrait masks (*imagines*) of politically important members of the family's ancestors.[16] Here also occurred some of the most important rituals of Roman family life.[17]

A key part of the open and public nature of the Roman house is visible in the ritual of the morning *salutatio*, where poorer or lower status members of society visited the homes of wealthier or politically significant contemporaries in order to receive the *sportulae*.[18] In the late Republic and early Empire, the *clientes* of a wealthy patron would be met in the *atrium* (Figure 4.1, A), which acted effectively as a reception hall during this daily ritual of patronage. Open accessibility from the street of this front part of the Roman house was significant in its role in the construction and portrayal of social identity as Wallace-Hadrill (1988, 46) emphasises, 'Except when closed as a symbol of mourning, the doors of noble houses stood open to all'. According to Wallace-Hadrill, views into the heart of the Roman house were carefully framed from the threshold of the front door.

One vivid sign of this lack of privacy is the visual transparency of the Roman house. The visitor standing in the *fauces* of the standard Vesuvian house is immediately presented with a vista that leads through the heart of the house (Wallace-Hadrill 1988, 82). The potential for a visitor to view the centre of a Roman house whilst standing at the threshold is important and shows a significant level of interpenetration of the public world into the domestic sphere. The placement of a clear view through the *atrium* and into the *tablinum* (Figure 4.1, B) with the backdrop of a *peristyle* garden (Figure 4.1, C) behind was aimed at emphasising prestige, for it was in the *tablinum* that the important records of the house and family were located (Flower 1999, 204–205).

Vistas into the interior of a Roman house played an important role in manipulating the image of its owner but to consider the Roman house only in terms of transparent and unobstructed views into the core allows us to understand only a fraction of the role played by the domestic realm in the construction of personal identity. Further investigations into both the visibility into the house from the street, as well as the level and nature of permeability between the house and the immediate vicinity outside together with a consideration of other sensory experiences from the street into the residence are vital. Such enquiries will help develop deeper understanding of the complex methods by which owners sought to control multi-sensory experience of the Roman home as part of the construction and display of elite identity.

Whilst the ritual of the *salutatio* was a way in which the public life of Rome's elite permeated into their domestic spheres, the proximity of house and street also encouraged the interpenetration of private and domestic into the public and urban realm of the city. The discovery of stone benches (Figure 4.1, A2) on the pavement outside houses has long been connected with the *salutatio* as scholars have suggested that they were used by *clientes* waiting to see their patrons.[19] When we look at literary descriptions of the Roman house, however, we can also observe the way in which the private

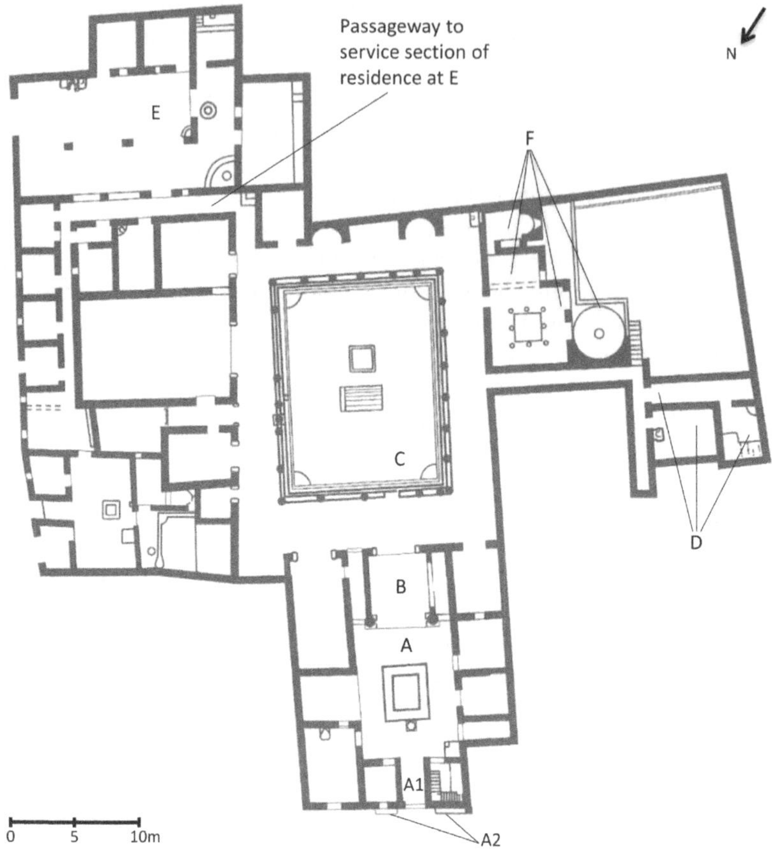

Figure 4.1 House of Menander, Pompeii

Source: Penelope M. Allison, *Pompeian Households: An Analysis of the Material Culture*, fig. A3, 2004, UCLA Cotsen Institute of Archaelogy Press. Permission agreed with copyright holder

world of Rome's elite was perceived to extend physically into the urban domain. In his investigation into the etymology of words, Aulus Gellius considers the meaning of the word *vestibulum*, which was connected to the entrance of the Roman *domus*.

> Gaius Aelius Gellus … says that the vestibule is not in the house itself, nor is it part of the house, but is an open place before the door of the house, through which there is approach and access to the house from the street … Those then in early times who made spacious houses left a vacant place before the entrance, midway between the door of the house and the street. There those who had come to pay their respects to the master of the house took their places before they were admitted, standing neither in the street nor within the house … the name vestibule was given to the great places left, as I have said, before the doors of houses, in which those who had come to call stood, before they were admitted to the house.[20]

What is significant here is the two-way relationship between the house and the pavement outside. The public realm of the urban sphere entered the house on a daily basis via the *salutatio*. Likewise, with the *vestibule* space immediately outside the house being where *clientes* gathered, the domestic realm appears to be spreading beyond the space of its built confines and out into the urban arena. The complexity of the relationship between urban and domestic, public and private can be observed through this mutual blurring of the boundaries. It was through the processes of public life filtering into private life and vice versa, and the urban realm seeping into the domestic sphere and vice versa, that Rome's elite sought to control the image they presented to others.

As well as considering in more detail the complex nature of physical interpenetration between urban and domestic space in the Roman world, we need to consider how far views into the heart of the Roman home were as transparent as have been previously suggested. It must be noted that whilst the doors at the entrance to the Roman house were left open regularly, thereby ensuring its accessibility, literary evidence shows there were times, beyond that of mourning, when the doors to a house were closed. In particular, house doors were closed at night and, during the late Republic, we hear of instances where the entrances to houses are closed at times of political unrest.[21] Wooden screens, which could be drawn to seal off the *tablinum* and the back section of the house from the *atrium*, further complicate our understanding of constant, uninterrupted and uncontrolled views through the entrance doors and into the heart of the Roman *domus*. Evidence for such partitions have been found in numerous dwellings from Pompeii and Herculaneum, including the House of the Wooden Partition and the Villa of Diomedes.[22]

Even when the doors of the house were open, and a view through the *atrium* into the *tablinum* and beyond could be had from the entrance, we must also consider the extent to which the owner contrived to maintain control over the image perceived by visitors or passersby. The structure of the standard Pompeian residence contained a passageway (the *fauces*; Figure 4.1, A1) which varied in length and which led from the house entrance into the *atrium* (Figure 4.1, A). The *fauces* presented an ideal way not only to frame the *tablinum* located on the central axis of the house, but to direct the gaze straight ahead and physically block views to the right and left to ensure that initial views into the house were focused upon the owner and the room where household records and evidence of achievements and public duties were stored. The right and left hand sides of the *atrium* would be hidden from the street entrance. Furthermore, no view into the interior would be possible except through the entrance since the only windows in the dwelling's street façade were small and well above eye level. There was no possibility of peering in through windows into other rooms of the residence, which thus ensured a high level of control from the owner of the image he permitted those beyond the household to experience.

Spatial coding through sensory immersion in the Roman home

The complex relationship between the Roman house and the street outside in terms of interpenetration between the two realms and the extent to which views in from the street were permitted, yet controlled, emphasises important ways in which social relations were organised and articulated in the Roman world. Further investigations within the Roman domestic realm suggest, however, that it was not just visibility and accessibility into the house that presented a means by which Rome's elite sought to construct and present their image to others.

As well as a means of controlling the vistas into the residence, the *fauces* of the Roman house served as a way of setting back the main body of the house from the busy street. Whilst the lack of large windows and the presence of doors (when shut) would act as a means of dampening some of the sounds from the street, it is important to understand that the further back from the road a residence was located the more seepage of external sounds into the dwelling would be reduced. When standing in a residence's *atrium*, it is likely that some noise from the traffic of carts and people would have been experienced. For every sound made, when the distance is doubled from the sound's source, there is a sound drop of 6 decibels (db). In the Roman period, few sounds above 75 db (a human shout at a distance of 1 metre) would have regularly been heard.[23] A sound of 75 db at 1 metre from the source (for example a human shout) will measure 69 db at a distance of 2 metres and at 4 metres, 63 db.[24] Consequently, the longer the *fauces* leading from the street into the *atrium*, the greater level of control the owner has as to the auditory experience of a visitor. Moreover, the deeper into the residence a visitor was admitted, the harder it would have been to hear the street outside, irrespective of whether there were sound-muffling obstacles, such as doors, walls or curtains, or other noise-emitting objects such as fountains between the shout in the street and the location of the visitor in the dwelling.[25]

Within the residence we can see further evidence of the articulation of domestic space not just via architecture and decoration but also through attempts at controlling a visitor's multi-sensory experience. The House of Menander at Pompeii presents an excellent example of the way in which not only the house's connection to the street could be manipulated by the owner but also the way in which the spatial structure of the dwelling could be organised to enable an owner's control of the visitor's sensory experience once inside (Figure 4.1). The distance from the entrance to the back wall of the *peristyle* garden was approximately 45 metres.[26] Thus, by the time a visitor reached the *peristyle* garden, the level of noise from the street would have been significantly reduced. Furthermore, when we consider the location of the kitchen and latrine quarters (Figure 4.1, D) and the service quarters and stables (Figure 4.1, E) within this residence, it becomes clear these are located at a significant distance from both the *atrium* of the house and the residence's *peristyle* garden. The entrance to the residence's kitchen and latrine section near the bath suite was situated at least 14 metres down a narrow 'L' shaped corridor. Likewise the stable and service quarters of the dwelling were situated in the back left-hand area of the dwelling, up a ramp and 8 metres along another narrow 'L' shaped corridor. Views of, and access to, both sections of the dwelling were restricted by the narrowness of the corridors as well as by the narrow entrance to the spaces themselves. Furthermore, movement from either of the narrow corridors into the open space of the *peristyle* garden would have produced an acoustic effect referred to as an *aural cut out*. This would have significantly diminished further any sounds from the kitchen or the service quarters being heard in the *peristyle* garden (Mongelluzzo 2013, 101).

It was not just the sights and sounds emanating from these sections of the residence that were carefully regulated by the location of these spaces down long, narrow corridors. The owner also aimed to manipulate the visitor's olfactory sensations by trying to diminish smells that emanated from these areas as well, employing the plantings in the gardens as a way of masking the remnants of any particularly strong odours.[27]

Through studying the organisation of the residence's spaces surrounding the *peristyle* garden together with the carefully mapped distance between the dwelling's service quarters and its reception area, which played so crucial a role in the construction of elite identity, we can observe how important a role multi-sensory responses within the home were perceived to be. Thus we can start to understand the complex way in which the spatial design of the Roman domestic realm could be used in order to articulate social identity.

Conclusion

Space-coding presents an important way of understanding how the domestic realm, in various eras, might communicate social and cultural information; and key to understanding the spatial codes of the home is the role of sight. However, within any experience of space is a crucial level of corporeal immersion and physical responses beyond that of vision. As Seneca's references to the noise of 'panting' and 'grunts', 'yells' and even splashing water highlight, the Romans were acutely aware of the role that other senses played in responses to the domestic realm.[28]

Current interpretations of the Roman home have concentrated upon the open nature of the front section of the house where the *atrium* was located. Emphasis has been placed upon the importance of views and accessibility into the heart of the dwelling and the way in which this reflected the public image of the owner. This chapter suggests, however, that the spatial layout of the Roman home was organised with a view to manipulating a visitor's multi-sensory experience: thus not only could vistas into and within the residence be carefully controlled, but likewise a guest's auditory and olfactory sensations could also be influenced whilst within the dwelling.

The relationship of the Roman house to the outside world was more complex than has generally been understood. Whilst the ritual of the *salutatio* brought the public realm into the domestic sphere, the residence's built structure, with the vestibule in front of the house, demonstrates that at times the domestic realm of the home physically spread beyond its built confines and into the public space outside. Interpenetration between the domestic and the urban realms could be mutual. Furthermore, that the spatial layout of the house lent itself to being visually transparent on occasion, enabling views into the heart of the dwelling, is key to understanding an owner's presentation of his image and identity to contemporaries. Once again, however, a rather more complex insight into visibility from the street and within the residence needs consideration in order to understand in more depth how owners sought to control their domain and the image it portrayed to others. The sightlines from the street into elite residences were not as uncontrolled as has previously been assumed: the *fauces* and the lack of windows on the front façade of the dwelling served to restrict, and thus focus, the views of passersby onto the *atrium*/ *tablinum*/ master accordingly.

Within a residence its spatial organisation served not just to affect a guest's visual responses but also aimed to manipulate other sensory responses, for, as we see from Seneca above, the Romans were well aware of the role that multiple sensory experiences had on an individual's reaction to the space in which they were located. By examining the spatial layout of the Roman home in terms not only of sightlines and décor, but in terms of the effect a building's structure and layout might have had upon the other physical senses, we can start to understand the highly complex role played by the domestic realm as a place for displaying an owner's social identity.

Notes

1 Seneca, *On quiet and Study*, LVI. 1–2.
2 On the analysis of settlements see Hillier and Hanson (1984, 82–142). For analysis of buildings, see Hillier and Hanson (1984, 143–197).
3 For archaeological studies into spatial organisation see, e.g. Hodder (1977); Clarke (1977). For archaeological focus on spatial organisation of the domestic realm, see, e.g. Kent (1990); Grahame (2000); Robertson et al. (2006).
4 'Elite domestic space' is interpreted here as residences of size or with evidence of opulent decoration including wall-painting and mosaics.
5 Seneca, *On quiet and Study*, LVI. 1–2.
6 See, e.g. Stöger (2011) on the analysis of the distribution of *scholae* using justified access-maps within ancient Ostia; Grahame (1993, 1997, 1998, 2000) on spatial analysis of Pompeian residences using justified access-maps.
7 Cf. Williams (2013, 208) citing Lefebvre (1991, 39). For criticisms regarding Hillier and Hanson's space syntax analysis see Lawrence (1990, 75). On criticisms of Grahame's space syntax analysis study of Pompeian residences see George (2002, 239).
8 The need to look beyond the architecture of a site and consider simultaneously its artefacts has been pointed out by Allison (1997, 81); cf. Ault and Nevett (1999, 51–52).
9 On views into the centre of a house from the street through its open entrance, see Wallace-Hadrill (1988, 46, 82–84; 1994).
10 Seneca, *On quiet and Study*, LVI. 1–2.
11 See, e.g. Wallace-Hadrill (1988, 1994); Platts (2011); Platts (2017b).
12 Cicero, *De Domo*, XXXVII. 100; cf. Vellerius. Paterculus II. 14. 3. On the importance of residences being located in visible areas, see Wallace-Hadrill (1988, 46); cf. Hales (2003, 41–46).
13 Vitruvius, *De Arch.* VI. I–VIII.
14 Vitruvius, *De Arch.* VI.V.1
15 The Roman house, and the *atrium* in particular, has recently been subject to attempts to escape the rigid interpretations of room use that have been imposed upon the Roman domestic realm. That the *atrium* was multifunctional has been demonstrated by Allison (1993); cf. Berry (1997).
16 On *imagines* in the *atrium* see, e.g. Polybius, 6.53.4; Cicero, *Phil.* 2.26; Valerius Maximus, 5.8.3; Seneca, *Ben.* 3.28.2. That *imagines* were a status symbol is emphasised by Cicero, *Ver.* 5 who notes that they were granted when individuals achieved the office of *aedile* on Rome's political career ladder, the *cursus honorum.*
17 For rituals centred around the *atrium* see, e.g. Polybius, 6.53–55; Suetonius, *Nero*, 6.2; Macrobius, *Sat.* 1.16.36; Propertius, 4.1.131–2; Perseus 5.30–1; Dionysius of Halicarnassus, 4.15.5; Juvenal, 10.334. Cf. Flower (1999, 91–95, 200–202).
18 A gift generally of money or food. In return *clientes* would offer wealthy patrons support, often of a political nature. On the *salutatio* see Goldbeck (2010).
19 Kroll (1933); Mau (1902); but see Hartnett (2008, 2011) on additional ways in which these benches can be interpreted.
20 Aulus Gellius, *Attic Nights*, XVI.6.1–12, cf. Macrobius, *Sat.* VI 14.-23. Whilst later Gellius gets confused as to the location of the *vestibulum*, his emphasis within the bulk of the text that the vestibule is located outside the entrance to the house is reflected in other authors, c.f. Varro, *Ling.* 7.81; Statius, *Silv.* 4.4.41–2; *Digest*, 10.3.19.1.
21 On the closure of house doors at night see, e.g. Juvenal, *Sat.* III. 127. Cf. Flower (1999, 188), who refers to doors being open in the day, but makes no specific reference to doors being closed at night. Wallace-Hadrill (1988, 46; 1994, 5) emphasises mourning as being the only time doors were shut. On the closure of house doors at times of civil unrest, see Asconius 43C.

22 On the ability to close off the *tablinum* and back section of the Villa of Diomedes and the effect this might have had on multi-sensory responses to the residence see Platts (2017a). We must also be aware that curtains instead or as well as wooden partitions might have been used to temporarily segregate areas of the house from views at the entrance/*atrium*.

23 Betts (2011, 125). The loudest sound experienced in the Roman period would have been thunder (120 decibels). Typical speech measures about 60 (db), whilst rainfall is around 50 db. Measurements from http://ele.aut.ac.ir/~wind/en/tour/env/db/dbdef.htm, accessed 1 November 2014.

24 Information on the inverse square law from http://www.nonoise.org/hearing/noisecon/noisecon.htm, accessed 1 November 2014.

25 See Platts (2017a) on the possible auditory experience of visitors to the Villa of Diomedes at Pompeii.

26 The *fauces* was approximately 4 metres long.

27 According to Hamilton and Whitehouse (2006a, 178) the smell of meat cooking in an enclosed outdoor area travels a minimum of 17 metres and a maximum distance of 122 metres. Taking the kitchen area of this residence as an example, whilst we cannot be sure what was cooked here, any smells produced in this area would be diminished by the architectural layout of this area of the house.

28 Seneca, *On quiet and Study*, LVI. 1–2.

5 Space for neighbourhood

Social identity and the built environment in medieval York

Gareth Dean

Introduction

For those studying the development of urban space the development of a 'spatial turn' by archaeologists and historians has primarily focused either on the classical world or the period after 1600 (for examples see Hicks and Beaudry 2006; Paliou *et al.* 2014). The development of approaches to medieval towns and the influence of scholars in a range of disciplines has been discussed in detail elsewhere (e.g. Classen 2009; Palliser 2000), highlighting the broad research agenda for towns by scholars in history, archaeology and historical geography. Topics commonly studied include the process of urban development and specific themes relating to politics, the economy and decline (e.g. Slater and Higgins 2000; Scholkmann 2012, 379). Within these studies neighbourhood and neighbour are used freely with little consideration of their complexity and there is therefore scope for archaeologists to develop further the concept of neighbour, neighbourhood and our understanding of the development and use of urban space. This chapter proposes that through careful consideration of approaches to the study of urban space, the built environment, and changing definitions of neighbourhood through interdisciplinary study it is possible to examine the characteristics of medieval neighbourhoods and the wider social networks of city residents.

Studying urban space

Neighbourhood is a spatial concept; it is defined by the interaction of groups of people within a given area and is therefore closely linked to the development and uses of spaces within any given environment. These interactions range from the private to the public and the consideration of the different spaces, ranging from streets to buildings, is necessary to provide the context for the study of neighbourhood. To this end, when exploring the medieval city, it is necessary to draw on a wide range of approaches and disciplines to understand the topographic layout and the built environment because with the medieval city there are no true maps, but the representations we have are drawn from documentary and cartographic sources which show urban space defined by walls with ordered streets, lined with uniform buildings. As Keith Lilley (2002, 18–21) has argued, our imaginings of medieval cities drawn from these depictions may not reflect the realities, nor the views of contemporaries who would no doubt have understood literary and visual representations for what they are: an idealised depiction of the city, perhaps illustrating how the city would like to be perceived and its connection to wider symbolic references. As in the modern city, the spaces of late medieval urbanity were far from equal to the idealised vision of the city, but were an active element in developing its character and form; spaces were socially produced and socially productive (Howell 2000, 17–19). We also have to consider Edward Soja's (2000, 8) argument that urban space included the mental or ideational field, a conceived or 'mental map' of city space carried in people's heads. Such local knowledge is an important factor in how we consider the use of urban space for neighbourhoods.

The study of urban space and its social uses can be traced to the social reform movements of the nineteenth century (for an overview see Pacione 2005, chapter 2). Subsequently, characterising and defining the social and physical elements of space has been mainly developed by scholars in social sciences. For example, Henri Lefebvre (1991, 31, 83) has shown that space is a product, created by social relationships determined by hierarchies of power which dictate how it is organised and given meaning. Hillier and Hanson (1984, 27) argue that people relate themselves through their relationship to one another but also through the patterning of space determined by the built environment. Rappoport (1990, 1993) argues that the built environment can be seen to be made up of a number of different elements to which people respond at a subconscious level, eliciting different modes of behaviour depending on an individual's understanding of those cues. He states that the built environment is constructed of a number of elements which help us to shape our responses to it. This ranges from fixed features such as walls, streets and buildings, which change slowly, to semi-fixed features comprising a wide range of elements including street signs, window displays, clothing, gestures and language and street furniture which could change more rapidly.

Drawing on these approaches we can begin to consider the complexity of the urban environment of the medieval city. An interdisciplinary approach is essential and such studies are commonplace for archaeologists working in the period after 1600. However, these studies rarely consider the built environment in greater detail although there are exceptions (e.g. Karskens 1999; Rimmer 2011). It is this work on the more recent past of our cities that has the potential to be applied to the study of the medieval city where the focus has tended to be on economic, political/religious and functional development (e.g. Palliser *et al.* 2000; Schofield and Vince 2003). Whilst there has been a growing interest in the role of space in a medieval urban setting, studies have tended to focus on individual buildings rather than city space (e.g. Graves 1989; Giles 2000). Indeed, John Schofield (2011, 113) argues that delineating neighbourhoods or groups in society by archaeological methods remains an object not fully realised or only in the broadest terms.

Arguably, it is historians who have developed broader approaches to understanding the urban landscape, and the use of the built environment through documentary evidence to consider the uses and perceptions of medieval city space. For example, Vanessa Harding (2002, 550) argues the varied street frontages of the medieval and early modern town were the result of individual histories of ownership and development. Anne Lester (2010) has used documentary sources to provide an insight into the 'mental' map of medieval urban residents through the order in which bequests of property are set out in wills. Equally important has been the work of urban geographers such as Keith Lilley (2009) exploring the influence of Christian cosmology on the medieval urban environment. In contrast, archaeology is still influenced by a focus on the descriptive discussion of the evidence and the development of patterns of streets and urban form which have been a focus for many studies influenced by the work of urban geographers (e.g. Barley 1975; Hall and Hunter-Mann 2002).

Historical approaches to urban space have shown the importance of signs and buildings as landmarks have been used by historians examining documentary sources to explore how they were used to map and shape the experience of the medieval resident of the city (e.g. see papers in Hanawalt and Kobialka 2000). The street therefore has to be seen as more than a means of linking urban spaces, but as a space which had multiple functions acting as a public forum, a place of trade and the venue for multiple social contacts. These footways marked the daily comings and goings of the resident, forming a line of communication for pedestrians, although travel on foot could be supplemented with the use of carts or pulled wagons (Mumford 1989, 308; Rothschild 2006, 126). However, the space of the street has received little attention despite the fact it is and was as meaningful as the monuments and buildings that line them (Celik *et al.* 1994, 1; Hartshorne 2004, 15–18). The buildings lining the streets were public, ecclesiastical, institutional and private spaces within which a mix of devotional, domestic, commercial and industrial activities were carried out, or indeed sometimes spilled out into the street. Camille (2000, 4, 11, 23) argues that the signs that adorned these buildings to advertise goods and services (a picture of a loaf of bread for a baker or a shoe for a cobbler) and other forms of visual decoration were for the inhabitants of the medieval city part of the texture and negotiation of everyday life.

Many of the buildings were home to the residents of the street, constituting its neighbours. However, the evidence for properties, drawn from archaeological and historical evidence, has traditionally been used to describe the form of the urban environment, or to focus on specific elements of social interaction of the groups represented (for example see Baker 2010; Burch and Treveil 2011). The activities within buildings were as important as the activities within the street in creating and shaping neighbourly relations and neighbourhood character. There has been an increase in studies of urban housing, but these have focused either on the surviving standing buildings, documentary sources or the below ground archaeology (e.g. Grenville 1996; Pearson 2004). As yet there have been few attempts to integrate the evidence from these different approaches, often being treated as separate sources of evidence with limited consideration of the social use of urban space or structures (e.g. Bowsher *et al.* 2007; Hall and Hunter-Mann 2002). Combining the evidence for the use of streets and buildings provides the means to consider the sounds and smells of the urban environment. This noise would have spilled out from buildings, mixing with other sounds typical of the general noise of everyday life (Woolgar 2006, 66, 68) and would have formed an integral part of the mental map of the city's inhabitants. Streets and buildings are therefore more than the representation of movement or urban development; they form the spaces that facilitate and have an active role in social engagement.

Research into medieval urban housing shows a complex network of spaces with ground floors used as a workshop/shop with living spaces on the first floor, and some buildings would even have been subdivided between different groups on different floors. At ground floor, buildings were a liminal space between the public access from the street for commercial functions and the domestic (private) (e.g. Pearson 2007, 47–50; Schofield 2011, 68). This would have an impact on perceived levels of neighbourly interaction through the extent individuals could permeate this liminal space. This is reflected in court cases; witnesses often report seeing what was going on in an adjacent house or comment on what has been heard through the walls (Capp 2003, 185). Research by Jane Rimmer (2007) and Jeremy Goldberg (2008) into probates and wills have shown how these provide valuable insights into the furnishings and multi-functional uses of rooms and how they were a reflection of status, conveying a message to residents and visitors. The historical sources in conjunction with excavated and surviving medieval buildings in towns has the potential to shed important light on construction, health, hygiene, daily life and a wide range of craft and industrial processes, which if integrated together provides the evidence for a consideration of social interaction amongst residents.

The social interactions of residents however are rarely confined to the buildings or the streets within which they reside. Despite this, cities are often broken into manageable areas when being studied, with the focus either constrained by documentary evidence, an excavation or formal boundary. Such boundaries will invariable impose a false limit on the area which constituted a social network or neighbourhood, and Craig Cessford (2009, 312) has argued that to understand the development and use of urban space it is necessary to move beyond arbitrary boundaries to allow an analysis that relates to entities that existed in the past, such as individual properties, blocks of land or the whole town. This reflects a tension within archaeology regarding scale of analyses from the location of a single artefact to a whole landscape (Lock and Molyneaux 2006). Studies of neighbourhoods need to move between these scales to examine the urban landscape in detail whilst adding wider ranging methods of study. While interdisciplinary studies are becoming commonplace for archaeologists working in the classical world or the period after 1600, they still rarely consider the built environment in great detail (see Hicks and Horning 2006) although there are some exceptions (e.g. Karskens 1999; Laurence 2010). These studies highlight the adoption of many approaches in identifying the space for neighbourhood, but they rarely define 'neighbourhood' which is equally important to the study of urban space.

Defining the neighbourhood

The study of social space and the built environment has therefore shown how they are intertwined, each influencing and shaping the other. A question that has to be asked is how do we define these communities/neighbourhoods that used and shaped the built environment within which they lived and worked? The definition of community and identity is a source of debate in a range of disciplines, and is arguably shaped by the traditional models proposed by Ferdinand Tonnies of community as *Gemeinschaft* or *Gesellschaft*; the former characterised by intimacy, kinship networks and stability, the latter by ego-focused, discontinuous relationships and social tension (Tonkiss 2005, 12). These reflect the perceived breakdowns in community in the late nineteenth century as a result of industrialisation. Other approaches developed in human geography from the late nineteenth century built on the work of sociologists such as Weber and Marx and explored the particular spatial configuration of a city and the social repercussions of its spaces on the inhabitants of large, dense, nucleated populations (Blake 2004, 237). Urban/social geographers have explored the spatial elements of neighbourhood (e.g. Pacione 2005, 32–34), as well as the work of historical geographers looking at cities in the past, such as Keith Lilley (2002) who draws attention to the relationship between the built environment and social structure. Anthropological and ethnographic approaches to communities and neighbourhoods have also stressed the importance of shared interests, a shared social system or network, and a shared locale (e.g. Agbe-Davies 2010). Ethnographic studies show that spatial and social boundaries are often porous and that spaces may be seen in very different ways by different factions within the community (cf. Agbe-Davies 2010, 379; Blake 2004, 243; Gosden 2004, 163–164).

Theories for conceptualising neighbourhoods and the interaction of people have been developed through the social sciences. For example, Henri Lefebvre (1991) explored the concept of neighbourhoods (or quarters) and has been highly influential. Soja (2000, 10) builds on these approaches to consider the real and imagined representations of space and place as not being mapped into specific and fixed geographies, but filled with movement, change, tension, conflict, politics and ideologies, passion and desires. Anthony Giddens (1984) emphasises that all human action is carried out by knowledgeable human agents who structure the world through their actions but whose actions are also constrained by that world. Soja (2000, 9) suggests structuration allows the linking of dynamic production and reproduction of city space to studies of social life such as family, community or government. The study of these smaller scales through *habitus* has gained remarkable interest within archaeology (cf. Bourdieu 1990; Hodder 2004, 34). *Habitus* has been widely adopted among medieval archaeologists exploring parish churches, public, institutional and domestic buildings, and the street (e.g. Giles 2000; Graves 1989; Hartshorne 2004; Johnson 2010). The importance of space and temporality has been explored through *The Annales School* which has had a significant impact through its conceptualisation of three temporal scales (short-term events; medium-term cycles; and the very long-term or *longue durée*) to examine historical developments, by understanding their geographic and temporal scale (Johnson 1999, 150–151). These frameworks have been highly influential in providing a context for understanding the characteristics, spatially and socially, of space.

Recent studies by sociologists, influenced by these approaches, of modern neighbourhoods highlight how people perceive neighbourhoods. Anna Agbe-Davies (2010) has shown that anthropological and ethnographic approaches to communities and neighbourhoods have stressed the importance of shared interests, a shared social system or network, and a shared locale. These themes have to be considered when exploring neighbourhoods in the past, and highlight the fact that neighbourhoods are not easily spatially defined but are bound by a sense of community spirit, trust, reciprocity, organised activities and through people who come into the area for specific services providing links with the wider area (Mumford and Power 2003, 31–32, 35, 38). Durrschmidt's (2000) examination of London explored the loss of the concept of neighbourhood in transient populations. This was contrasted with remembrance of more stable patterns of residency associated with people living in a street or area for generations; these communities were galvanised by almost exclusively local relations, perpetuated by shared conduct of daily practices such as washing and controlled by strict standards that were necessary for the use of shared facilities. It is also important to remember that with ideas of identity, each of the inhabitants would have a different experience of the urban environment (Borsay 1990, 17).

The approaches and definition developed in the social sciences have influenced approaches by historians and archaeologists studying a range of periods and locations from the ancient world (e.g. Laurence 2010, 39–40), the later medieval and early modern period (e.g. Wrightson 2007) and the eighteenth and nineteenth centuries (e.g. Garrioch 1986; Rothschild 2008, 18–20, 67–68). However, there is an assumption that neighbourhood and neighbour is an understood given, comprising a spatial area, social contact and community. This is not reflective of the multifaceted character of urban space and the organisation and interaction of the various human inhabitants as outlined above. When consideration is given to neighbourhood and community they can be seen as complex and ambiguous concepts defined through the management and regulation of membership, social and spatial boundaries and interaction with other groups (Nevola 2010, 351). Indeed, scholars are beginning to show medieval cities and the varieties of urban experience they fostered are more complex than previously thought (see Goodson *et al.* 2010; Halvorson and Spierling 2008, 3–10). The studies of neighbourhoods discussed here provide a framework for consideration of the characteristics that made up the neighbourhood in the medieval city, highlighting the binding characteristic of a neighbourhood is its residents' sense of identity manifested through a friendly atmosphere, a sense of trust and reciprocity that relates directly to where they live, alongside which there is the interplay and tensions between different social levels that help to form, shape and influence social identities.

Defining the medieval neighbourhood

To understand neighbourhood in the past, it is therefore necessary to consider how people in the past might have perceived their social world, not imposing our contemporary understanding of boundaries, and to recognise the historically particular uses and character of the built environment. Naomi Tadmoor (2010, 23) argues neighbourhood was a key concept in early modern England as most people lived their entire lives in small communities where interaction took place first and foremost among neighbours. To frame these communities, some scholars have examined the role of the parish as a focus of community (French 2001; Kumin 1999), although late medieval society was not contained, immobile, within fixed parochial blocks (Rosser 1988, 29–30), and however neighbourhoods were defined, they may or may not have had clearly defined geographical parameters (Wrightson 2007, 22–23). In Norwich many parishes were small and their residents could be expected to know one another. The form of the urban landscape provided other points of contact and interaction that were important in the social lives of the townspeople and people may have distinguished themselves through the 'micro-geographies' of individual streets and lanes, which cut across formal ward and parish boundaries (King 2006, 71–73). John Schofield (2011, 108–113) suggests that within these spaces identity is established and reinforced by differences in buildings, clothes or eating habits, with distinctions of ethnicity, although rich and poor lived in a mixture in every street.

Neighbour and neighbourhood are associated with Christian teaching and began to adopt their modern meanings through the translation of the bible into English in the fourteenth century. The original sense of the Anglo-Saxon word was one who dwells nearby; the first usage of neighbour as denoting fellow man dates from the 1300s (Tadmoor 2010, 26). The *Oxford English Dictionary* (2012) shows that from the late fourteenth to seventeenth centuries, the noun 'neighbourhood' was associated with individuals living in or near a particular place or in proximity as neighbours.

Keith Wrightson's (2007, 31) study of neighbourliness in the Elizabethan and Jacobean period shows it involved a combination of 'place, personal knowledge, active reciprocity, the avoidance of conflict and aspirations towards Christian charity'. The early uses reflected an ideal with a multi-dimensional meaning combining religious, utilitarian and sentimental elements. In contrast, modern understanding focuses on divisions and organisation of space within a town or city, reflecting changing attitudes of the eighteenth and nineteenth centuries. How we use these terms must therefore reflect the period that is being studied.

Neighbourhood arguably represented an ideal, but one which has an important role in urban space, and this chapter argues that a working definition for neighbourhoods can be based on three broad themes. First, spatial, with neighbourhoods forming within an area defined not only by streets or buildings but also by residents' perceptions of space and points of social interaction. Second, boundaries are not constrained or easily defined, despite scholars' desire to ascribe boundaries, either official or arbitrary. The ties to a given area were the principal form of support for residents, but ideas of Christian teaching on neighbourly values were an important tool for the residents to keep the peace through church courts, and ward courts as a forum for complaints and issues to be aired relating to the upkeep of the streets and buildings (Barron 2005; Hartshorne 2004). However, documentary sources also shed light on the neighbourhood as perceived by residents whose boundaries and spaces were important to them and whose 'mental map' connecting them to the wider city was able to transcend formal boundaries. Finally, neighbourhoods are multi-layered, made up of overlapping and intersecting spaces and activities: private versus public, daily routines, civic routines, visitors, differences of age, sex or profession. The consideration of evidence from York through an interdisciplinary approach has the potential to reveal this complexity of neighbourhood and to demonstrate the nuanced and subtle use of urban space in the medieval city, and the extent to which these can be considered neighbourhoods.

Space for neighbourhood in the medieval city

To begin to look for the spaces for neighbourhood in the medieval city, this section will draw on an example from York, to show how documentary and archaeological evidence enables the identification of potential neighbourhoods. In York, fourteenth- and fifteenth-century information for the management of the urban environment comes from the Wardmote Court books which show disputes between neighbours and within neighbourhoods which ranged from the built environment including the disposal of rubbish, the maintenance of ditches, streets and the maintenance of buildings, to concerns over social issues from arguments between neighbours to accusations of prostitution (Hartshorne 2004). These documents show it was the street, or very localised areas within a long street, that formed the neighbourhood. Further information for urban space in the late eleventh to sixteenth centuries can be traced in other documentary evidence concerned with the disposal of rubbish; the division of burgage plots into smaller tenements (the burgage was a originally a large land holding forming part of an urban estate); regulation of crafts and trades; the construction, rents and maintenance of buildings, tenure and the organisation of households (e.g. Swanson 1989; Rees Jones 2008). These sources, in conjunction with the archaeological evidence for the detailed evidence for activities within tenements, can shed insights into the multi-layered nature of urban neighbourhoods.

The archaeological and documentary evidence from York shows that the street-front buildings were used for workshops/shops (Grenville 2004; Rees Jones 1987[ii], xii–xiii). Indeed, documents from other cities indicate these buildings would have had signs on the street indicating services on offer (e.g. Camille 2000) and the archaeological evidence allows consideration of what the signs might have been and how they would have added to the noise, smell and character to the street. These workshop/shop houses lining the streets were to some extent liminal spaces and archaeology allows consideration in conjunction with the documentary evidence for the division between private and public space within buildings. The permeability of the space between streets and buildings is highlighted in defamation cases (see above).

The area behind the properties could vary in size and this influenced the activities that were carried out behind the buildings. Regardless of the size of the available plot, the area behind the street-front buildings was used for the disposal of waste and rubbish. In the tenements with larger plots areas behind the street-front buildings could also be used for industrial purposes. The documentary records for tenants in York show that alongside those working in properties lining the streets, others worked away from home such as builders and labourers. How were those not working on the street, or new to the area, integrated into the neighbourhood of the street? The answer is possibly through building social relations with the neighbours, literally those who dwelt nearby, as well as more formal associations through craft guilds or parishes.

The study of a group of adjoining tenements in Petergate, York provides a useful study for the use of space and concepts of neighbourliness. Excavations in tenements on the corner of Petergate and Hornpot Lane (see Figure 5.1) produced evidence for a range of activities (metal, bone, horn and leather working). A consideration of the activities represented by the waste materials indicates that the tenements were producing items associated with the cutlery trade, and this suggests the level of interaction between the residents within the tenements, the possible social composition of the household and wider networks associated with craft guild affiliations. Each of the tenements was making components that formed part of the finished knife – sheathes, handles and blades. Understanding the process carried out on each tenement allows consideration of the impact these crafts would have had on the sounds and smells of the street.

The evidence for different artisanal groups within adjoining tenements engaged in the production of items integral to a finished item raises questions relating to potential tensions and social interactions. This grouping of residents could transcend the craft guild affiliations and implies regular and frequent communication through visits to each other's workshops. Different perceptions of craft importance or social standing may have led one group to seek to dominate or control the others, and such tension between collaborative craft groupings has been shown by Derek Keene (1996, 98–99) amongst the residents around Cheapside identifiable through disputes over payment. These tensions might also be reflected in who held which tenement. In Petergate, the metal workers were on the street frontage whereas those making sheathes and bone handles were in properties along a side alley.

Understanding the craft activities carried out at a tenement can be used in conjunction with the historical sources to consider the social makeup of the household.

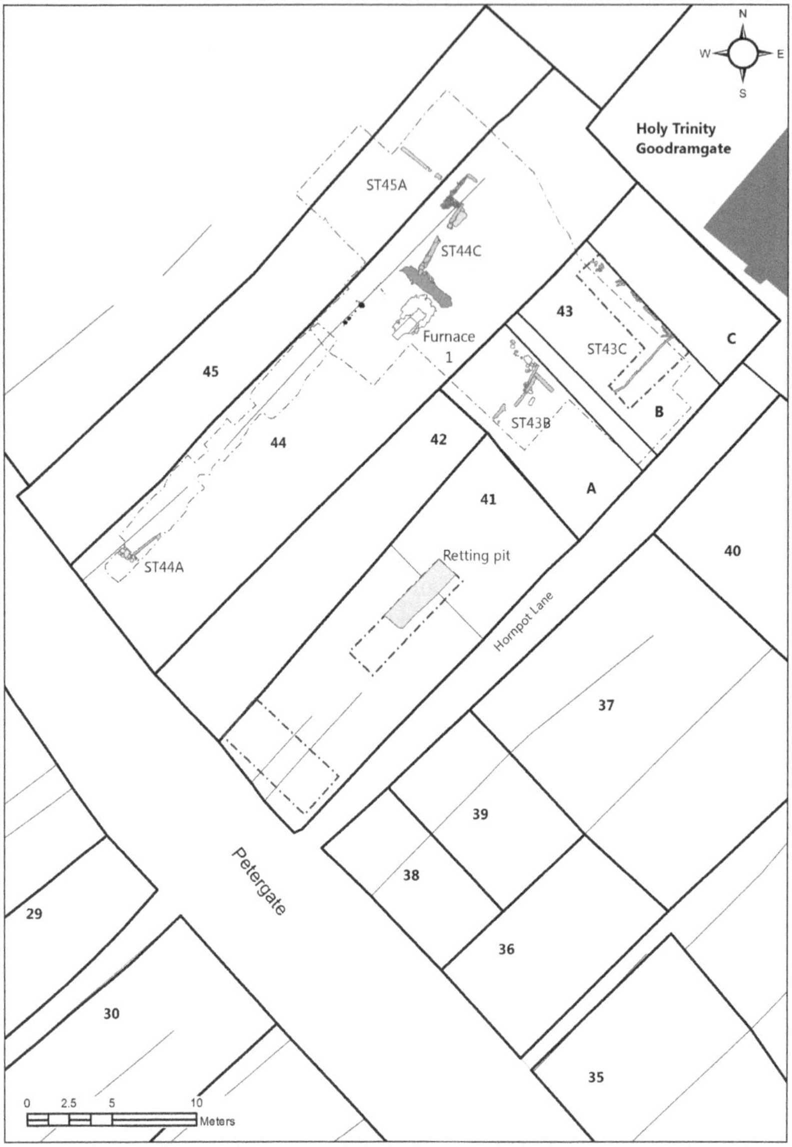

Figure 5.1 A medieval neighbourhood? Tenements (thick black line and numbered) identified from the historical records and the archaeological information (shaded within dash lines) that suggest the presence of a cutlery trade. Tenement 41 produced evidence for hornworking, Tenement 43 the working of leather and Tenement 44 for metal working

Source: author

Jeremy Goldberg (1996) has shown that York workshops employed the immediate members of the master's family (his wife and children), male and female servants (including apprentices, who lived with their employers as part of the household), as well as men and women who were employed within the workshop on a daily basis. Each of these residents/workers would have constituted a neighbour, but each would have had a different perception and understanding of the neighbourhood within which they lived and worked. The residents of the street were connected to the wider community of the city through parishes and craft affiliations and by the daily routines of purchasing raw materials or purchasing goods from the markets. Within the established community there would also have been transient visitors and uses of the neighbourhood space, from the poor/vagrants to those passing through either for work, shopping or social visits. The wider social affiliations through craft guilds, parishes and wards would have played a part in shaping and determining social relations, regulating urban space, and a means for advancement in the city with the parish and ward forming important forums for tensions to be aired and settled (Hartshorne 2004, 35; Phythian-Adams 1979, 115–116).

There are therefore no clear and easy solutions to identifying the spaces for neighbourhood within the medieval city, but it is important to consider the uses and character of the built environment with the historical evidence for crafts, households and Christian values in shaping the concept of neighbour and neighbourhood identity in the medieval city. Such a study requires a flexible and interdisciplinary approach, allowing scholars to move beyond the description of the form and uses of the built environment, to consider the quotidian character of the city which determined its appearance and character, the relationship between households and their activities and the links this provided within specific localities and the wider city.

Neighbours, neighbourhoods and the medieval city

The argument presented here is that interdisciplinary approaches to urban space provide insights into how it was shaped by social interactions, and in turn how social interactions gave character to urban spaces; a consideration of the built environment and neighbourhood are closely linked. Studies of neighbourhood also show that arbitrarily assigned study areas such as a group of streets or official boundaries do not necessarily represent neighbourhoods, either in the modern world or in the past. For the medieval city, it is equally important to understand the contemporary meaning of neighbour and neighbourhood; the Christian social rules which shaped and bound the residents through networks of overlapping but distinct social groups, transcending the ties of the parish or official groups or institutions such as wards. This is not to underestimate the importance of these institutions; they were part of the system for managing, controlling and diffusing tensions as well as the means for social advancement.

Studies of contemporary neighbourhoods, as well as those from the more recent past, show that a broad range of factors shaped a neighbourhood ranging from the built environment to the long-term residents and the more transient element of the population. Within these groups the neighbourhood will be perceived differently depending on age, gender, profession and social standing. The physical form of urban space gave meaning and a character to areas through repeated social activity, either by association with particular trades or particular forms of behaviour that built on the social cues that dictated what was appropriate. This was not a static process; rather, it was subject to constant reshaping by individuals and above all by the repeated practices of everyday life (Hartshorne 2004, 200). It also has to be remembered that neighbourhoods identified themselves not purely as groups with shared priorities; they were also identifiable by whom they excluded (Swanson 1999, 89–90, 128). This sense of belonging to a social group needed to be continuously negotiated because communities consist of overlapping groups defined by kinship, political institutions, economic or social status, gender, age or religion (Spierling and Halvorson 2008, 7).

To achieve a fuller understanding of the links between urban space and neighbourhood an interdisciplinary approach is needed. Scholars working across the humanities are developing approaches to the examination of the form, character and social interaction within cities across a broad range of temporal and spatial levels from the classical word to the present day, although they have not been widely adopted within studies of the medieval city. By moving beyond a descriptive discussion of archaeological excavations to consider the evidence for daily life, crafts and trades represented, it is possible to begin to draw upon documentary sources to consider the social composition of the household represented. In turn this allows examination of the evidence within the wider context of surrounding properties to begin to identify potential social groupings and their relationship with the built environment. It is proposed that such an approach can lead to a more nuanced examination of the urban environment, and in turn facilitate more detailed consideration of the character and form of medieval urban neighbourhoods. However, urban space and concepts of neighbourhood are fluid and never stop developing and this has to be recognised. It is equally important to minimise the imposition of modern concepts of community, or focused consideration of specific elements of the urban landscape such as individual buildings or administrative units such as ward or parish which can mask the complex networks of social interaction that defined communities. The challenge for researchers is to develop a flexible and interdisciplinary approach to find the space for investigating the idea of neighbourhood in the medieval city.

Part II

Spatial cultures in the long nineteenth century

6 Thirdspace?

Historians and the spatial turn, with a case study of political graffiti in late eighteenth- and early nineteenth-century England

Katrina Navickas

Introduction

Historians have been slow to consider space as a framework for understanding historical action and change. Their main frameworks have been based around time, and, more recently, text. Space is becoming a current buzzword, but there are potential problems with the way historians have embraced the 'spatial turn'. They risk the danger of treating space as just another form of representation in the same mode as culture and text. They also tend to neglect or blur the relationship between space, place and time. This chapter explores the methodological problems for historians taking the spatial turn, particularly in the limitations of the primary sources. It offers a case study of political graffiti and placards in late eighteenth- and early nineteenth-century English towns. How can the spatial turn enlighten historians' understanding of this form of popular protest? I suggest that other features of cultural geography offer useful ways of thinking about both space and place. There is more to space than its semiotics: its materiality, its relational nature, its role in shaping human emotions, and how humans embody space through performance and action.

Until the 1970s, historians usually regarded time as the most important force in history. Whig, Hegelian and Marxist approaches to understanding historical change were centred on teleological notions of progression and stages of development over time (Warf and Arias 2008, 3). Space and place formed merely the background to events. Historical geography was (and to some extent remains) a separate field. Historical geographies were empirical and quantitative in method, particularly keen on collating and analysing data from the census and other demographic records. For studies of industrial England in the nineteenth century, they examined the effects of industrialization upon tangible features of urban life and society, especially the structure of the family, the geography of work and home, and the physical separation of classes (Dennis 1984; Baker and Billinge 1982). Marxist historians were keen to chart the extent of residential segregation of the working from the middle classes and, in turn, make some broader conclusions about social relations and the potential for class conflict caused by industrialization (for example Foster 1977). But both Marxist and historical geographical approaches led to geographic determinism: an assumption that working-class inhabitants were powerless against the changes in their urban environment, or indeed were conditioned by it to proletarianize. Both considered *place* within a Cartesian or a socio-economic framework but paid little attention to *space*.[1] Space was presented as a neutral medium in which action and social relations operated: place was the 'bare stage on which the historical drama was enacted' (Gunn and Morris 2001, 2–4).

The advent of post-structuralism from the late 1980s onwards shifted historians' attention away from attributing structural causes to events. They increasingly focused on words and cultures as sources of agency and change. To generalize, first came the 'linguistic turn', with its emphasis on the multi-layered semiotics of text, and then followed the cultural and material turns, with their focus on representations in culture in its various forms. Words, culture and material objects, in this formulation, give subaltern groups agency against hegemonic elites. Space has followed as the next point of study and method. Since the 2000s historians have increasingly sought to understand how historical actors used and represented space as tools of power, agency and resistance. Historians of the eighteenth and nineteenth centuries in particular have been keen to spatialize culture and politics, with London in the 1790s being a particular focus. James Epstein, John Barrell and Christina Parolin have argued that the radical London Corresponding Society and its republican successors operated in political and gendered spaces that shaped both activists' behaviour and the attitudes of the authorities towards them. They have shown how the prosecution at the trial of the radical John Frost in 1793 sought to change the definition of meeting places in taverns and coffee houses from private to public spaces. In doing so the prosecutors thus sought to prove that Frost had committed treason by uttering seditious words 'in public'. They also illustrated how the court was in itself a non-neutral space, as the spatial hierarchy of the placing of the judge, prosecutors and defence shaped the words and actions of the theatre of the trial (Barrell 2006, 4; Epstein 2003, 109; Parolin 2010). Jon Stobart, Andrew Hann and Victoria Morgan have examined what they term 'spaces of consumption', representations of politeness and commercial exchange that produced and shaped the spaces of English towns in the eighteenth century (Stobart *et al.* 2007).

A central influence upon spatial approaches has been the tripartite model of space developed by Henri Lefebvre and refined by Edward Soja (Lefebvre 1991; Soja 1996). They posit that first or perceived space is material, topographical and Cartesian; second or conceived space is abstract, represented in plans or the imaginary. Third or lived space is a combination of perceived and conceived spaces, the space between the material and the representative, as practised in everyday life. Lived space allows the formation of 'counterspaces', spaces of resistance to the dominant constructions of the material and the representational by elites (Soja 1996, 68). Linked to this notion is Michel Foucault's concept of heterotopia, 'counter-sites, a kind of effectively enacted utopia in which the real sites ... are simultaneously represented, contested and inverted' (Foucault 1986, 24).

Stobart *et al.* apply Soja's model directly to their spaces of consumption. They develop Peter Borsay's influential notion of an 'urban renaissance' occurring from the late seventeenth century onwards, a physical expression of and facilitator for the 'polite and commercial people' of Enlightenment Europe (Borsay 1991; Langford 1998). The eighteenth century was an 'age of improvement', in which the centres of towns were redeveloped and redesigned by local elites, facilitated by private and parliamentary acts of improvement. Patrick Joyce applies a Foucauldian framework to the history of urban improvement. He argues that local elites instituted a 'political economy of detail', in which the new gridiron street plans were designed for the uninterrupted flow of people, traffic, and underneath, sewerage, with all obstacles and restrictions removed (Joyce 2003). This form of urban improvement reflected the spirit of the time, moving away from customary and medieval parochial forms of government (represented in the tangle of medieval streets and chaotic markets) to laissez-faire free trade liberalism (represented by the new public and civic spaces). Similarly, for Stobart *et al.* (2007, 92):

> central to many of the improvements made to the street was its reconceptualisation as a holistic architectural unit. The street was conceived less as a succession of plots and buildings, and more as a unitary space, drawn together by the design and layout of the thoroughfare and the buildings with which it was tied.

But whereas Borsay and Joyce's vision of the town was predominantly first and second space, in which the wealthy elites constructed and represented their vision of a 'polite and commercial people' in architecture, Stobart *et al.* show that consumers constructed their own thirdspace by negotiating these identities in the performance of consumption, negotiating and everyday life. They argue that the performances of shopping and the lived experience of such acts enabled eighteenth-century consumers to challenge elite conceptions and perceptions of the street. Consumers 'emerge as active agents who experienced their environment as lived space, at once social and practical, and attached significance to the knowledge-making process that came about through, for example, shopping' (Stobart *et al.* 2007, 194).

Thirdspace offers a way of thinking about spaces in which social movements protest. As social movement theorist William Sewell has argued, protesters often demonstrate in places with politically salient meanings and uses (Sewell 2001, 62). They also attempt to claim or subvert those meanings and uses.

Geographers of resistance argue that strategies of domination by elites and resistance by subaltern groups occur around spatial practices and the control over space. Elites usually have hegemony over space, but struggles can lead to new forms of space, identity and spatial practice (Murdoch 2006, 18). Subversion can be achieved by occupying a square at a time not determined or expected by its owners, by sitting in the wrong seats in a meeting, or by giving alternative meanings by using new names, symbols and rituals in places. Spaces of power, therefore, as sociologist Fran Tonkiss argues, are 'not merely locations in which politics take place, but frequently constitute objects of struggle in their own right'. Buildings, squares and streets provide sites for political action but are themselves politicized in contests over access, control and representation (Tonkiss 2005, 59). Protests can also create heterotopias. The idea of the world-turned-upside down, *charivari*, in riot or demonstration, is a major feature of the history of popular protest in the early modern period, notably explained by E.P. Thompson (1991) as when established rules are overturned, if temporarily, in a particular place and time as a means of enacting community justice against its transgressors.

The problem with many historians' adoption of all things spatial, however, is that for many it is just the natural (and easy) follow on from the previous turns rather than a focus of theoretical and empirical enquiry in itself. So the semiotics and representations of text, culture, material objects, and now space, are essentially one and the same. And if every space and action within that space is categorized as symbolic or representative, it runs the risk of rendering all spaces meaningless. Borsay's (1991) notion of the urban renaissance in itself is based on cultural representation. By focusing on the spaces of leisure, consumption and display in the eighteenth-century town, the urban renaissance defines space as static, monumental and architecturally symbolic. Cultural geographers debated these issues over twenty years ago. Denis Cosgrove's (1984) influential approach to landscape as a way of seeing was then criticized for its over-emphasis on cultural representation, which privileges the views of the elites as observers and accords the inhabitants no agency. The spatial turn is useful in understanding the contests over representations of power, but it says little else about the wider forces (and indeed structures) composing that power, particularly in relation to change over time.

Historians risk repeating the approach of Cosgrove and other geographers of thirty years ago. As Leif Jerram has warned, historians' take on the spatial turn has been merely to categorize space in the same way as culture and text, as a symbol and representation, rather than interrogating what space actually is. Part of the problem is the assumption that the spatial turn is the same as the linguistic or cultural turns, and that therefore space is just another form of representation. This misunderstanding is compounded by a reliance on Lefebvre and Soja's tripartite model (for example, Stobart *et al.* 2007, 22). Jerram argues that Lefebvre and Soja's tripartite model of space as a lived, represented and representative one actually ignores the materiality of space, the very 'things' that space is supposed to be 'living' or 'representing' (Jerram 2013, 404, 410). Space, according to Jerra, is matter, not representation. It shapes physical action by its materiality not its symbolism. When human ideas, codes and practices are associated with a location, it is called a place, not a space. This may seem like semiotic pedantry, but the elision of terms has implications for methodologies and understandings of the effects of space on human action. James Epstein, for example, described the Percy coffee house in London in which the London radical John Frost was alleged to have uttered seditious words in 1793: 'This was predominately, if not exclusively, a male space; it was also space marked off, at least ideally, not only from the controls of royal government but also from the street, the fair and the low tavern culture of the metropolis' (Epstein 2003, 109). This is not to criticize the rest of Epstein's work, which is highly sensitive to the nuances of human behaviour and cultural symbolism in nineteenth-century popular politics. But Jerram would no doubt take issue with this use of the word 'space'. He powerfully asserts that such terms as 'male space' or 'sacred space' are misleading because the materiality of the spaces in themselves do not possess human qualities such as gender or religious belief (Jerram 2013, 404). A historian examining the space of the Percy coffee house cannot tell from descriptions of its physical matter that it was confined to men or reflected and shaped male identities, or that it was 'marked off at least ideally' (Epstein 2003, 109) from other forms of behaviour of the court, the fair or the pub. They infer such information from other knowledge about the space as a place: the actions of the people who occupied it and how they represented it in other arenas (notably in the texts of the main sources upon which historians rely: newspapers, trial documents, correspondence). This is not to deny that buildings and streets are culturally constructed by people, who themselves are defined by underlying economic and political power structures. But it is place that has associations and representations, and it is the performances of its inhabitants that should be studied to understand those spaces and places. Or in the words of Nigel Thrift (2003, 105): 'Place refers to the process whereby spaces are ordered in ways that open up affective and other embodied potentials.'

It takes a leap of the imagination to wander round a modern streetscape trying to work out how historical actors may have behaved in the contemporary space. It also complicates the question of agency. Within the spatial turn, space is posited as a tool that empowers subaltern groups with new tactics and identities. Structural restrictions on oppositional groups' use of space often counteract this agency. The tension between the agency and power in space is difficult to resolve in historical research. Can space cause events? If subaltern groups challenge the hegemonic symbolism associated with a place, or devise their own symbolism, does that give them agency? Is space a form of power? The following case study illustrates the possibilities of this approach for historians, but also the continual tensions with the historical sources and limitations of the spatial turn.

Graffiti and placarding

Graffiti is a complex type of spatial tool of human expression, with a multitude of types, meanings and intents. Ironically it is much easier to study graffiti in the ancient world than it is for later periods. There are abundant studies of graffiti that examine ancient Greece and Rome because of the wealth of evidence. Scholars have recorded over 11,000 pieces of text on the walls of the two excavated towns of Pompeii and Herculaneum (Milnor 2014, 5). Historians of eighteenth- and nineteenth-century urban life, by contrast, have much less material to work with. Streetscapes and their buildings in Britain have been razed and altered continually since the early modern period. Moreover, different forms of graffiti have different timespans. Archaeological studies examine graffito etched into stone that is much more likely to survive than the chalked words or pasted paper common in the early modern period. Archaeologists also emphasize the materiality and performance as an integral part of the action of etching the stone, whereas studies of modern graffiti are more likely to focus on the symbolism and meaning of the text or pictures drawn.

The meaning and intent of graffiti also fluctuated over time. It is not a static symbol or act and occurred in a variety of forms. Modern definitions of graffiti consider it to be vandalism or a tool of resistance. It subverts hegemonic spaces built by elites; it disrupts because it is an individual act, performed by an anonymous individual. It is considered as offensive because it is 'understood as a triumph of one person's individual expression over another's', and a violation of the property which it defaces (Milnor 2014, 4). But this definition did not always apply to the ancient world. Graffiti could be legitimate in religious rituals in a temple or commercial dealings in a marketplace. Writing on interior walls was not 'unusual or undesirable in the ancient world' and not the preserve of the 'vulgar' sections of society (Baird and Taylor 2010, 3–4). Some elements of this alternative meaning of graffiti still existed in eighteenth-century Britain, including the chalking of notes on the interior of pubs for gambling and authorities fixing public notices on walls.

Eighteenth-century urban improvement sought to define the 'proper' place for writing: that is, on paper and on officially sanctioned inscriptions such as on memorials or carvings on buildings. This reflected attachment to private property that infused political ideology. Guides to towns lauded the 'polite' bourgeois terraces and villas, grand commercial exchanges and new straight streets constructed by improvement acts, and denigrated dirty, untidy and dark early modern and medieval streets. Palladian architecture idolized the clean and uncluttered style of the Roman Empire (Green 2003, 70), although its uptake and style varied across the country and ignored the fact as evidenced above that the ancient buildings it emulated must have at the time been covered in graffiti. As Lisa Forman Cody has commented about London, graffiti and placards 'contradict our image of what the eighteenth-century public sphere actually looked like' (Forman Cody 2003, 83). That is, this was not a polite rational debate within the confines of the coffee houses and the newspapers, but on the contrary, the politics of the street continued, and it was messy and dirty and visual as much as textual.

Chalking was common for example during the popular campaigns for the outspoken politician John Wilkes in the 1760s. His supporters used the number 45 to represent their movement in symbol and propaganda. Wilkes was originally arrested for publishing seditious libel in the forty-fifth issue of his newspaper, the *North Briton*, but the number also implied the movement's anti-Scottish sentiment by alluding to the failure of the Jacobite rebellion in 1745. The number was chalked and daubed on doors and carriages across London during the tumultuous Middlesex election campaign of 1768 (Crosby 1849, 233). During the anti-Catholic Gordon Riots that swept across London in June 1780, 'no popery' was the most common slogan chalked on the doors of suspected Catholic sympathizers. Furthermore, it was reported that men went from door to door demanding 'mob money'. Following the practice of beggars, they chalked various circles and lines on the doors as shorthand to indicate the extent of generosity they received from the occupants. This act was not merely a communicative sign to others to instruct whether it was profitable to knock on a particular door, but drew from a wider culture of vagrants and the poor, a strategy for survival when excluded from parish welfare (Hitchcock 2004, 77). E.P. Thompson has argued that chalking on walls and pavements, 'which had certainly been known in London from at least the time of Wilkes, became a far more serious means of propaganda during the French Wars' (Thompson 1975, 281–282). Arguably, therefore, the political potency of graffiti increased as the 'urban renaissance' and improvement emphasized the clean and blank facades of buildings as an ideal. Political graffiti became 'more serious' during the French Wars because it became associated with the radical ideas emanating from the French Revolution. The republican followers of freethinker Thomas Spence, for example, used chalk and charcoal to write such phrases as 'Spence's plan and full bellies' across London. It is difficult for historians to surmise the purpose of these chalkings: did the Spenceans believe that their slogan would convert the general populace to follow Spence's revolutionary plan of land redistribution? Or was this a mere act of subversion, anonymous protest against the owners of the property on which they scrawled? Certainly it was interpreted by its viewers in different ways. The writer and publisher William Cobbett later wrote to the radical orator Henry Hunt, 'We have all seen, for years past, written on the walls, in and near London, these words, "SPENCE'S PLAN," and I never knew what it meant' (Hunt 1820, 381).

Graffiti could be a form of spatial agency, allowing supporters of Wilkes, Gordon and Spence to express their voice in an arena in which they felt they were excluded. They produced thirdspace by subverting the representative spaces of 'improved' London. On the other hand, thirdspace cannot solely describe the meaning of such actions. In spite of its emphasis on performance, it also ignores the physicality of the act of writing or sticking a paper on a wall (amongst other acts of resistance). John Wylie has indicated how cultural geographers maintain 'persistent anxieties' over the issue of cultural representations versus materiality. He argues that a solution to this tension can be found in a conception of space as being produced by embodied practices (Wylie 2007, 11). Performances of actions like walking, or constructing buildings, connect the physicality of space with human conceptions of place in a tangible but also constantly changing way. Scholars of ancient graffiti underline its performative aspects, the 'material remains as an object which has been created by this event' (Baird and Taylor 2010, 6). The number 45 was not just chalked on the walls, but was a physical act on human bodies as well as on inanimate spaces. Suspected opponents of John Wilkes were attacked in the disturbances of 1768; a trial report claimed that John Percival attacked Richard Capel, 'took him by the collar and said damn you I'll mark you and accordingly he did mark him [with a] large figure number 45 on the cape of his great coat' (Shoemaker 2004, 120). In chalking 45 on a person's back as well as on a wall, therefore, the spaces were embodied.

Graffiti and placarding should be considered beyond their demonstrative spatial semiotics: as emotional and invoking an emotional response, as physical and material, as performance in space of an embodied geography. Placards and graffiti instigated emotional reactions among their viewers. Political groups used them to inspire to action or to invoke fear or anger about an issue. Thompson described graffiti and anonymous threatening letters as tools of the 'crime of anonymity' and common methods of protest in the eighteenth century (Thompson 1975, 281). Whereas the threatening letter was generally private, chalking or posting a handbill on the wall was public, a warning to all who saw it. But it could also be intensely personal, designed to induce fear in the person whom it targeted. As Carl Griffin (2010) has shown in his study of the maiming of trees and animals, such threats were designed to invoke 'disembodied pain', transferring the action of violence on the human body to another body, thus amplifying the threat. Threatening graffiti changed public space into an arena of potential or imagined danger. For example, 'Vengeance for the Blood of the Innocent' was chalked on many doors in the woollen manufacturing villages around Halifax in Yorkshire in 1812. The slogan appeared after the funeral of Samuel Hartley, a woollen cropper who was shot by soldiers during a massive Luddite machine breaking attack on the largest mill in the area (Thompson 1968, 642). 'Liberty or Death!' was perhaps the most common slogan chalked or painted on walls during times of disturbance, simple in its direct threat but general enough to be malleable to any cause. During political agitation in March 1817, the Manchester postmaster reported to the Home Office that 'The streets were yesterday much placarded and there was every reason to apprehend the Intention was Riot. P.S. Written about the Streets – "Be ready – be steady – Liberty or Death" '.[2] In July 1819, as tensions mounted with mass meetings calling for universal suffrage in Carlisle in Cumberland, a board painted with 'Liberty or Death' was found fixed to the market cross.[3] Again it is difficult to assess the direct impact of such signs. Graffiti and signs alone could not cause a revolution, and indeed most reformers in this period sought peaceful change. But the writers of such placards knew that by affixing such signs in public places, and like the market cross, at sites where the authorities placed official declarations and announcements, they assured that the authorities would take their threat seriously. Often the authorities found out about planned meetings via placards, and they often had to retaliate by posting up their own placards announcing that the proposed meeting was illegal. Posting placards formed a battle between authorities and their opponents over news in the public sphere. Like rolling news, in Liverpool during political agitation in 1820, 'The walls were, every hour, covered with placards respecting their [the radical societies'] various meetings and arrangements'.[4]

Thirdspace also shifts the historian's attention away from the materiality of graffiti and placards as well as its spatiality. Unlike the heterotopia of a demonstration, deliberately temporary, the unwashed wall or lingering scraps of a placard, as Cody suggests, 'contributed to a lasting political landscape in the city, long beyond the political moment itself' (Forman Cody 2003, 94). Surviving posters, often seized by the authorities as evidence and kept in their files or sent to the Home Office, are material evidence to the way in which they altered the physical and visual environments. Usually printed on the cheapest thin paper but in several colours, dirty with soot, plaster or brickdust, the papers sometimes survive with layers of previous posters underneath, a palimpsest of previous protests. For example, there is a poster in the collection from Huddersfield, Yorkshire, railing against the imposition of the new poor law of 1834, especially against its punitive regime and workhouses. It was torn down from the wall by the magistrates, who sent it to the Home Office.[5] The original has remnants of older posters stuck to it. This is where modern history meets archaeology, uncovering the layers of protests in the archives.

If material items could change the appearance of a space and the meanings associated with place, so could sounds. Focusing on the visual and symbolic elements of space, the tripartite model thereby neglects the soundscapes of the town. Performances of sound could be a form of aural graffiti. Brass bands were an integral customary part of all processions and parades, reflecting the local identity of the groups involved as well as drawing attention to their cause by drowning out the usual humdrum sounds of everyday life by the distinctive sound of the tunes played. Crowds also performed their politics aurally in space as they processed, hissing or booing outside the houses of their opponents and cheering and 'huzzaing' outside the premises of their sympathizers. In disturbances, by contrast, insults were designed to invoke disembodied fear or at least annoyance. Directed at individuals, often soldiers and the police, swearing and insults disturbed the 'polite' design of public space, and altered the soundscape. After the military put down striking weavers' disturbances in Barnsley in Yorkshire in 1829, for example, the magistrates reported:

> the town is quiet this morning, but the gibes, the threats, the sarcasms, which the Mob throw out now that the soldiers are departed fill the Inhabitants, especially the females, with great terror; and the Inhabitants are exceedingly anxious that we sh[oul]'d have a few soldiers quartered in the town.[6]

Aural graffiti subverted the ideal of a polite and quiet urban centre, creating a heterotopia, but it was also a bodily performance by those shouting or singing, embodying their space with emotions and engendering fear among those targeted.

Conclusions

Space can shape events and human actions, but in its material form and only also in combination with its role making up place, with all the latter's representations and associations. This leads to something of a chicken-and-egg situation in that we must also account for the fact that spaces are themselves constructed by human behaviour; but then, as Jerram argues, there is a 'historical tipping point' whereby the building or other space once constructed then begins to shape the behaviour of its users and occupants (Jerram 2013, 418). The problem therefore for historians, particularly of pre-twentieth-century periods where no oral histories or the original spaces survive, is that often all we have are textual and pictorial representations of space. The symbolic representations in maps, for example, have to be interpreted through text, in the form of their toponyms and legends. It raises questions not just for how historians should understand the material experience of space – the phenomenology of its dimensions, smell, tactile characteristics – but also for examining how space, for example the layout of a building, can shape human behaviour and relationships. There remain some difficulties for historians, particularly in identifying human actions that did not as consciously change the landscapes or who did not leave sufficient evidence of their relationship with space.

The most important lesson for historians is not to treat space as bounded or one-directional – which cultural representations and the tripartite model of space tend to do. Influential geographer Doreen Massey argues that space is relational, 'a product of practices, trajectories, interrelations' in everyday life (Massey 2005, 4; Mouffe 2013, 25). Spaces are sites of connection, where different groups meet or contest and ideas converge or disagree at particular points in time, and thus those spaces are constantly in flux. At times of political agitation or great socio-economic change, those spaces change more rapidly than at times of social stability (Murdoch 2006, 16; Thrift and Dewsbury 2000, 415). Spaces are constructed and produced by the effects of those interactions and connections, though they also serve to shape further actions and events as a result. The form of spaces and the associations of the places they create are shaped by human performances, embodying space in action. The relational model of space also offers great possibilities for understanding the relationship between space and time, a connection arguably even more important for historians than it is for geographers. The layers of placards on a wall provide a palimpsest of particular challenges at specific points in time, connecting different activists, movements and ideas in one space.

Notes

1 For more on Cartesian space, see Chapter 16 in this book.
2 R.H. Whitelock to Francis Freeling, 3 March 1817, HO 33/2, fo.13, Post Office correspondence. The National Archives, Kew, London (hereafter TNA).
3 *The Times*, 27 July 1819.
4 *Manchester Observer*, 25 November 1820, 'Public Notice' against an anonymous placard; *Liverpool Mercury*, 24 November 1820.
5 Handbill 'Men of Paddock', enclosed in Armitage to Home Office, 17 June 1837, HO 52/35, fo. 80, Home Office papers. TNA. TNA, HO 52/35/80.
6 J. Beckett to Spencer-Stanhope, n.d. 1829, SpSt 3/26, Spencer-Stanhope muniments. West Yorkshire Archive Service, Bradford.

7 From lines on maps to symbolic order in the city?

Translating processional routes as spatial practice in nineteenth-century Sheffield

Sam Griffiths

Introduction

This chapter considers the emergence in the English city of Sheffield of what the historian Gunn (2000b, 231) has called the 'processional culture' of the nineteenth-century city. It takes as its starting point the fact that processional routes have rarely been subject to systematic treatment despite their importance in accounts of processional topography (for example, Howe 2007; Vandeweghe 2011). Scholarly interest in processional activity as a mechanism for asserting symbolic order on the built environment (Gunn 2000b, 230) proposes the processional route as a legitimate object of 'topographical hermeneutics' (Lünen 2013, 118, after Sombart 1992). As such the task of decoding processional routes as mapped materialities of social practice does not appear secondary to that of deconstructing the symbolism of the procession itself. The effect of a too-exclusive emphasis on processional symbols, it is argued, is to close down critical consideration of the historical relationship between the emergence of symbolically privileged urban regimes and broader patterns of changes and continuity in the quotidian life of the city. Such an historical elision is especially unhelpful during a period in which rapid urbanization created new possibilities for spatial practice.

For the majority of historians principally concerned with the study of human subjects in the past, sources pertaining to the built environment are more likely to have illustrative or contextual value than to form a starting point for their investigations, the alleged 'spatial turn' in historical studies notwithstanding (Kingston 2010; Gunn 2001). When the built environment is directly addressed the critical effort is typically engaged in deconstructing the sources (for example, maps, photographs and art works) as representations of particular cultural discourses, as much as towards assessing what they can reveal about *what actually happened* in the past, an epistemological swerve that has been criticized by Ethington (2000). While many important questions of urban history are necessarily addressed in representational terms the difficulty of conceptualizing the agency of inhabited space in giving spatial form to social practices such as processions, increasingly constitutes a limitation on research into why the materiality of social life should matter at all. As Jerram (2013, 410–411) has argued, even Lefebvre's otherwise useful notion of 'spatial practice', that is the unreflexive bodily performance of everyday life, struggles to articulate a sense of material agency that extends beyond the holism of bodily experience to become socially meaningful (Lefebvre 1991, 38). The consequence in research terms is a tendency to collapse the materiality of spatial practice into the representational domain (Griffiths 2013).

Whyte (2006) has proposed that built environment agency can be revealed by a method of *translation* from the architectural form to manifold meanings that arise in different social and historical contexts. Yet exactly what this process involves is unclear since spatial forms do not, in themselves, connote meaning in any straightforward, referential, sense (Gieryn 2000; Hillier and Hanson 1984, 48–49). The danger is that such translation works only in reverse to render the built environment 'as a text'. In fact, as Tschumi (1996, 20) has argued it is the very *disjunction* of spatial forms with their intended use ('programme') and actual use (practice) that asserts architecture's particular claim to agency. The 'translation' of processional routes offered in this chapter, therefore, is not intended as a neat mapping of meaning onto movement but rather to draw attention to the material contingency of processional culture as it emerged in a single nineteenth-century industrial city.

Transposing and translating processional routes

From the late eighteenth century the routes of officially sanctioned public processions were usually published in advance in the local press. Once the procession had taken place newspapers would use the route as the organizing motif for their often exhaustive coverage of the event. Readers would be invited to relive the procession on a street-by-street basis, identifying those stages at which they had been present and catching up on anything they had missed. They could, for example, read the detailed reports of the illuminated displays set up in shop windows on patriotic procession days, such as the coronation of a monarch. Processions not sanctioned by civic elites but which were of sufficient scale to be newsworthy (for example some Chartist demonstrations) would also be granted coverage, the exact nature of the reporting depending on the political sympathies of the newspaper.

The extent of local press coverage means that for most of the nineteenth century it is possible to be fairly confident of the exact routes that processions took. Familiarity with contemporary cartographic sources and a comparative sample of processional routes is usually sufficient to make it a relatively straightforward matter to supply omissions of small or ambiguously named streets in the press reports. Figure 7.1 shows the processional route for the celebration of the end of the Crimean War that took place in Sheffield in May 1856, as it was announced in the *Sheffield and Rotherham Independent*. The researcher can transpose the list of streets into a sequence of lines on a cartographic base map dated as closely as possible to the event itself.[1] The result is a representation of the route inscribed on the map, as in Figure 7.2 showing the routes and orientations of major civic processions in Sheffield 1798–1856.

How might these 'lines on maps' be deployed in historical research to provide an account of processional activity? A common approach is to identify sites of symbolic significance (for example, the town hall or a principal religious building) that indicate a given processional narrative with a definable origin and destination. This conception of the 'route' is consistent with the definition given by the anthropologist Ingold (2007, 79) who contrasts *routes* with *trails*: while the former implies surface movement as a series of connections between discrete points, the latter implies a continuous flow of bodily movement through the material world. Yet if the line of a processional route is certainly a 'mapping' of symbolic authority onto urban space, it is equally a trace of bodily movement through a built environment that was not *terra*

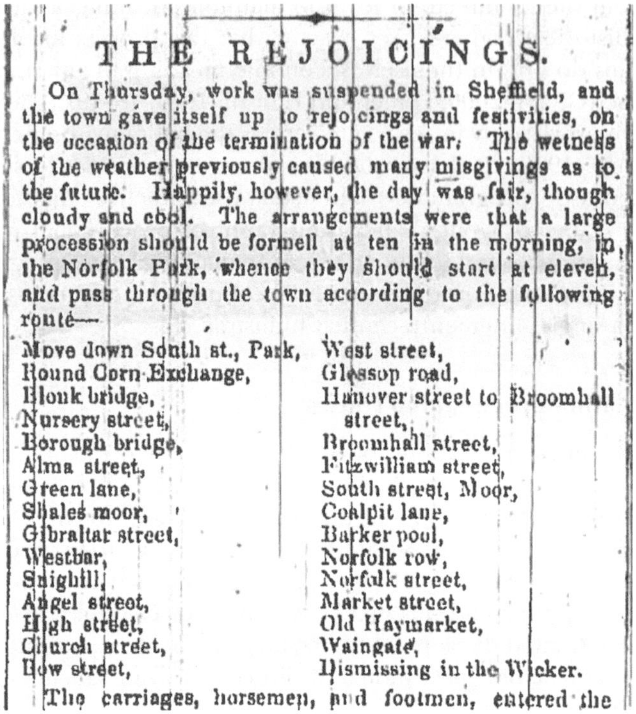

THE REJOICINGS.

On Thursday, work was suspended in Sheffield, and the town gave itself up to rejoicings and festivities, on the occasion of the termination of the war. The wetness of the weather previously caused many misgivings as to the future. Happily, however, the day was fair, though cloudy and cool. The arrangements were that a large procession should be formed at ten in the morning, in the Norfolk Park, whence they should start at eleven, and pass through the town according to the following route—

Move down South st., Park,
Round Corn Exchange,
Blonk bridge,
Nursery street,
Borough bridge,
Alma street,
Green lane,
Shales moor,
Gibraltar street,
Westbar,
Snigbill,
Angel street,
High street,
Church street,
Bow street,
West street,
Glossop road,
Hanover street to Broomhall street,
Broomhall street,
Fitzwilliam street,
South street, Moor,
Coalpit lane,
Barker pool,
Norfolk row,
Norfolk street,
Market street,
Old Haymarket,
Waingate,
Dismissing in the Wicker.

The carriages, horsemen, and footmen, entered the

Figure 7.1 Route of procession to celebrate the end of the Crimean War in Sheffield 1856
Source: Sheffield and Rotherham Independent, 31 May 1856

incognita but familiar on a day-to-day basis to most of the many people taking part in processions as participants or spectators.

Allowing that these two aspects of the processional performance, the mapped route and the embodied trace, are mutually constitutive in some sense highlights the epistemological danger of interpreting the 'meaning' of public processions in isolation from the materiality of the city itself. Whether viewed primarily as symbolically charged 'invented traditions' (Hobsbawm 1992, 1–15) or as ritual practices embodying a latent sense of social continuity, public processions in the nineteenth-century city served to ritually codify patterns and orientations of urban movement that were, in other contexts, already routine in terms of everyday street life. Connerton's argument that even highly programmed mass ritual activities can be regarded as 'incorporating practice', invoking the body rather than the text as the principal hermeneutical object, is significant in this light (Connerton 1989, 101–102). The anthropological emphasis on embodied agency, however, begs (perhaps somewhat paradoxically) the question of *extra-somatic agency* that extends beyond individual bodies in space and time to constitute the material domain of collective bodily action. Without such a notion it is hard to conceive how spatial practice can be realized as socially as performance. After all, the idea of per*form*ance invokes a concrete act of *making* in which the material conditions of spatial practice are implicated in realizing symbolic meaning. Indeed, the appropriation of what Hillier *et al.* (1976, 180) refer to as the

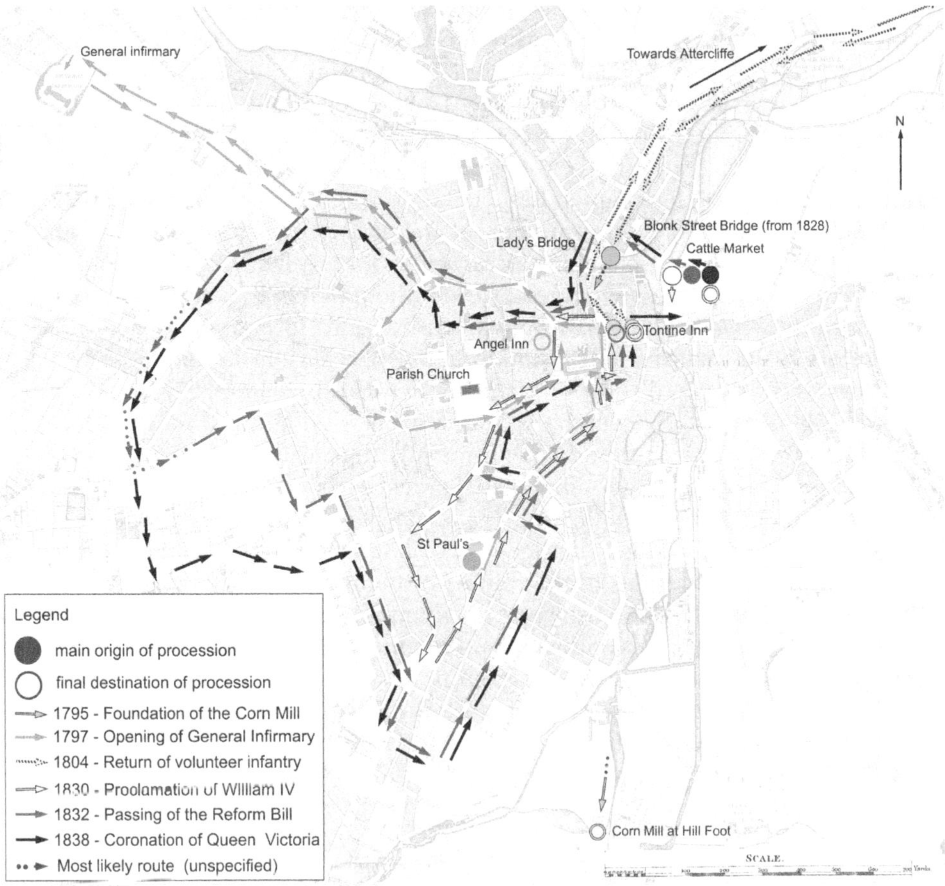

Figure 7.2 Routes of selected major public processions in Sheffield 1795–1855

Source: Sheffield Iris, 13 October 1797 and 19 April 1804; *Sheffield Independent*, and *Yorkshire & Derbyshire Advertiser*, 10 July 1830, 16 June 1832, 23 June 1832, 23 June 1838 and 30 June 1838

'profane', routine, social space of the street for ritualistic purposes, was a distinctive feature of nineteenth-century processional culture that enabled it to adapt, materially *and* symbolically, to the rapidly expanding built environments of industrial cities. An interpretative opportunity is lost, therefore, if the transposed processional route is deployed naïvely as a gazetteer of the procession. It is the absence of a sense of the material conditions of processional performance which means Harrison's interesting description of urban crowd activity as 'patterned' is not particularly well served by his account of processional routes as a linked sequence of symbolically significant sites (Harrison 1988, 166). The task of translating the symbolic topography of processional routes should be expanded to consider whether their symbolic articulation as *processional culture* can be explained in terms of incorporating practices with or against the grain of a city's quotidian *spatial culture*, with its emphasis on the everyday materialities of social practice.

That the processional route might also serve an explicit symbolic purpose is clear from the way in which urban groups with different social and political agendas sought to enhance their status through the temporary occupation of principal thoroughfares in the nineteenth-century city (Goheen 1993, 142). Identifying the symbolism of a given route in particular urban contexts (one immediately thinks of Orange order parades through predominantly Catholic areas of Belfast; see Bryan 2000), is an essential aspect of the topographical hermeneutic of processions but it is not the main focus of this chapter. Rather, the proposition is that processional routes represented the periodic appropriation of the agency of urban street network for the performance of the 'spatial infrastructure of the ritual' (Nejad 2013, 22). This 'spatial cultures' perspective implies a view of the nineteenth-century industrial city as a basically *intelligible* built environment with regard to everyday spatial practice. It therefore problematizes rather simplistic but pervasive representations of industrial cities as 'fragmented' (56) or 'amorphous' (Gunn 2000b, 227) places in need of 'ordering'. An exclusive emphasis on the procession as an ordering device, it is argued, works to negate, rather than engage with, the particular quality of built environment agency in catalyzing routine spatial practices that do not, in the first instance, possess representational significance. It effectively restricts the negotiation of symbolic forms to social groups (whether elite or popular) with power to assert 'order' on the city, and excludes the larger number of people whose role in the emergence of processional culture can hardly be accounted for in these terms. The question is then: how to arrive at translations of processional routes that do not categorically separate the materially practised city from the symbolically meaningful procession? The approach advanced here is to disclose the hermeneutical potential of formal descriptions of built environment structure in order to decode their appropriation for processional performance.

Space syntax as a spatial hermeneutic

This section reflects on conceptual and methodological insights from the space syntax perspective on architectural-spatial morphology developed by Hillier and Hanson (1984) and explains how these can contribute to the broader project of topographical hermeneutics raised by von Lünen. In space syntax research the formal spatial descriptions provided by analysis pertain not to pre-social material elements (which would be physical geography) but to material arrangements or *spatial configurations* that denote a pre-semantic (or 'non-discursive') (Hillier 1996, 38) relational domain through which quotidian patterns of movement and encounter realize a basic social intelligibility. Although, as will be explained, spatial configurations can take a highly symbolic form reflecting a strongly normative programme of social organization, Hillier *et al.* (1976, 180) hold that the agency of space is such that social life can never be entirely determined by socio-economic structures.

My argument is that a historicized interpretation of configurational descriptions can help to unlock the multiple dimensionality of the material world as an agent, though not a cause, of social meaning. This requires unlocking the latent temporality of space syntax theory, so far largely repressed in Hillier's work (Griffiths 2011). Space syntax analysis reveals how any given 'location' (imagine a dot on a map) is simultaneously implicated in any number of configurational descriptions extending in space and time. An historical perspective on this ambiguity reveals it as highlighting how descriptions of material arrangements are constantly being assembled and corroded, accumulated and dispersed, appropriated and displaced in accordance with changing social mores, in a non-linear temporal process. The agency of urban space lies in affording spatial form to performative practices in mediating material and representational semantics. It follows that while a given processional route reflects the agency of a particular social group to assert symbolic order, this assertion cannot be considered singular or absolute. At the least it is mediated by the degree to which the configurational possibilities of urban space affords symbolic emphasis to a given performance.

Axiality convexity

The fundamentally ambiguous qualities of any given location in real world space are set out in an important passage in Hillier and Hanson's *The Social Logic of Space*:

> Any point in the structure of space ... can be seen to be a part of a linearly extended space ... passing through [a] point, which represents the maximum global or *axial* extension of that point in a straight line. But the point ... is also part of a *fully convex fat space* ... that is, part of a space which represents the maximum extension of the point in the second dimension, given the first dimension.
> (Hillier and Hanson 1984, 91 [italics are the original authors'])

While 'axiality' and 'linearity' are fairly conventional terms, the notion of 'convex' space is peculiar to space syntax analysis. Formally, it can be defined as existing 'when straight lines can be drawn from any point in the space to any other point in the space without going outside the boundary of the space itself' (Hillier and Hanson 1984, 97–98). Importantly however, convexity, like linearity, is a 'scale free' spatial description in that it does not conform to an ideal (or mean) standard. Its definition can be extended to refer to any approximately circular area of space characterized by a high degree of inter-accessibility between all the different points, for example, where urban blocks are relatively small and densely packed together (Hillier 1999, 117). At whatever scale they are constituted, therefore, spatial descriptions that emphasize linearity are more likely to pertain to movement through an urban system and those that emphasize convexity are more likely to pertain to relatively static areas of social interaction within that system (Hillier and Hanson 1984, 17; Hillier 1996, 316). The fact that any given point in space has a spatial description pertaining to *both* its linearity and convexity and that these are also ambiguous with regard to the resolution of their description suggests how scales of spatial practice might acquire imbricated layers of hermeneutical significance when considered historically in particular social contexts (Griffiths and Quick 2005).

Spaces of production and reproduction

Hillier (1996, 215–232) distinguishes between two ideal types of descriptions of urban axes (linear extensions of space) – the symbolic and instrumental – which, he argues, relate to cities of 'production' and 'reproduction' respectively. In *cities of production* social life is said to be perpetuated through urban form functioning to generate probabilistic patterns of movement and encounters associated with the rapid exchange of goods and information. In cities of reproduction, by contrast, ease of movement is considered less important than symbolic emphasis for the purposes of ritual performance. Whereas an *instrumental axis* will prioritize inter- and intra-urban accessibility, *a symbolic axis* is likely to terminate at a right-angle to a building of symbolic significance constituted by a convex space (for example a cathedral on a square). If the category 'city of production' applies anywhere it must be to the 'shock' cities of the Industrial Revolution such as Sheffield. Certainly the first industrial cities did not invest resources in dedicated ceremonial sites. Even here, however, the evidence is that ritual forms of ceremonial performance developed at a rapid rate from the late eighteenth century. This raises the interesting historical questions of the extent to which instrumental and symbolic axiality co-existed within a single urban form, how far the balance between the two shifted over time, and how processional culture was enacted within largely 'functional' urban space. There is therefore ambiguity in 'instrumental' and 'symbolic' spatial descriptions. The relationship of spatial culture and processional culture can be conceived in these terms.

Distance concepts

Another source of ambiguity in spatial description derives from Hillier's theory of 'distance concepts'. Hillier argues that distance possesses three distinctive modalities: topological (directional change), angular resistance (relative straightness of path) and metric (units of geographical distance; cf. Hillier and Iida 2005). Hillier and Iida contend that 'our notions of distance are compromised by the visual, geometrical and topological properties of networks' (2005, 476). Their research, based on observational surveys of traffic and pedestrian flows in London, suggests that both topological and angular descriptions correlate strongly with vehicular and pedestrian movement whereas metric descriptions do not. On this basis they conclude that metric distance is predictive of movement flows only at the most localized scale but that when considering space 'above a certain threshold' people conceptualize distance through a mixture of topological and angular intuitions. Hillier (2005, 19) characterizes this as a 'phenomenology' of distance. The implication of this conceptualization (couched in largely cognitive terms) is far-reaching. By demonstrating how distance concepts differentiate probable movement patterns at three modalities of scale, Hillier hints at how these variables tap into a multiplicity of configurational arrangements that are materialized in the spatio-temporal world as descriptions of possible spatial practice.

Morphic languages as performance

It is its essentially ambiguous quality that realizes spatial description in the syntactic sense as a legitimate hermeneutic object. The intrinsic ambiguity of space means social programmes (i.e. desired normative outcomes achieved by assigning social 'meaning' to space) instigated by hegemonic groups and their configurational description can never be entirely conflated. Hillier and Hanson (1984, 48–49) have convincingly argued why there is no straightforward sense in which buildings or cities *can* mean anything. They differentiate between space syntax as a 'morphic' language and the 'natural language' of the spoken word to support their claim that the characteristic feature of natural language, that it refers to things beyond itself (the referent), does not apply to morphic languages which refer only to themselves in denoting the material organization of space. Yet, as Netto (2015) has argued, these two languages, though epistemologically distinct, can never be entirely separated from each other as social practice. It follows that the translation of processional routes implies a process in which morphic and natural languages, actions and representations, are brought into relation with one another through processional performance.

From routine journeys to processional routes: changing patterns of home and work in the Rockingham locale, Sheffield

An experimental phase of space syntax analysis was used to explore whether Hillier and Iida's distance concepts could provide useful spatial descriptions to inform historical research into emergent modes of spatial practice during Sheffield's eighteenth- and nineteenth-century urbanization (cf. Griffiths 2008, 2009). The idea was to constitute the spatial configuration of the city's road network as an hermeneutical object, emergent over time and practised at different 'modalities of scale' – equivalent to the three contrasting descriptions of distance in Hillier and Iida (2005). Specific spatial descriptions recoverable through formal analysis were deployed to decode persistence and change in Sheffield's processional culture by illuminating the broader spatial culture in which such developments took place.

Figure 7.3(a–c) presents space syntax 'segment-tulip analysis' using *Depthmap* software (Tuner 2000–10).[2] Tulip analysis calculates angles of incidence between road segments using a simplified algorithm that does not compute exact angles but places similarly sized angles into 1,024 'bins'. This analysis was used to calculate the space syntax measure of 'Choice'. Choice is a relational measure that calculates how often a given segment features on a path between all pairs of segments within a specified network radius.[3] This radius measure can be differently weighted according to the three distance concepts: angular (a), topological (b) and metric (c), representing angular deviation, number of turns and metric units respectively. The descriptions of network structure provided by Choice are useful to interrogate the different modes of network distance at which a street system emerged historically since it can identify, on an heuristic basis, the basic scaling properties exploited by subsequent urban expansion and developments in transport. On each of the maps in Figure 7.3 road segments are shaded on a gradient of white to black with the highest value segments in white visualizing the characteristic network structure of each distance concept.

In Figure 7.3a showing angular choice radius-*n*, it is the highly linear centre-edge network structure, representing the oldest roads through Sheffield Parish, as well as more recent large-scale interventions in the urbanized area, that are most evident. In Figure 7.3b topological choice radius-30 (b) indicates how multiple local grid intensifications constituted a web of lateral connecting roads that gave network structure to the emerging suburban neighbourhoods on the periphery of the mid-nineteenth-century town. Such lateral structures effectively connected local neighbourhoods with intra-urban accessibility structures, relating what Hillier (2009) calls the 'background' (i.e. residential) and 'foreground' (i.e. thoroughfares) structures of space. In Figure 7.3c metric choice radius-5 presents a large area of intensive griddy development that covers the late nineteenth-century suburbanization of Sheffield Parish. There is little consistent large-scale structure here and it is best thought of as representing multiple intensifications of local development or 'noise',

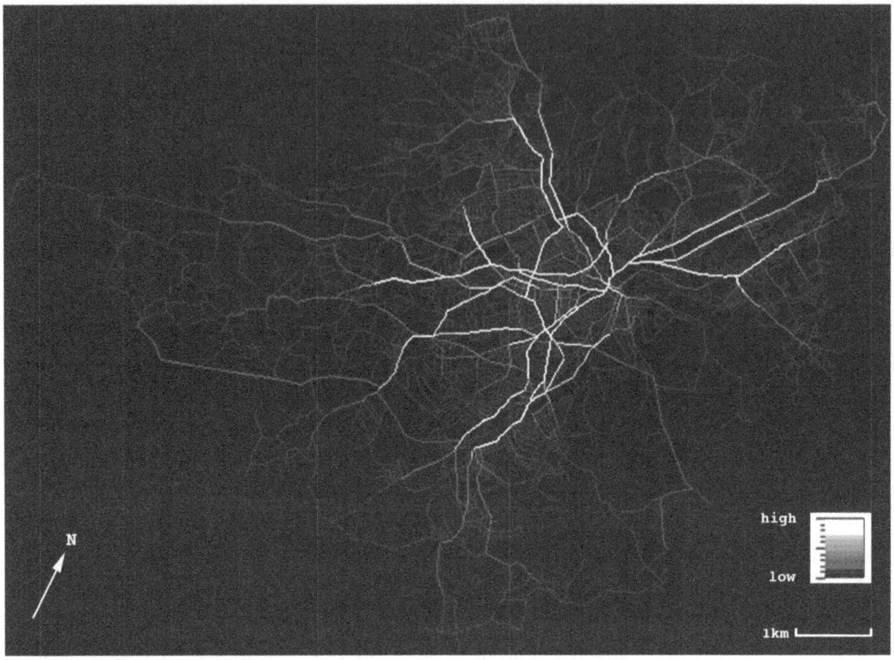

Figure 7.3a Distance concept (a) angular weighting, radius-*n*
Source: author

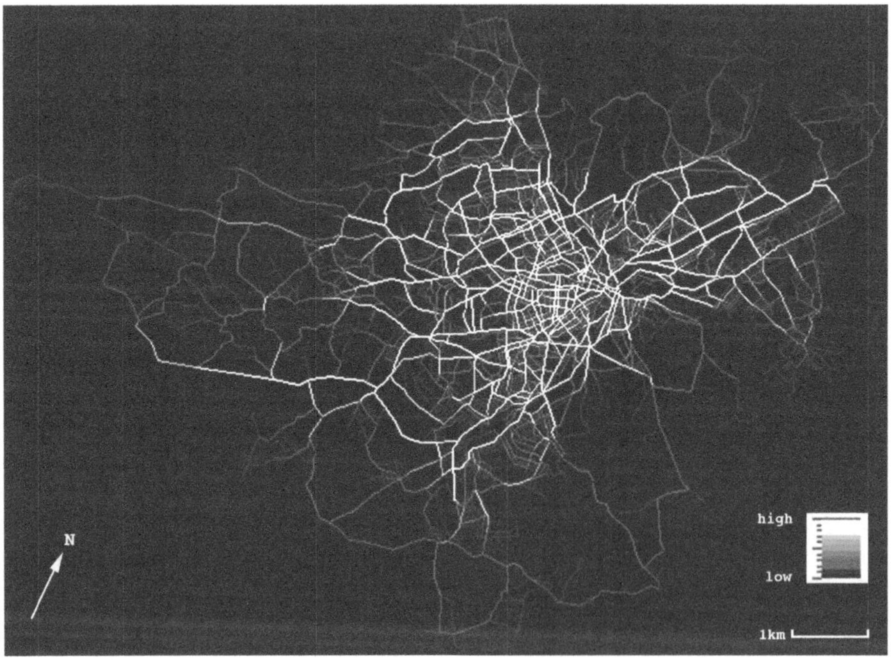

Figure 7.3b Distance concept (b) topological weighting, radius-30
Source: author

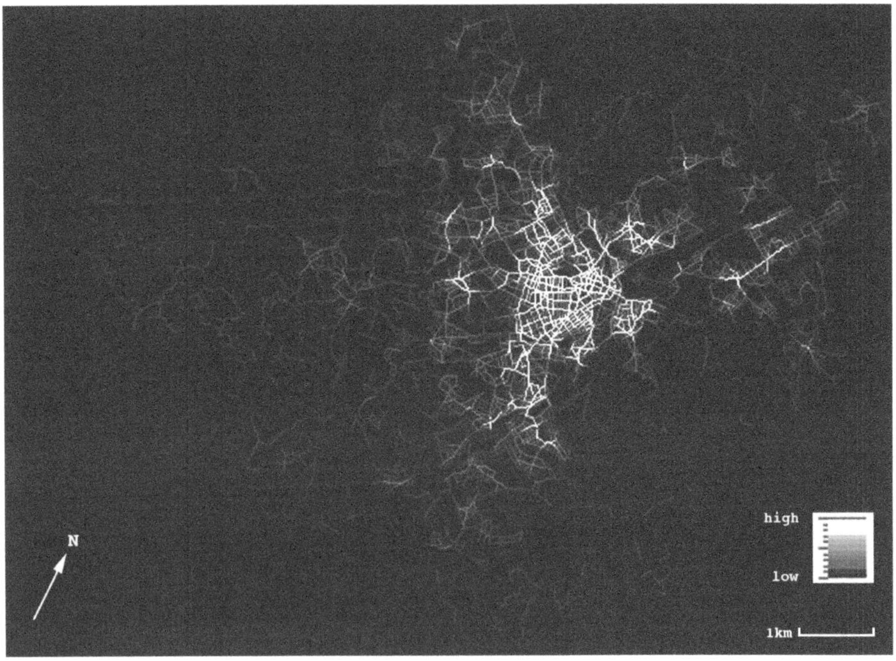

Figure 7.3c Distance concept (c) metric weighting, radius-5
Source: author

where urban space was most densely packed. The details of the formal analysis matter less here other than to establish the conceptual point that *all* elements of a road network can be differentiated in terms of these three distance concepts and at different network radii. These manifold numerical descriptions, it is argued, offer a starting point not only for morphological *heuristics* but also for a topographical *hermeneutics* in which complex network analysis is brought to bear on re-conceptualizing the material agency of the built environment in dialogue with established sources pertaining to urban culture.

The separation of home and work is one of the master narratives of urbanization. Soja (1990, 151) has argued that it is an index of modernity itself because of its key role in structuring the human experience of space. In nineteenth-century studies the differentiation of home and work denotes the rise of suburbia (Thompson 1988, chapter 6), the separation of public and private spheres (Dennis 1984, chapter 4) the extension of intra-city urban transport connections (Kellett 1969) and the rapid development of the urban periphery (Dyos 1966). Alongside the appearance of new transport technologies (for example, the electric tram and the train) these developments gradually transformed the structure of 'walking and horses' cities during the second half of the nineteenth century (Dennis 1984, 113). Routine journeys lend themselves to being conceptualized performatively as 'place ballets' (Seamon 1979) or 'choreographies' (Pred 1977), in the spirit of Jacobs (1993) rather than through their association with any explicit codes of 'meaning'. Identifying shifting patterns of home and work following Sheffield's suburban expansion in the second half of the nineteenth century provides a vital key in translating formal syntactical analysis of

Table 7.1 Changing patterns of home and work in the Rockingham locale 1841–1905

	Employment							
Year	*Artisan/skilled labour*	*Manufacturer*	*Merchant*	*Merchant and manufacturer*	*Professional*	*Retailer*	*Other*	*Home/work total (%)*
1841 (home and work in)	10	9	3		3	1	4	30 (25)
1841 (home in – work out)	6	11	2		5	10	1	35 (30)
1841 (home out – work in)	10	23	3	9	6	1	1	53 (45)
Total 1841	26	43	8	9	14	12	6	118 (100)
1905 (home and work in)	6	1						7 (4)
1905 (home in – work out)	2	2					1	5 (3)
1905 (home out – work in)	70	57	5		5	8	14	159 (93)
Total 1905	78	60	5	0	5	8	15	171 (100)

Source: White's Directories of Sheffield (White 1841, 1905)

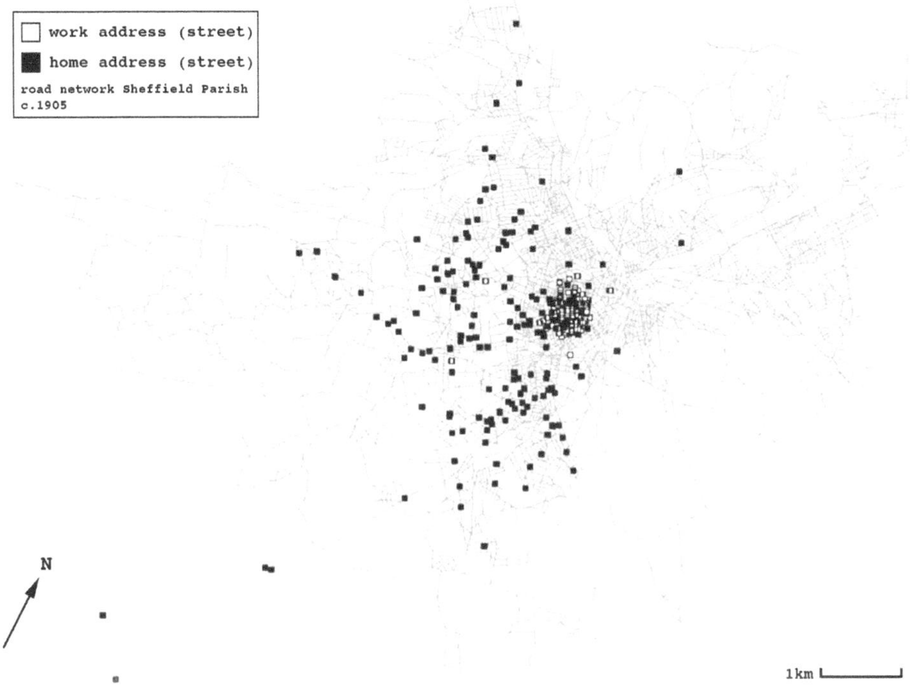

Figure 7.4 Differentiation of home and work in the Rockingham locale 1905
Source: White's Directory of Sheffield (White 1905)

spatial structure into an account of the quotidian life of the nineteenth-century city. These shifts are said to indicate a transformation in the scale of the city's spatial culture, intelligible as time-space practices emerging at different modalities of scale in which larger scale movements emerged as symbolically privileged.

Table 7.1 presents data on the location of home and work for everyone listed in *White's Directory* for 1841 and 1905 as either living or working in the Rockingham locale (a neighbourhood in the heart of Sheffield's cutlery industry) and who gave a separate address for both. Trade directories are notoriously imperfect sources but they are useful in conveying a sense of trends in the changing pattern of home and work for tradespeople and the professional classes. Before 1841 the number of those listed in directories who gave separate addresses is negligible, indicative of how people lived and worked in close proximity – often in the same premises. The proportion of people listed as both living and working within Rockingham locale declined dramatically between 1841 and 1905. The relative absence of residential addresses listed in 1905 suggests how the residual residential population, the majority belonging to the poorer working classes living in back-to-back houses, would have become increasingly socially segregated from the professional and business-owing middle classes as suburbanization progressed. Figure 7.4 maps the data presented in Table 7.1 for 1905. It is clear how the dense socially heterogeneous occupation of the Rockingham locale, characteristic of the early nineteenth century, was displaced by multiple spatialities of distance, redefining the home–work relation at different modalities of scale.

The change in the relationship between home and work in the Rockingham locale 1825–1905 was characterized by the emergence of an urban place ballet that increasingly took the form of linearized movement at the maximally urban scale. The argument is not that localized forms of movement within and between proximate locales did not persist throughout the period but that the modality of scale that described linearizing movement was increasingly privileged over more local, circulatory forms by a dynamic, suburbanizing middle-class culture that asserted home and work as distinct origins and destinations. This middle-class separation of home and work increased social differentiation but also enabled new forms of sociability to be realized in a considerably expanded area of urban space.

From spatial culture to processional culture

The increased differentiation of home–work place ballets at different modalities of scale in nineteenth-century Sheffield produced novel time-space materializations of urban form that became available to the city's elites to appropriate for symbolic purposes. Routine and ceremonial practices in Sheffield's urban milieu were not therefore, categorically distinct but realized different representational potentials in spatial practice.

Figures 7.5–8 present a visual summary of systematic analysis of processional routes in Sheffield in the late eighteenth and nineteenth centuries. These routes (especially

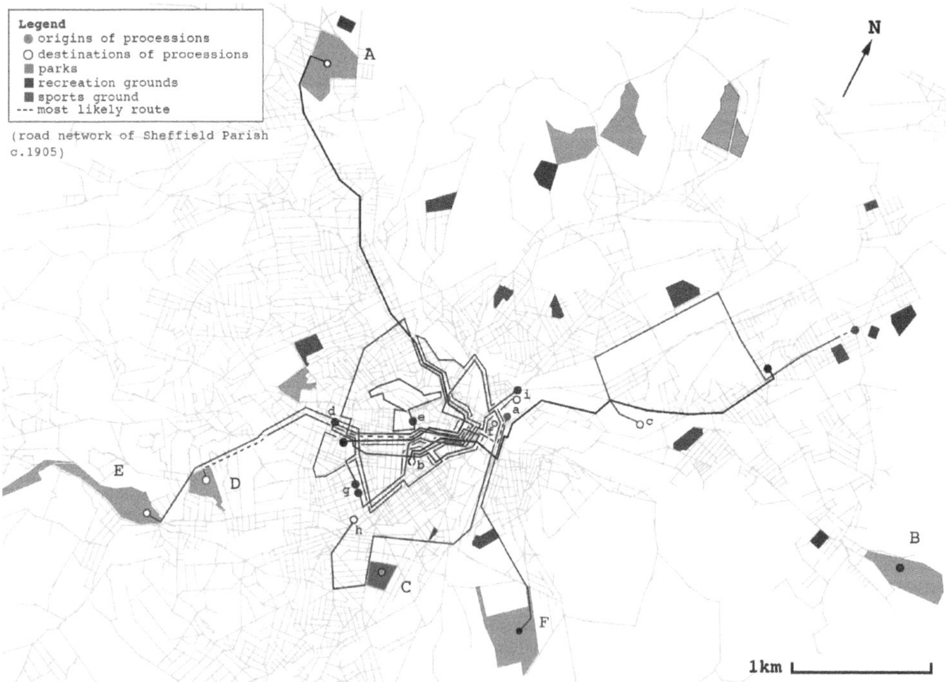

Figure 7.5 Routes of selected major public processions in Sheffield 1856–1908

Source: Sheffield and Rotherham Independent, 31 May 1856, 12 July 1864, 6 March 1883, 21 October 1884, 16 January 1893, 21 October 1884, 25 June 1898 and 27 June 1898; *The Sheffield Daily Independent,* 12 June 1905, 12 March 1906, 25 June 1906, 6 May 1907, 21 June 1907, 7 May 1907, 5 May 1908, 22 May 1908 and 25 May 1908

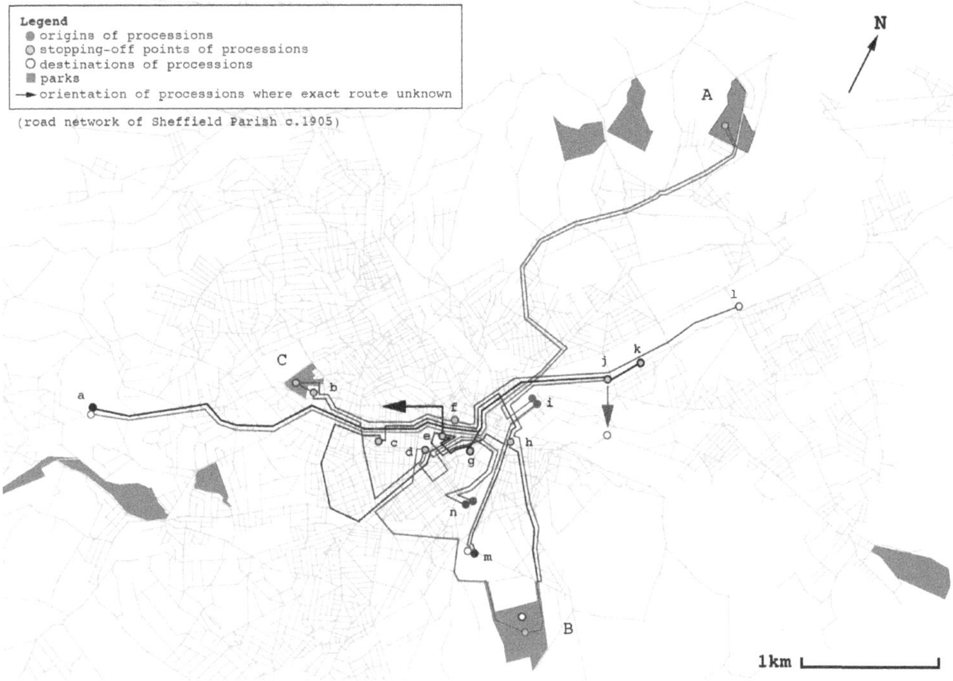

Figure 7.6 Routes of royal processions in Sheffield 1875–1905

Source: Sheffield and Rotherham Independent, 16 August 1875, 11 May 1895, 13 May 1895 and 22 May 1897; *The Sheffield Daily Independent,* 4 July 1905, 8 July 1905 and 13 July 1905; *Pictorial World,* 21 August 1875; *Official Programme* (1875; 1897)

those in Figures 7.5 and 7.6 which are most directly comparable) can be usefully compared to those in Figure 7.2 with regard to the changing orientation and destinations of public processions. The intention here is not to give an exhaustive account of this research but to sketch some preliminary conclusions about the relationship of processional routes, spatial culture and processional culture arrived at through the method of translation that has been outlined here.[4] Five points pertaining to the emergence of Sheffield's processional culture as a material, embodied and representational phenomenon are highlighted.

1. The dominant circulatory orientation of major civic processions gradually extended to encompass a greater area of urban space endured until approximately the mid-nineteenth century when long linear routes and shorter routes restricted to suburban neighbourhoods became more typical. This shift in processional practice can be characterized in terms of urban-scale orientations of 'departure and return' giving way to routes of well-defined 'origin and destination(s)'.
2. The early nineteenth-century orientation of processions of all types towards the historic urban centre was gradually reversed by an increasing orientation towards the suburban fringe and distributed neighbourhood centres.
3. The deployment of parks, recreation grounds, sports grounds and cemeteries as the point of ritual dénouement is increasingly evident in Sheffield from the

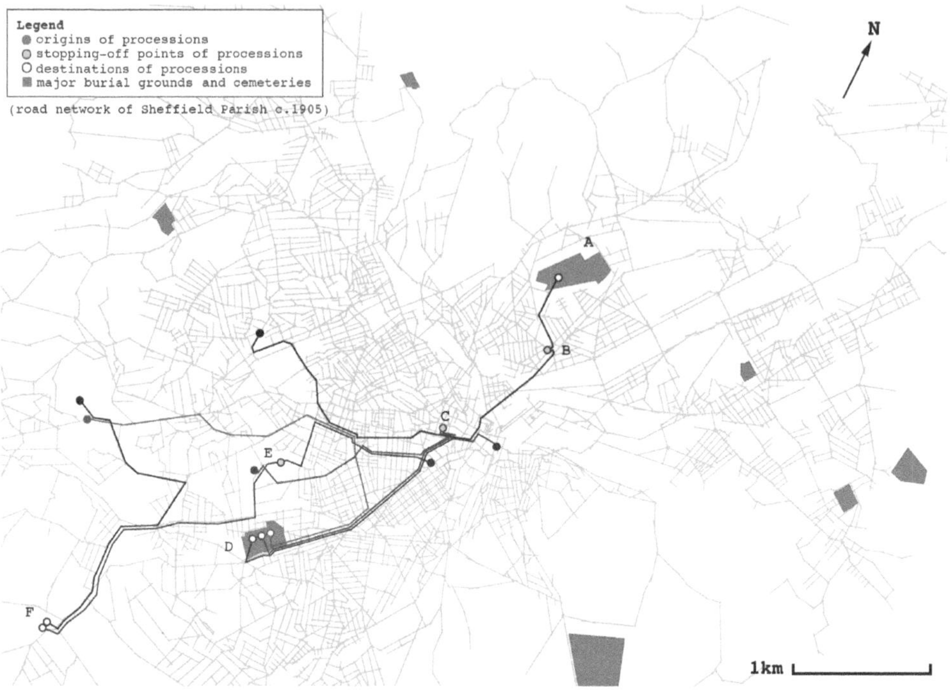

Figure 7.7 Routes of selected public funeral processions of Sheffield notables 1856–1910

Source: Sheffield and Rotherham Independent, 28 April 1865, 27 September 1873, 2 November 1880, 7 November 1885 and 5 October 1898; *Sheffield Daily Telegraph*, 6 October 1898; *The Sheffield Daily Independent*, 24 March 1910

mid-nineteenth century. These destination venues were often situated proximate to historical access routes into the urban centre from the Parish and/or tram-route termini. The impression is one of movement along centre-to-edge routes practised at an increasingly linear modality of scale (because over a larger geographical area) until arriving at open spaces, realized as distinctive ceremonial sites.

4. The linearization of processional routes to destination sites entailed an active process of widening and straightening Sheffield's historical road network. This process was the consequence of transport engineering works, most notably to accommodate tram lines, and the removal of encroachments (Griffiths 2009). In a more fundamental sense, however, it can be explained as a materialization of the rescaling of the spatial practice of distance to a more linear mode through the separation of home and work and the suburbanization of nineteenth-century Sheffield.
5. There is significant continuity as well as change in processional routes over time and across different categories of ceremonial type. This indicates not only how some urban elements – notably those belonging to the early-modern town – sustained symbolic significance over time but also that they did so by functioning as processional spaces at different modalities of scale, enabling them to accommodate various changes in the spatial practice of processions.

Figure 7.8 Orientation of Whitsun processions of the Sheffield Sunday School Union, 1 June 1903

Source: The Sheffield Daily Independent, 2 June 1903

While the symbolic significance of Sheffield's early-modern centre endured throughout the nineteenth century, perpetuated through elite participation in social events such as the annual Cutlers' Feast at the Cutlers' Hall, mass participation processions increasingly re-orientated the city's population outwards along linearized concatenations of historical roads towards bounded convex spaces located at the city's periphery. At the urban scale public parks, sports grounds and especially the newly landscaped cemeteries possessed high symbolic potential for the performance of organized ritual practices (see Figure 7.7). At the neighbourhood scale the popularity of Whit walks (see Figure 7.8) and sings that took place simultaneously in parks and recreation grounds across the city was indicative of the emergence of a thriving suburban culture with its network of interlinked locales (see Entwistle 2012).

While spatial practices at different modalities of scale were not mutually exclusive it was the larger-scale movements to suburban locations that signified social as well as bodily mobility that became symbolically encoded in processional activity. Early- to mid-nineteenth century attempts to re-present the essential unity of Sheffield's eighteenth-century urban community at a larger geographical scale, particularly evident in the great procession to celebrate the passing of the Reform Act in 1832 (Figure 7.2), proved unsustainable once ongoing suburbanization decisively transformed the spatial culture of the city, clearly signifying the social distance between the middle classes and working classes who remained in relatively central locations. The possibility for symbolically privileged mass movement to focus in and around the central Township was increasingly undermined by an emergent urban-scale place ballet that went against the grain of the highly localized 'shuffle' of home and work that sustained the socially heterogeneous spatial culture of Sheffield's workshop-based cutlery industry in its early nineteenth-century pomp (Griffiths 2015a). Thus even within a fervently generative 'city of production' such as Sheffield, socio-spatial mechanisms of ceremonial 'reproduction' gestated, and were realized at the transformed urban scale of the mid-to-late nineteenth-century city. These two images of Sheffield were, in fact, intriguingly combined in royal visits, such as that of Queen Victoria in 1897, during which inspections of major industrial works were combined with a high degree of processional spectacle.

Conclusion

Nineteenth-century urban historians have tended to view processions as impositions of symbolic order on unintelligible industrial urban environments. This argument is premised on the assumption that the built environments of rapidly growing manufacturing towns possessed no agency of their own to sustain meaningful forms of social practice. This chapter has argued to the contrary, that nineteenth-century urbanization enabled, indeed necessitated, the discovery of new performative possibilities in urban space that was appropriated for a range of symbolic purposes. The danger of approaches to civic identity that allege their 'invention' or 'imagining' (after Anderson 1991) in symbolic terms is that the aspatial and totalizing *concept* tends to escape the time-space concreteness and mutability of spatial practice. The idea that spatial practice is articulated in morphic language with a material – that is a social – form problematizes the idea of 'Invention' since from this per*form*ative perspective it is more accurately described as an historical process of making and re-making. The anthropologist Parkin notes that the very 'doing' involved in the performance, even of highly formulaic ritual practices, inevitably introduces 'innovation' (Parkin 1992, 18). It is the intrinsic ambiguity of spatial forms that demands their continual fabrication as spatial practices that extend temporally and materially beyond individual bodies and, in that extension make continual demands on our social capacity to translate meaning into making and making into meaning. As Netto (2008, 359–360) has argued, the agency of space has to do with enabling individualized action to become genuinely social *or communicative* as practice. This poses the historical relationship of morphological forms, performance and representation as a key area for further enquiry.

Appendix: Notes to accompany Figures 7.4–7.7

Figures 7.4 to 7.7 are useful mainly to get a visual sense of the orientations of processional routes across the different categories of processional activity, consistent with the main purpose of this chapter. More information about the sample of processions represented in the figures is given below to aid interpretation.

Figure 7.4

1856 – from Norfolk Park (F) to the Wicker (i), to celebrate the end of the Crimean War
1864 – Oddfellows procession from the Cattle Market (a) to the Botanical Gardens (D)
1883 – Salvation Army procession from Thomas Street (g) and to the Albert Hall (b)
1884 – Franchise demonstration from Attercliffe to Mar's Hill (c)
1893 – parade of the Sheffield Engineers Corps to and from the Drill Hall (d)
1898 – parade of the Lifeboat Saturday movement from the Wicker (i) to Endcliffe Park (E)
1905 – parade of the Grand United Oddfellows from Attercliffe to the Albert Hall (b)
1906 – Temperance Mission procession from and to the Temperance Hall (e), through the Crofts area
1907 – Socialist demonstration terminates at the Castlefolds market (f)
1907 – procession to Hillsborough Park (A) for horse show
1908 – Empire Day pageant at Bramall Lane Sports Ground (C)
1908 – unemployed parade from Thomas Street (g) to London Road (h)
1908 – Socialist demonstration at High Hazels Park (B)

Figure 7.5

Sunday Schools from different areas of the city were organized by the Sheffield Sunday School Union into different divisions for their various processions which converged at different parks and recreation grounds for the annual Whit Sing. In 1903 these included Firth Park (A), Attercliffe Recreation Ground (B), High Hazels Park (C), Norfolk Park (D) and Crookesmoor Recreation Ground (E).

Figure 7.6

The figure shows six processions associated with four royal visits to Sheffield. The first two royal visits (1875 and 1895) lasted two days and featured processions with distinctive itineraries on both days.

1875, day 1 – procession of the Prince and Princess of Wales from Victoria Station (i) to Oakbrook (a) via Firth park (A)
1875, day 2 – procession of the Prince and Princess of Wales from Oakbrook (a) and back via Cyclops Works (j), Norfolk Works (k) Rodgers and Sons (g) and the Cutlers' Hall (e).
1895, day 1 – procession of the Duke and Duchess of York from Victoria Station (i) to the Farm (m)
1895, day 2 – procession of the Duke and Duchess of York from The Farm (m) to Norfolk Park (B) via the Corn Exchange (h), Royal Hospital (c) and the Cutlers' Hall (e)

1897 – procession of Queen Victoria from the Midland Station (n) to Cyclops Works (j) via the Town Hall (d) and Norfolk Park (B)
1905 – procession of King Edward VII from the Midland Station (n) to Vickers' Works (l) via the Town Hall (d), University (b), Weston Park (C) and Parish Church (f)

Figure 7.7

1865 – funeral procession of Wilson Overend from New Haymarket to the General Cemetery (D)
1873 – funeral procession of the Rev. Canon Sale from his residence to the General Cemetery (D) via the Parish Church (C)
1880 – funeral procession of Mark Firth from his residence to the General Cemetery (D)
1885 – funeral procession of Alderman Robert Leader from his residence to Burngreave Cemetery (A) via Wicker Congregational Chapel (B)
1898 – funeral procession of Chief Constable John Jackson from the Town Hall via Collegiate Crescent (E) to Ecclesall Churchyard (F)
1910 – funeral procession of Sir Frederick Thorpe Mappin from his residence to Ecclesall Churchyard (F)

Notes

1 Such research would typically take place in a GIS; see Chapter 16 in this volume.
2 *Depthmap* is an open-source software project at University College London that can be accessed via http://varoudis.github.io/depthmapX/ (accessed 7 April 2015).
3 This radius is calculated in terms of the network structure, where n extends to the entire system, rather than 'as the crow flies'.
4 Further information on individual processions can be found in the Appendix.

8 Weaving patterns in the suburban fabric

Carnival procession routes, mapping place and experiencing space on London's changing periphery, 1890–1914

Dion Georgiou

Introduction

This chapter explores how residents of London's burgeoning late nineteenth- and early twentieth-century suburbs related to their rapidly transforming surroundings. It does so through the prism of routes for carnival processions held in Enfield, Tottenham and Finchley in Middlesex, and in Ilford in Essex, between 1890 and 1914, to raise money for local hospitals. It interprets the procession route as part of a wider process of mapping London's suburbs, considering how carnival organisers identified and amended routes through their districts as they underwent rapid physical change, as well as pondering the question of how processionists and spectators experienced the carnival route. It argues that suburban development, mapping and processions were connected procedures through which new spaces were envisaged, embodied and experienced as places. Place, in this reading, is a multi-authored text, continually redrafted through adjustments to the physical environment, new mappings of it, and new movements through it. My research draws primarily on local newspaper reportage on carnivals and their build-up,[1] including listed procession routes, which I traced against Ordinance Survey (OS) maps to illuminate continuities and changes over time. This analysis is contextualised through reference to scholarship on suburbanisation, mapping, and processional culture, as well as my own readings of late nineteenth- and early twentieth-century maps and books representing suburban histories and geographies.

Contexts: suburbanisation, maps and processions

Suburbanisation, cartography and processional culture were facets of and responses to modern western urban expansion. Reformers, concerned by cities' physical and population growth, and by perceived ills like poverty, disease and vice, sought to render urban space transparent and orderly. Richard Dennis has highlighted measures carried out to this end in London, New York and Toronto, such as slum clearance, traffic improvements and street lighting (Dennis 2008, 113–144). Yet in practice, urban spaces were regularly contested as part of power struggles between different groups, as Krista Cowman's (2001, 153–157) work on gender and class contestation of public spaces in Liverpool, for example, and Kate Hill's (2001, 193–197) on class contestation of museums and art galleries, illustrate.

The middle classes increasingly responded to urban problems by departing for suburbia. 1,405,852 people lived in the London's 'Outer Ring' – the County of London's hinterland comprising Middlesex and parts of Essex, Hertfordshire, Surrey and Kent – by 1891, rising to 2,995,678 by 1921 (General Register Office (GRO) 1922, 1). This was facilitated by speculative developers acquiring former agricultural land around the capital, and new transport networks serving emergent suburbs, but also owed much to growing enthusiasm for residing in greener, less crowded, more select areas, as work by historians such as F.M.L. Thompson (1982, 8–16) and Jerry White (2007, 78–81) has shown. However, social segregation, although aspired to, was never wholly achieved and suburbs accommodated varying assortments of occupational groups, including working-class populations (White 2007, 95–96). Analyses by Gail Cunningham (2000, 52–55), Matthew Ingleby (2010, 128–132, 135–137) and others of period writing have demonstrated that many contemporary commentators were scathing about suburbia as a cultural and physical phenomenon, decrying it as rootless and sprawling – echoing earlier criticisms of urban growth. Yet positive images of suburbs abounded in developers', local authorities' and railway companies' promotional materials (Heller 2014, 7–9, 13–15; Ward 1998, 110–136), as well as in suburban-set popular fiction (Cunningham 2000, 67–68; Hapgood 2005, 39–43), casting them as semi-rural idylls offering escape from and easy access to the city, and personal success and contentment.

Urban expansion also prompted diverse attempts to map the city. John Pickles has stated that:

> at the heart of the territorializing practices of modernity is a cartographic impulse, comprising mapping practices that contribute to and draw upon the coding of nature, space and social life in terms of metric and parametric models of space and the emergence of capitalism and the state.
>
> (Pickles 2004, 68–69)

Studies such as Pamela K. Gilbert's (2004) work on mid-Victorian public health maps, Lucy E. Hewitt's (2012) on the 'Civic Survey of Greater London' undertaken during the First World War, and Richard Hornsey's (2012) on Harry Beck's 1933 London Underground map have highlighted this governmental tendency within maps of metropolitan spaces. Yet as Pickles (2004, 91) has also argued, mapping goes beyond standard cartography, and maps contain multiple meanings. Maps are intertextual, produced through compromises and conflicts over time. Moreover, as Hornsey (2012, 675–693) has noted, historians of cartography now recognise that maps should be considered not merely as representations of space, but in terms of negotiations between map, user and terrain, through understanding 'dynamic environments', and how historic subjects are likely to have mobilised the map.

A third outcome of urban expansion was a revival in processional culture. Work by Simon Gunn (2000a, 175) on Britain, Mary Ryan (1989, 136) on the United States and Peter G. Goheen (1990, 237–238) on Canada has highlighted how processions were used to communicate social hierarchy within expanding, increasingly chaotic-seeming cities. Yet, like maps, processional culture was not the preserve of one group. The working classes, ethnic and religious minorities could assert their own legitimacy and synonymy with the locality by participating in large-scale municipal processions, or else by staging their own, as Gunn (2000a, 167–174) and Ryan (1989, 138) have also argued.

Urban and suburban expansion, mapping and processions were therefore connected procedures through which spaces were envisaged, embodied and experienced as places. Changes to the built environment inspired new mental and physical maps, which enabled visualisation and execution of new construction projects. Maps were also precursors of processions, enabling organisers to plot courses through built environments for participants to proceed through. Directing this guided and facilitated occupation and usage of the built environment, in turn adjusted individual and collective mental mappings of it. The remainder of this chapter will examine suburban carnival procession routes as a paradigm of this relationship within the context of broader efforts to map the suburbs.

Making and mapping the suburbs

The urban districts (UDs) of Tottenham, Enfield, Ilford and Finchley had varying lifecycles of expansion.[2] The population of Tottenham, north of London, grew dramatically from 14,262 in 1871 to 102,703 by 1901, and although the pace of enlargement slowed thereafter, it still reached 146,711 by 1921 (GRO 1872, 120; GRO 1922, 3). Further north, Enfield's population trebled from 19,104 in 1881 to 60,738 in 1921 (GRO 1883, 245; GRO 1922, 2). Growth was the most drastic in Ilford, northeast of London. In 1891 its population stood at 10,922; this quadrupled over the next decade, doubled again in the 1900s, and by 1921 Ilford had 85,194 inhabitants (GRO 1912, 648; GRO 1922, 2). The population of Finchley, to London's northwest, did not climb at quite this rate, but still doubled from 11,191 in 1881 to 22,126 in 1901, and then again to 46,716 in 1921 (GRO 1883, 245; GRO 1922, 2).

Despite their different rates of growth there were commonalities in patterns of suburbanisation in these districts. In each case, development initially focused on existing points of settlement and long-standing roads – their pre-eminence enhanced by the locating of railway stations near them from the mid-nineteenth century onwards – before extending as nearby rural estates were sold off and built upon (see Figures 8.1–8.4 for patterns of suburban development in each district). In Tottenham it centred upon the main network of roads, almost fully in place since the early seventeenth century (Baggs *et al.* 1976, 309–324). In Enfield suburban growth took place around the existing market town of Enfield Town in the centre of the district, and along the Hertford Road on its eastern flank (Baggs *et al.* 1976, 208–224). In Ilford suburbanisation concentrated around the original village of Ilford, in the southwest of the UD (Powell 1966, 249–266). In Finchley, development initially took place along a triangle of long-standing roads – Ballards Lane, East End Lane and the Great North Road – with concentrations of built up area at each corner (Church End, North End and East End), as well as further north along the Great North Road at Woodside Park and Whetstone (Baggs *et al.* 1980, 38–55). Pre-suburban physical environments therefore played a primary role in shaping suburban morphologies. As swift as suburbanisation often was, developers and builders did not begin with a blank page. Nor did they work alone on the redraft, for they had to anticipate and respond to changes in transport networks and population movements.

Further adaptations came from those who sought to represent these suburbs in word and image, reflecting and stimulating a desire for spatial knowledge about them. In some cases this process was explicitly cartographic; in other instances it involved relaying information whose attainment and interpretation was interwoven with formation of mental maps. Suburban areas were repeatedly mapped by the OS, for example (as evident from Figures 8.1–8.4). Maps of the case study districts were

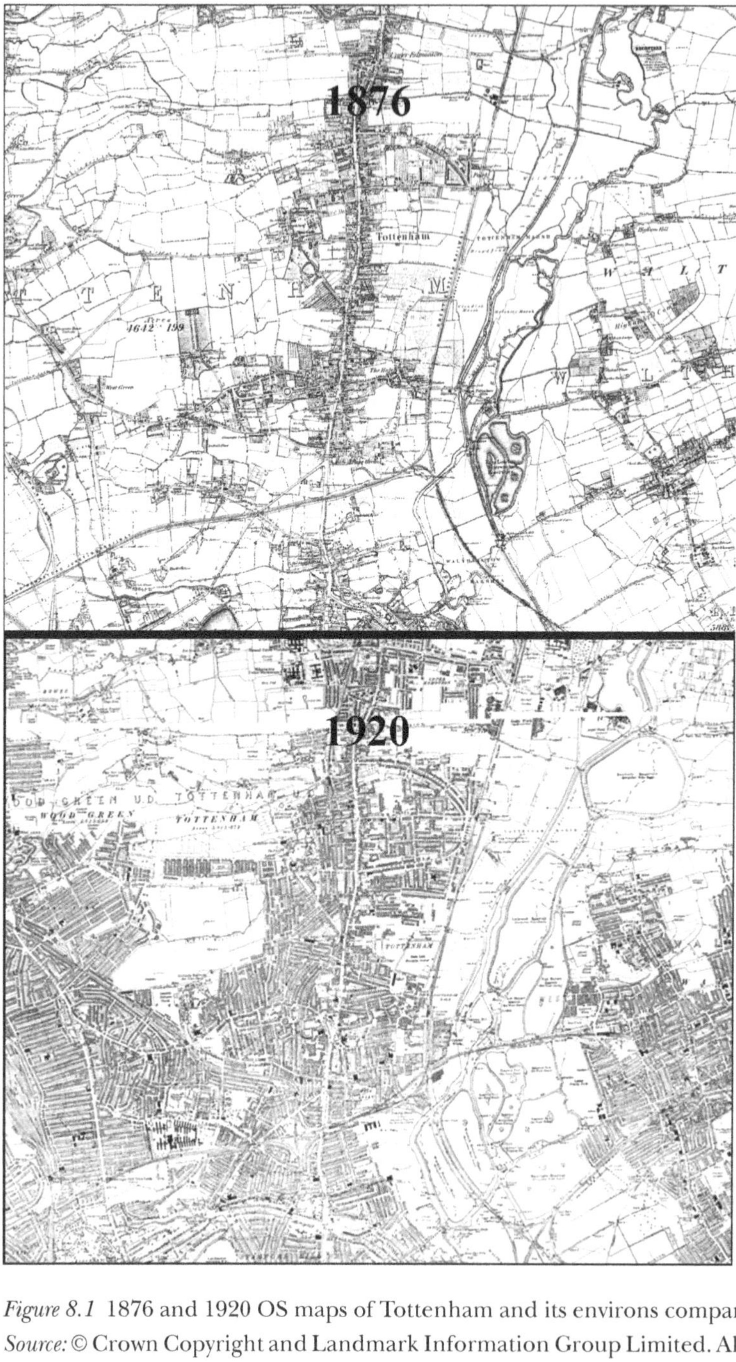

Figure 8.1 1876 and 1920 OS maps of Tottenham and its environs compared

Source:

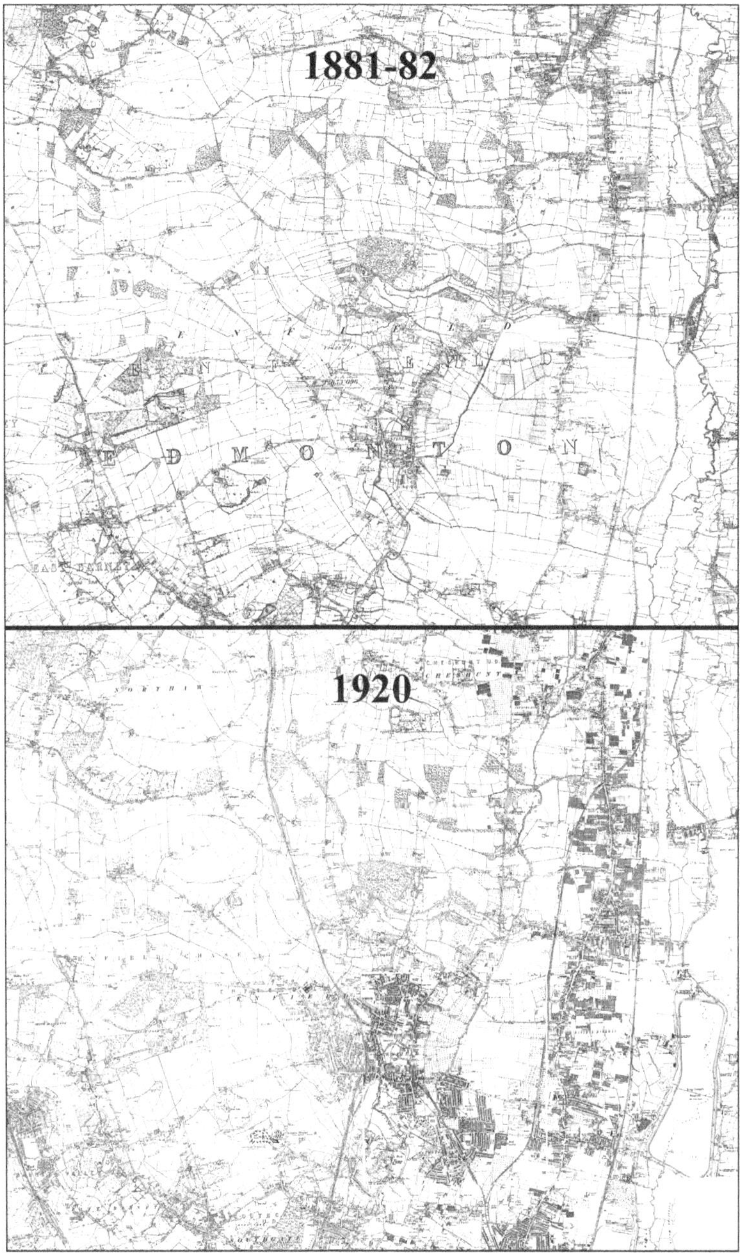

Figure 8.2 1881–2 and 1920 OS maps of Enfield and its environs compared

Source:

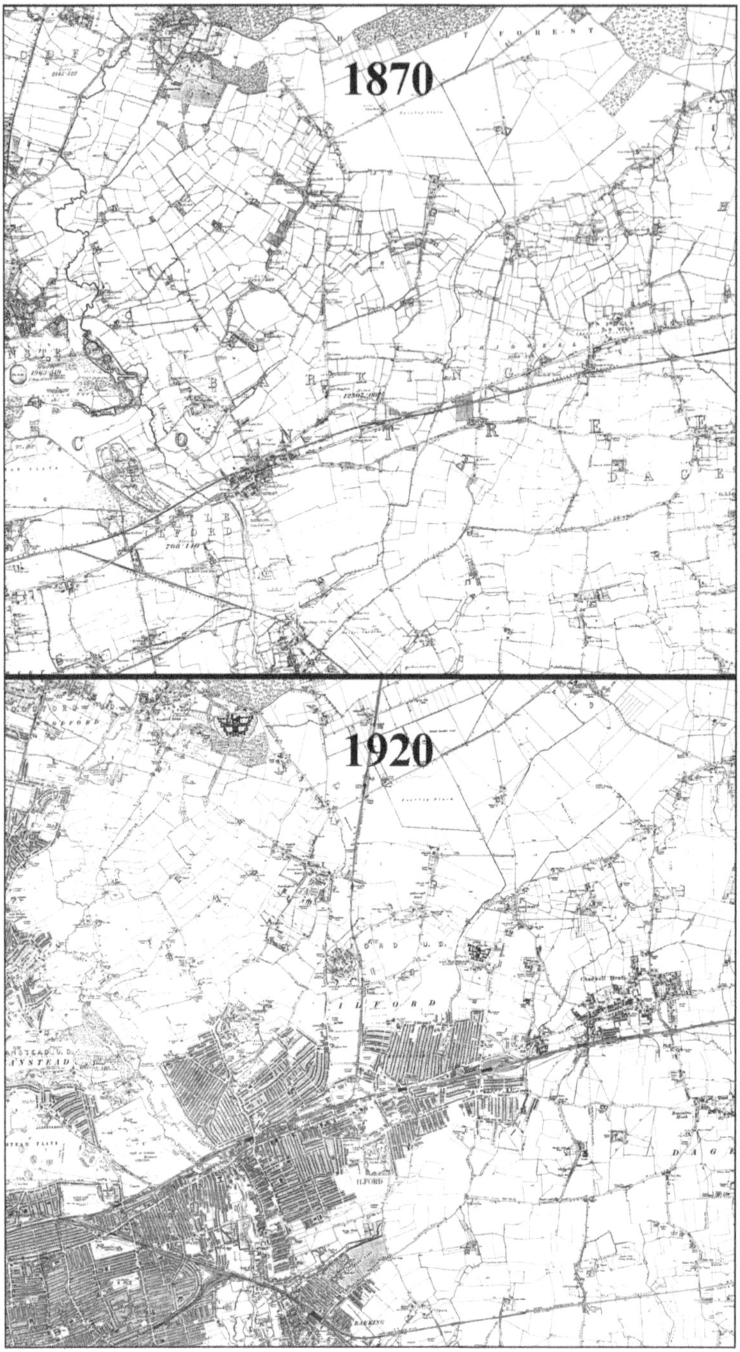

Figure 8.3 1870 and 1920 OS maps of Ilford and its environs compared

Source: © Crown Copyright and Landmark Information Group Limited. All rights reserved

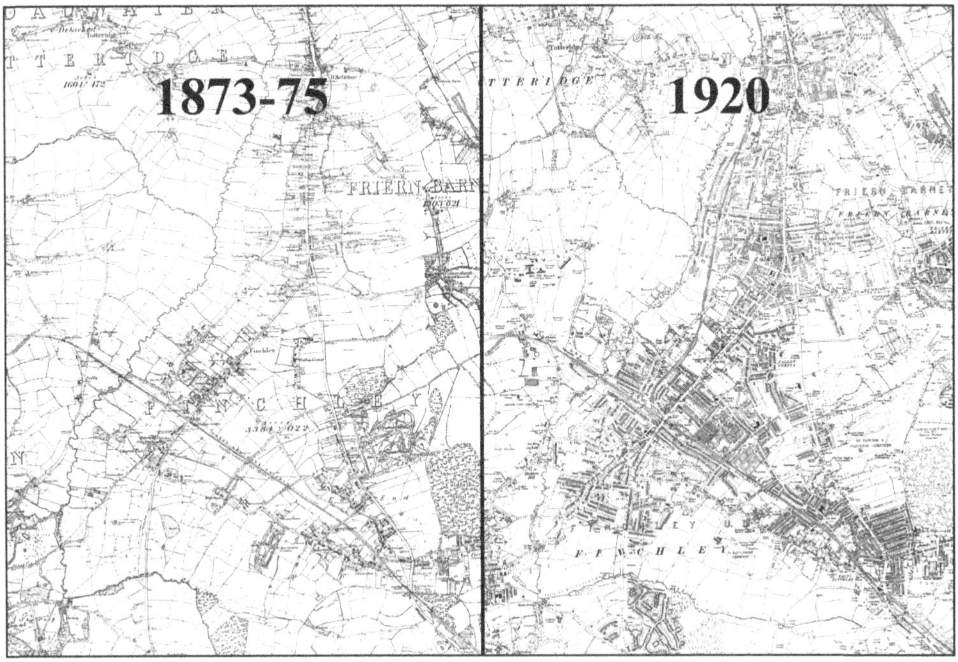

Figure 8.4 1873–1875 and 1920 OS maps of Finchley and its environs compared
Source:

also included in suburban directories published annually by London-based Kelly's from the 1880s onwards, along with other information such as lists of addresses and details of local institutions (Shaw and Tipper 1988, 121–122, 209–213). Builders and developers also tended to use estate maps and bird's-eye view images in their marketing (Ward 1998, 110–112).

There were also guides for suburbanites who wished to explore their district and its environs. *Country Rambles by Field Path and Road in and around Ilford* and *Rambles around Enfield*, for example, stressed the pleasantness of these suburbs' rural surroundings, provided itineraries for walks and cycle rides taking readers through the local countryside, and included local maps relating key roads and landmarks to assist their progress (Round 1905; Tasker 1910). Histories of suburban districts by long-standing residents were also common (Allemandy 1914; Fisk 1913), as were personal reminiscences (Couchman 1910; Hill 1899). Suburban histories sometimes contained historic and contemporary maps too (e.g. Whittaker 1911). More generally, histories and reminiscences took a micro-geographic approach to their subject, dwelling on the significances of particular buildings and natural features and adorned with illustrations and photographs of them.

Surveying this material highlights aspects of how and why the suburbs were mapped, mentally and literally. First, new maps and histories were often based on older ones, while books frequently reproduced visual sources derived from elsewhere. Second, they were often drawn from personal experiences of movement through suburban space, as well as acting as cues to further movement. Third, they historicised local geography and reanimated the pre-suburban past in the suburban present. Fourth, they emphasised suburban districts' rurality, albeit while also engaging with their expanding built environments. Fifth, commercial motives stimulated mapping, which in many cases was undertaken by stakeholders in suburbanisation effectively advertising their product, while publishers also sought to profit from producing guides. Yet these co-existed with other causal factors, such as writers seeking to educate residents about their district or encourage healthy perambulation of it. Sixth, surveys of suburbs were often undertaken by centrally based agents with requisite resources, such as OS and Kelly's, but also by locally based writers and organisations. Finally, these various, continual attempts to render the geography of the suburb reflected, and were prompted by, the rapidity with which it changed. OS maps soon became outdated; directories each year updated lists of streets and residents; reminiscences and histories captured a physical environment now partially lost; and rural rambling routes disappeared beneath bricks and mortar.

Street carnivals in suburban London

During the 1890s and 1900s carnival processions proliferated in London's suburbs, raising funds to support existing and establish new hospitals in response to local population growth. Fancy-dressed individuals and decorated vehicles paraded through streets to raise money for charity, while collectors, usually also in fancy dress, solicited donations from watching crowds. Businesses, local public sector bodies and an assortment of voluntary associations including sports clubs, youth organisations and friendly societies all took part in the processions. I have discussed the festive function of carnivals elsewhere (Georgiou 2014); suffice to say here that suburban carnivals served as occasions for both ritual and inversion, implying a break with the everyday but simultaneously recurring (with varying frequency and consistency) within a broader set of temporal cycles.

From 1890 to 1901 a group calling themselves the Bonfire Boys organised a Guy Fawkes Night carnival in Enfield Town primarily in aid of the Enfield Cottage Hospital. Each year processionists undertook a relatively unchanging loop around Enfield Town, where the Bonfire Boys generally resided and worked, culminating in the lighting of a bonfire on Enfield Chase; the route of the 1901 procession is shown in Figure 8.5.[3] From 1896 until 1913 Enfield Tradesmen's Cycling Club also held an annual carnival and fete; this included a street procession, covering similar ground to that of the Bonfire Boys', in tandem with various other entertainments held originally in the grounds of Enfield Court, but relocated to the new Town Park from 1903. In eastern Enfield another carnival was put on in Enfield Highway and Ponders End between 1900 and 1902 in aid of Tottenham and Enfield's hospitals; again, the 1901 route is exhibited in Figure 8.5.[4] The procession began at the Bell inn, proceeded along the Hertford Road northwards to Freezywater, which it circulated, before travelling back down the Hertford Road and turning off into Ponders End, where it concluded. From 1912 an Eastern Enfield Carnival was organised in aid of both the Ponders End, Enfield Highway and District Nurse Fund and Enfield Cottage Hospital. This commenced in the recently opened Durants Park, toured Ponders End, then travelled northwards up the Hertford Road to circulate Freezywater, before travelling back south and finishing at the Bell.

Tottenham held its first carnival in 1898 when two cycling clubs organised a parade in aid of the local general hospital. In 1900 the organising body was expanded to include tradesmen too, with committees subsequently being set up in each of Tottenham's six wards. The carnival's scope was extended to include neighbouring Edmonton, which the hospital also served. The carnival was organised annually along these lines with growing success until 1904 with an increasing range of separate entertainments supplementing fundraising. The carnival was then discontinued due to a dispute between the carnival's organisers and the hospital's board of management, but resumed in 1909. However, from 1911 its size and successfulness dwindled, despite Tottenham's eastern neighbour Walthamstow participating in the 1912 event, with 1913 witnessing the final carnival before the outbreak of war. There was a fairly typical structure to the Tottenham Carnival procession, as exemplified by the 1904 route, displayed in Figure 8.6.[5] Its first half involved parading through southern Tottenham, taking in mostly main roads like St Ann's Road, Green Lanes and Seven Sisters Road, although the exact pattern varied year-on-year. The procession's second half primarily involved a long passage northwards along Tottenham High Road into Edmonton. There were variations within this: in 1900, for example, the procession diverted eastwards from the Tottenham High Road to circle the built-up area in Lower Ward in the north of the district, before recommencing its route northwards.[6] The 1912 Carnival meanwhile undertook an entirely different route, covering none of northern Tottenham and instead journeying east into Walthamstow before returning.[7]

In 1905 Ilford held its first carnival to raise funds for establishing a local emergency hospital. The carnival was held on the first Saturday in July annually, with the hospital eventually opening in 1910. In its latter years the procession attracted estimated crowds of 250,000 people and the attention of the national press, while large metropolitan businesses also sent cars; funds were supplemented as in Tottenham by auxiliary entertainments. However, the procession was deemed unsuitable in wartime and abandoned in 1915. The Ilford Carnival's organisational system also involved committees being established in all of Ilford's seven wards, each with responsibility for organising their own procession contingents and selecting routes through the ward for them to pass along. The different contingents then joined together for the final procession route later in the evening along which most spectators gathered. The main procession route initially led along Ilford High Road, then northward along Cranbrook Road into Central (later Valentine's) Park. As of 1907, under the local council's orders, the procession turned off Cranbrook Road into Melbourne Road, entering the park through a different entrance. There, sporting entertainments took place, followed by a concert. Figure 8.7 shows the ward and main procession routes for this year.[8]

In 1905 and 1906, tradesmen in North Finchley organised a cricket match, followed by an illuminated parade through the ward, to help raise funds to set up a local hospital. This was superseded in 1907 by a carnival covering all of Finchley, held again in 1908 – the year the hospital opened – and then in 1910, 1912 and 1914. Although there were small variations in the route, the procession usually commenced in Whetstone, travelling down the Great North Road to North Finchley, southwest along Ballards Lane to Church End, eastward along East End Lane to East Finchley, and finally northwest along the Great North Road to North Finchley again; the 1914 route, demonstrated in Figure 8.8, was fairly typical.[9]

Figure 8.5 1901 Enfield carnival routes plotted on 1896 OS map of district

Figure 8.6 1904 Tottenham Carnival route plotted on 1896 OS map of district
Source: © Crown Copyright and Landmark Information Group Limited. All rights reserved

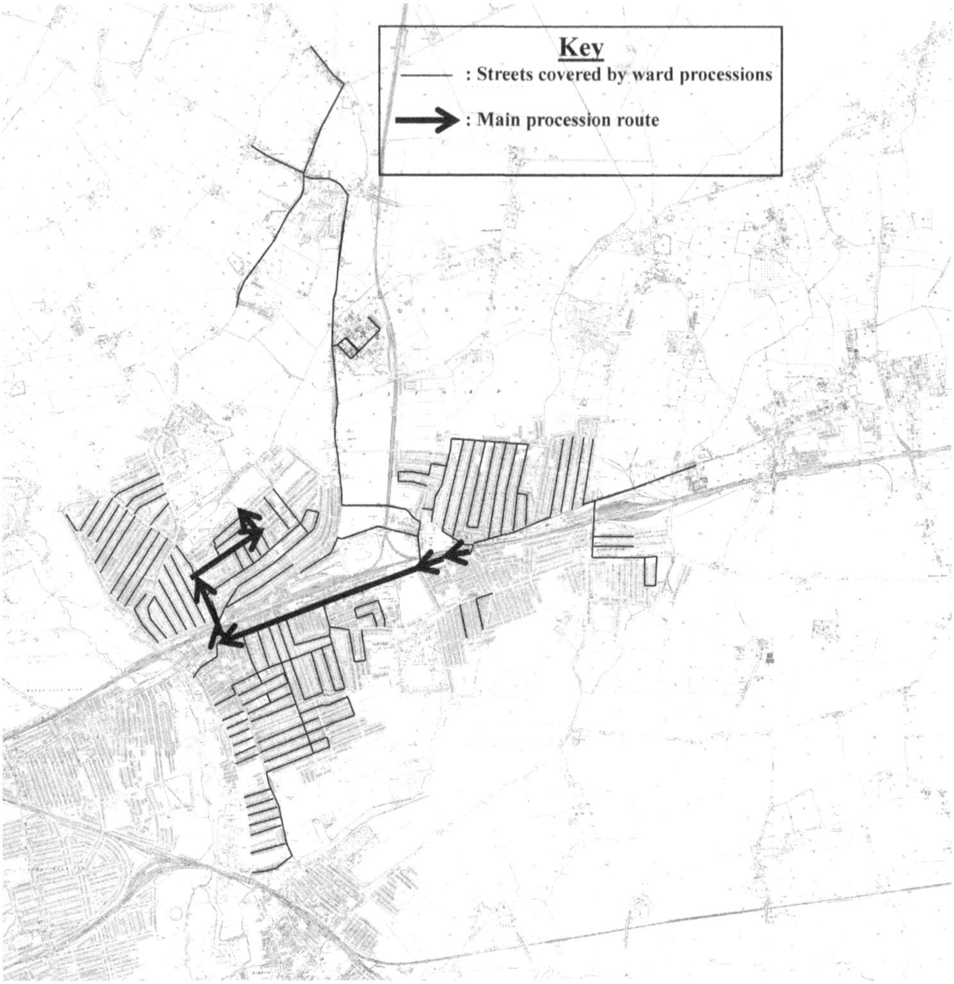

Figure 8.7 1907 Ilford Carnival routes plotted on 1920 OS map of district

Source:

Procession routes and the mapping of suburban place

In compiling these procession routes, carnival organisers composed narratives of their district by identifying primary roads and sites for inclusion, while omitting others, thereby making meaning from and imposing order on suburban 'sprawl'. This was the by-product of carnival organisers' more prosaic ambition to guide processions through areas where large crowds were likely to gather to watch, so money could be collected through them, thus aping cartographers and commercial leisure providers in monetising space. Carnival committee membership tended to consist mostly of white-collar workers and small business owners, accompanied by some professionals and skilled working-class members. Those responsible for compiling routes were also mostly male, although the Ilford Carnival's ward committees included a minority of female members.[10]

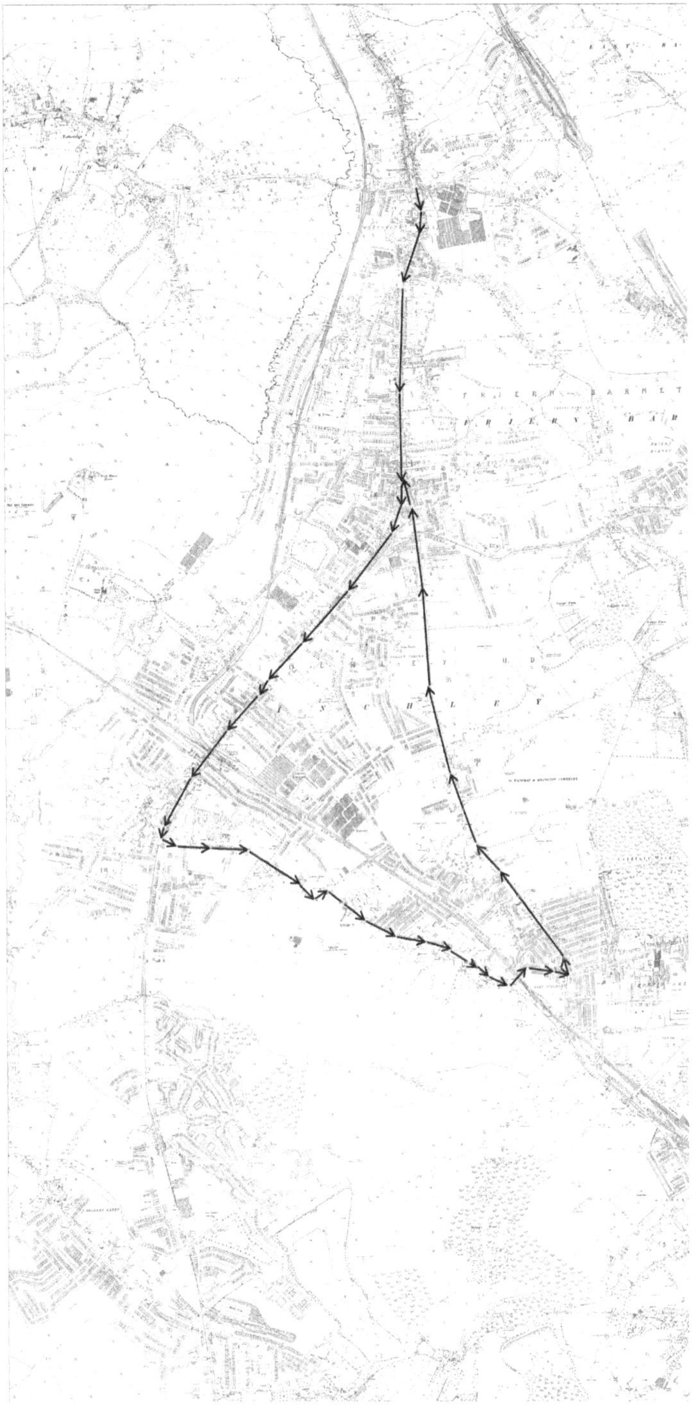

Figure 8.8 1914 Finchley Carnival route plotted on 1912–15 OS map of district

Source: © Crown Copyright and Landmark Information Group Limited. All rights reserved

Organisers drew up procession routes in communication with both the suburban physical environment and mapping processes discussed earlier, composing, adjusting and manifesting broader mental maps of their districts. This intertextuality highlights hidden layers of authorship within procession routes. If committee members did not compose them with local maps to hand, then some must at least have seen such maps before, helping to mould their local geographic knowledge. Yet limits to what maps could tell them, and their being quickly rendered out-of-date by rapid suburban development, meant committee members would have had to rely on experienced local geographic knowledge too. Many employed locally – like council employees, estate agents, builders, postal staff – would have dealt with maps and travelled the district extensively for work purposes. However, the procession route's composition must also be understood as a specific recurrent mapping practice in itself, reflecting the uniqueness of the carnival context, and the ritualisation of its organisational aspects.

Suburban developers, and the natural and manual forces that fashioned the pre-suburban environment before them, were also creative forces within these texts' evolution, providing raw materials carnival organisers could employ, but were simultaneously constrained and challenged by. Procession routes frequently ignored more recently added residential streets in favour of long-standing roads that long preceded suburbanisation. For example, High Road, integral to the Ilford Carnival's final stretch, possibly dated back to Roman times. Similarly, the Hertford Road central to eastern Enfield carnival routes dated back to at least the thirteenth century. This was pragmatic and financially sound practice: these roads were generally the best known and widest in the district, and therefore best suited to carrying processions. Yet continuity in procession routes was also a product of institutional memory. Despite turnover in membership, committees served as retainers of geographic knowledge, and on balance kept to older narratives of place rather than engaging with new suburban neighbourhoods. This perhaps also reflected these streets' prominence within the lives of the mostly male, lower middle-class committee members, who often worked along or commuted via them. The Ilford Carnival, by contrast, contained no single 'master narrative' of Ilford, but rather several more localised narratives prior to a unifying conclusion. Concentrating on smaller vicinities, Ilford's ward processions covered far more residential streets, and were eventually amended in response to new developments like Newbury Park in North Hainault, the Uphall estate in Loxford, and Seven Kings estate in Seven Kings.

The Ilford Carnival's structure also exemplifies tensions inherent in selecting procession routes. In 1906, senior figures in the Clementswood and Loxford ward committees advocated that ward processions be scrapped as most residents did not come out in numbers to watch them, instead gathering to see the full procession at the finale.[11] Nonetheless, ward processions must have had their supporters too as they were retained until 1914; financial considerations came second to administrative inertia, and maybe committee members' wish that processions visit their streets too. Tottenham also offers instances of procession routes as negotiations between pragmatics and local identity politics. The 1900 procession's deviation from the High Road section owed to Park Lane tradesmen contributing a decorated vehicle in exchange for the route being altered to pass through their street.[12] The following year, the Lower Ward committee promised to raise £10 if the procession passed through its vicinity; it failed to do so, and the procession therefore controversially did not travel the ward as advertised.[13] Such disputes emphasise street and ward-based identities' existence alongside districtwide ones.

Though procession routes sometimes passed through short stretches of countryside in travelling between centres of development, they generally marginalised rural areas (in contrast with the rambling routes discussed earlier). This is unsurprising, given potential donors there were far scarcer and roads lesser in quantity and quality. Yet it still illustrates the complex position the rural occupied within suburbia: central as an abstract notion to ideas of place, but sometimes more peripheral as a physical space within personal geographies.

By contrast, parks were frequently processions' starting and/or ending points, often hosting associated events too. These relatively democratic sites epitomised new suburban councils' self-confidence in acquiring private land for public health and leisure. As Carole O'Reilly (2013, 114–117) has argued, provision of municipal parks from the Edwardian era reflected local authorities' desire to take responsibility for local inhabitants' wellbeing and inculcate them with new ideas about citizenship. The park also typified suburbia's relationship with the rural: substituting agriculture and wilderness with tamer, demarcated green spaces for new residents' pleasure.

Suburban carnival processions reiterated these spaces' symbolic centrality and values within their narratives of place. The transferral of the Enfield Tradesmen's Cycling Club's fete from Enfield Court, granted access to them by a General Somerset, to Town Park embodied a shift from local elite patronage to collective ownership and rights to access (through municipal gatekeepers). Likewise Central/Valentine's Park in Ilford: former manorial grounds; opened as a public park in 1899; subsequently the finale for the Ilford Carnival's own narrative of (physical, cultural and philanthropic) progress (Powell 1966, 249–266). In combining these recently established sites with historical streets, procession routes also encapsulated the suburban built environment's hybridity of modernity and tradition.

Procession routes and the experience of suburban place

David Crouch (1999, 162–165) has argued that streets are conduits through which people organise their ideas of their locality, dotted with sites representative of practices and memories for individuals, as well as spaces of interaction. Alan Mace (2013, 87–95), focusing on London's present-day suburbs, has illustrated how suburbanites have invested their built environments with 'dreams' of an idealised future and 'ghosts' of past trauma and loss, with the passing of time helping individuals develop a sense of home, but also of alienation in the event of too much change. Drawing on these interpretations of relationships between communities and their built environments, and on insights gleaned earlier from fin-de-siècle maps and guides to suburbia, we can make three assumptions about experiences of suburban carnival procession routes. First, they would have differed between processionists and spectators, and between residents and visitors. Second, they would have been amalgamations of collective interpretations of place and personal meanings rooted in individual lives, influenced by wider social, cultural and economic forces. Published personal reminiscences of suburbanites hint at this, discussing particular sites' significance within the locality and their own lives (Hill 1899; Couchman 1910). Third, sites would have had connotations of past, present and future, relating particularly to their suburban context.

Processionists' and spectators' movements on carnival day illustrate how plotting routes functioned as cues to action. The former's trajectories were largely preordained, enacting place as a medley of streets, experienced from start to finish. By contrast, spectators' movements were more idiosyncratic. A route's planners delimited where the procession could be seen, and the built environment itself (along with the positioning of fellow carnival-goers) determined the best vantage points, but spectators ultimately chose specifically where to watch the procession, as part of their own personal itinerary for that day.

Procession routes' associations for residents would have related to their own ordinary routines. For those in paid employment (therefore mostly men, but also increasingly some women), routes would have intersected with their daily commutes. Perhaps proprietors of businesses along these streets had a particularly strong sense of ownership and security within them. In Enfield, meanwhile, where a relatively high proportion of residents were locally employed, procession routes would have cohered comparatively closely to their personal geographies.[14] Other sites along procession routes, such as shops, parks, schools, pubs and churches, would also have held particular meanings for residents, although their precise nature would have depended upon class, gender and age, as well as personal factors.

For processionists specifically, longer routes like those of the Tottenham and Finchley carnivals probably took them to parts of the district they had never or rarely visited before, giving them a broader sense of local place than daily routines normally allowed. By contrast, Ilford's ward routes took processionists along more concentrated tours of their own areas, through familiar streets they may have associated with friends and acquaintances residing along them, or else that they lived along themselves and associated with home. However, Ilford's ward processions would often have introduced processionists to newer, less familiar local neighbourhoods too.

Length of residence would also have influenced how people responded to their surroundings on carnival day. For longer standing inhabitants, particular sites might well have had particularly strong past associations, especially as routes usually encompassed districts' oldest roads. Yet it is also possible that for some of them the built environment was haunted by absences, physical and human, due to rapid social and physical changes caused by suburbanisation, and to the passage of time within their own lives. By contrast, for far more recent arrivals – a group particularly plentiful in Ilford – the procession route would have been more of a blank slate, either to project their dreams of an idyllic life onto, or else that they as yet lacked attachment to.

The carnival context is integral to understanding the procession route as an experienced space. It converted selected streets into spaces for individuals, groups and organisations to publicly project, seriously and playfully, various social identities. Some people also decorated homes and businesses for the occasion. This sense of the suburb on carnival day as an enchanted space was picked up on by the press; the *Ilford Guardian*, for example, said of the 1909 Ilford Carnival that 'The drab suburban streets were metamorphosed into a veritable fairyland'.[15] Moreover, participating businesses envisaged routes as marketing spaces and contributed cars advertising their products, echoing the quasi-commodification inherent in the route composition process. Yet utilisation of these streets for social performances and advertisement was hardly restricted to carnival day; rather, they marked peaks within longer term patterns of usage.

The carnival framework would also have affected individual experiences of procession routes. Processionists traversed some of these same streets regularly, but not usually in fancy dress, while huge crowds looked on. Likewise, spectators usually used those streets to pass through, rather than congregating to watch processions. As multisensory experiences involving multiple sights and sounds, not to mention long walks for those processionists on foot and heavy crowding for spectators, carnivals would have altered how suburbanites related to their surroundings, perhaps rendering them less familiar, or less noticeable. Isolated reports of drunkenness, placing of objects other than money in collectors' boxes and (during the Enfield Bonfire Carnival) throwing of fireworks in the streets suggest carnivals emboldened some attendees to behave in public spaces in ways that subverted everyday ideas of suburban respectability. Time of day and year would also have shaped responses to the physical environment, given variations in weather conditions and natural light levels: the Ilford Carnival, running from afternoon to evening in July, and the Enfield Bonfire Carnival, held on a November evening and illuminated by torchlight, were likely to have been markedly different experiences in this respect.

Conversely, carnival day may have affected suburbanites' long-term engagement with their surroundings. Perhaps it turned space into place by imbuing streets that, for many residents, were relative memorial vacuums, with positive specific temporal significances, providing subtexts to future narratives of place. Moreover, given that these events often occurred annually and seasonally, and involved fairly consistent routes, it is likely that, with time, processions came to function not as atypical experiences of the built environment, but as seasonal ones, offering local residents reassuring continuity and circularity as time passed and suburbs matured.

Yet many who watched and participated in carnivals were not local residents. Processions regularly featured contributions from individuals and organisations from neighbouring districts, and sometimes from further afield. Moreover, carnivals also often attracted large numbers of spectators from elsewhere. Reports on the Enfield Bonfire Boys' carnivals in the late 1890s made reference to thousands of people arriving at local stations to watch the procession, while reported crowd sizes for the Ilford Carnival suggest most present were outsiders. Carnival day therefore turned networks of suburban streets, ordinarily accommodating fairly localised social and economic activities, into tourist sites. Perhaps carnivals in this way helped market suburbs to potential new residents.

Not all visitors were welcome, however. The *Ilford Guardian* claimed that as the 1906 Ilford Carnival had only been advertised locally, 'The gangs of London's undesirables, who delight in visiting crowded fetes, did not know of the opportunities offered by watches and ladies' purses in that crowded High Road and Broadway at the time the procession was to pass along'.[16] Yet as Ilford's and other carnivals grew in renown, local papers increasingly reported arrests for pickpocketing at them. Listed perpetrators were working-class and usually either from more inner-lying London districts, or of no fixed abode. These instances highlighted limitations to suburbia's social segregationist capacities, with carnivals encouraging their transgression.

Conclusion

This chapter has demonstrated how socially middling, geographically peripheral groups compiled procession routes through their district as part of wider mapping practices and in conversation with a changing built environment. Interpreting place-making as perpetual adaptation, it highlighted how pre-suburban networks of roads retained their centrality in the suburban text, first through development and subsequently through praxes like processions. Recognising that the adaptation procedure continued beyond the composition of the route, the chapter subsequently considered the question of how individuals present on carnival day might have produced their own personal narratives of place in relation to their surroundings and the occasion, seeking to at least understand the processes by which these were produced, even if their actual nature can only ever be guessed at.

Moreover, the chapter highlighted the temporal specificity of place-making through carnivals, and its connection to more everyday imaginings and enactments of place; and the alternative usages and experiences of suburban space carnivals facilitated, some of them transgressive. Above all, the chapter has hopefully exhibited the synergistic insights the fields of urban (and suburban), cartographic and leisure history can offer to understanding historic and contemporary relationships between people and their built environments, as well as the importance of these relationships within the dynamics of festivals and ceremonies.

Acknowledgements

All maps used in this chapter are courtesy of Crown Copyright and Landmark Information Group Limited (2015). All rights reserved. The author and the editors would like to thank Landmark Information Group Limited for giving permission to use the maps here, and Edina Geoservices, University of Edinburgh, for their support in this.

Notes

1 I examined *Meyers's Observer* and its successor publications *Meyers's Enfield Observer*, *Enfield Observer* and *Enfield Gazette and Observer* for Enfield; the *Tottenham and Edmonton Weekly Herald* for Tottenham; the *Finchley Press* for Finchley; and the *Ilford Guardian* and *Ilford Recorder* for Ilford. All information regarding suburban carnivals is gleaned from these publications, although I will reference specific articles in the footnotes where appropriate.

2 Established under the 1894 Local Government Act, UDs were subdivisions of administrative counties, complete with their own councils.

3 Route taken from *Meyers's Observer*, 1 November 1901.

4 Route taken from *Meyers's Observer*, 18 October 1901.

5 *Tottenham and Edmonton Weekly Herald*, 24 June 1904.

6 Route taken from *Tottenham and Edmonton Weekly Herald*, 6 July 1900.

7 Route taken from *Tottenham and Edmonton Weekly Herald*, 26 July 1912.

8 Route taken from *Ilford Guardian*, 19 July 1907.

9 Route taken from *Finchley Press*, 8 May 1914.

10 Committee members' names were provided in local newspaper reports on carnivals, which I cross-checked against census returns to identify their occupations, using the online genealogy resource Find My Past, http://findmypast.co.uk (accessed between June 2011 and April 2012).

11 *Ilford Guardian*, 13 July 1906.

12 *Tottenham and Edmonton Weekly Herald*, 15 June 1900.

13 *Tottenham and Edmonton Weekly Herald*, 4 October 1901.

14 In 1921 – the earliest date at which census data for workplace location was available – 24 per cent of Enfield residents were in paid employment in their own district, compared to 17 per cent for Finchley, 15 per cent for Ilford and 14 per cent for Tottenham (GRO 1925, 6, 12).

15 *Ilford Guardian*, 16 July 1909.

16 *Ilford Guardian*, 20 July 1906.

9 Plasticine cities

On young people and historical urban morphology

Simon Sleight

> *The city belonged, now, in its limited size and dreams, to us ... All our plasticine buildings may not have been mathematically accurate, but they were, let there be no doubt about this, emotionally precise.*
>
> (Carey 2003, 141–4)[1]

Introduction

Young people do not just live in cities; they also mould them. In this chapter, it is my intention to outline the theme of historical reciprocity between youth and urban space. Following Henri Lefebvre and scholars of the 'spatial turn', space and spatial practice are commonly regarded within the academy as interdependent: spaces and spatial conventions presuppose behaviour, while the users of space shape its form and manipulate its meanings (Sleight 2013, 7). Cities and their inhabitants, it is held, are hence mutually constitutive. Yet for young city-dwellers, the urban environment has often been seen as a container, a determining feature that constrains anything much in the way of youthful agency. This assumption is questioned here with recourse to conceptual literature, scholarship on city childhoods and an excursion to Victorian-era Melbourne, a prime location for exploring the spatial cultures of youth. By way of a more playful preamble, I turn first to my choice of chapter title and to the literary conceit that informs both the epigraph beneath it and the metaphor I wish to sustain throughout what follows.

Cities in miniature

'Real' plasticine cities, I was surprised to learn in marshalling evidence for my analysis, are not uncommon. From humble arrangements depicting just a few streets to sprawling creations containing thousands of individually modelled buildings, plasticine has sponsored some highly creative urban dreams. Perhaps most spectacularly, Martin and Nigel Langdon's efforts – inspired by a trip to London for the then school-age brothers in the early 1960s – have resulted in an astonishingly intricate plasticine metropolis some 4.5 metres long and 1.2 metres wide. Around 30 years in the making (and featuring over 200 kilograms of plasticine), the city bears traces of urban clearances for later expressways as well as plazas and skyscrapers inspired by overseas tours.[2] In Kyrgyzstan, by contrast, Nikolai Tarasenko's smaller plasticine port city of Goldenburg – hosting 17,000 miniature buildings – has been under construction for over seven years (*AKIpress* 2014),[3] while in Melbourne, Pete Shand (2009–10) took to 'digital plasticine' to design his urban spaces, part of that city's Children's Voices project.

Most intriguing among such evocations of cites past, present and future is the compelling depiction in Edward Carey's novel, *Alva & Irva*, introduced above. The story sees adolescent twin sisters begin to model the fictional Mitteleuropean city of Entralla, the setting for the novel and their home territory. In undertaking this task they hope to understand both the city and themselves more deeply. An earlier plasticine creation by the twins is suggestive of a desire to shape reality towards their will, with their detested schoolhouse featuring in miniature hundreds of surface pinpricks, a sign of fictitious damage. This 'monument to wish-fulfilment' (Carey 2003, 65) is also exemplified by the far larger project to fabricate their whole city to scale, an endeavour which sees Alva setting out each day to measure and photograph city streets for subsequent modelling by her reclusive sister. Through this mammoth undertaking, Alva hopes to overcome Irva's crushing shyness and encourage her sister to trust outsiders as she explores the world beyond the modest family home (Carey 2003, 137). When towards the end of the novel the mayor of Entralla bestows gifts upon the twins in recognition of the utility of their creation following a devastating earthquake, Alva and Irva express surprise. 'We were given the keys of the city', Alva notes, 'and with them came something called the "freedom of Entralla", which we had thought all along was already ours' (Carey 2003, 198).

These examples of literal or literary plasticine cities suggest the shaping of the protagonists as much as the metropolitan spaces in which they dwell, or about which they dream. But so much for the fictive attic-based cities of Goldenburg, Entralla and elsewhere. How were their brick-built historical counterparts produced by (and producers of) the young people within them? And how might the idea of the city and the child as somehow bound together assist in the quest to better understand urban morphology? Like the book in which it sits, what proceeds in this chapter is intended to initiate cross-disciplinary discussion, rather than serve as a finished artefact. Speculative, but informed by wide reading and fine-grained source analysis, the hope is that the ideas introduced here will be reshaped and worked over for other historical settings and in other disciplinary contexts. As befits some of the practices of my historical actors, the theme of youth and the city requires a little of the 'undiscipline' and boundary-breaking often associated with coming-of-age (Figure 9.1).

Writing the city

Scholarly assessments of the effects of spatial conditions upon social interaction have an extended heritage. As long ago as 1903, Georg Simmel theorized 'the sociology of space', proposing that social formations and behaviours are grounded in specific territories (Simmel 1997, 137–70). Working in parallel, folklorist and ethnographer Arnold van Gennep observed five years later that 'the passage from one social position to another is identified with a *territorial passage* ... In fact, the spatial separation of distinct groups is an aspect of social organization' (van Gennep 1960, 192, original emphasis). Such thinking would from the 1920s influence the Chicago-based 'urban ecology' of Ernest Burgess and Robert Park, replete with its projections of the concentric rings and 'zones of transition' that were argued to arise from the conditions and associated desires of city living (Burgess 1967). Elsewhere, what became known as 'environmental determinism' – the notion that the (urban) physical environment

Figure 9.1 Building anew – a Chris Ridley photograph of early 1970s British children
Source: Ward (1972, 7)

circumscribed human activity – ebbed and flowed in popularity among geographers across the twentieth century, and only recently has the discipline reinvigorated the concept in altered form after a period of prolonged unpopularity (Creswell 2014, 7, 18–19). Architects and scholars of 'space syntax' have also reworked the idea that the spatial layout of cities influences social behaviour, underscoring further the broad-based interest in the relationships between spatiality and modern urban life. Bill Hillier (2007, 155, 160), for instance, has examined the patterns of movement in British housing estates, finding that poorly designed space can nurture a pathology pertaining to its subsequent use, but that nevertheless 'space is neither necessary nor sufficient for social decline'. Instead, he (2007, 156–9) maintains, one must carefully unpick architectural from social effects, the latter including acute overcrowding or negative perceptions of particular locations.

This emphasis on everyday practice and belief as significant factors in sponsoring the urban morphology of cities resonates with the earlier work of Henri Lefebvre, the French philosopher and sociologist. Especially influential for cultural geographers and urban historians like myself is Lefebvre's *The Production of Space* from 1974, which hosts an extended examination of the ways in which '(Social) space is a (social) product' (Lefebvre 1991, 26). Lefebvre aimed to reconcile lived space – or what would now be termed 'place' – with the mental space of imagining, opening the way for a wholesale re-theorization of how people use cities and produce (by literal construction or physical occupation) the spaces within them. Eschewing the looming postmodernist fascination with language and representation, Lefebvre (1991, 35) declared that 'Man does not live by words alone; all "subjects" are situated in a space in which they must either recognize themselves or lose themselves, a space which they may both enjoy and modify'. From writings on women's historical entanglements with space (Ryan 1990) to consideration of the gay city (Chauncey 1994) or Indigenous peoples' access to colonized territories (Byrne and Nugent 2004), Lefebvre's firm reassertion of agency has helped prompt extensive and ongoing scholarly enquiry.

Inspired by Lefebvre, it is the notion of youth as both shaping and being shaped by city spaces that I explore in the remainder of this chapter. Such an exploration is important because the dominant flow of cultural traffic concerning city youth has usually inclined historically towards a terminus of victimhood, with cities positioned from the Victorian era onwards as intrinsically detrimental places for young people, a cohort cast as passive. 'As the twig is bent, so the tree shall grow': thus ran the common English proverb at that time, its logic inspiring early adult intervention into the lives of the late nineteenth-century urban young (Sleight 2013, 47). Western reformers and 'child savers' depicted the urban scene in stark terms in this era, a projection supported by artistic depictions and literary evocations (Sleight 2014; Morrison 1896) as well as sensational charitable exposés (Hillel and Swain 2010). 'It seems at times as if a great city almost deliberately increases its perils', wrote Chicago-based activist, Jane Addams, in her seminal 1909 book, *The Spirit of Youth and the City Streets*. 'The newly awakened senses', Addams continued, 'are deliberately appealed to by all that is gaudy and sensual … [A] fundamental susceptibility is thus evoked without a corresponding stir of the higher imagination, and the result is as dangerous as possible' (Addams 1972, 27). Note here the emphasis on the city as *purposefully* harmful: a categorization heightening a sense of moral principle and widening the space for potential intervention. Curfews, the interception of street children and the initiation of adult-organized activities and associations for urban youth all stemmed from such perceptions.

In calling into question the one-sidedness of these accounts, it is also my intention at the same time to tackle two further issues. For too long, I maintain, urban geographers, anthropologists, educationalists and phenomenological child psychologists have confined their interests in city youth to the contemporary period; by the same token, historians have overwhelmingly failed to apply to young people's historical lives the insights of writers such as Doreen Massey (on the spatial aspects of contemporary youth cultures, 1998), Roger Hart (on young people's intimate connections with physical space, 1979), Kim Rasmussen (on the difference between 'places for children' and 'children's places', 2004) and Kevin Lynch (on children's perceptual understandings of the city, 1977b). Despite some recent forays by scholars including Gutman and de Coninck-Smith (2008), Pascoe (2011) and Sleight (2013), youth as a category of historical analysis is still inadequately 'spatialized', especially so by contrast to work on other social groups. Interdisciplinary thinking pertaining to a longer *durée* offers a way out of this impasse. By applying some of the lines of enquiry and methodologies generated outside the discipline of history to particular historical settings, it becomes possible to repopulate the cities of yesteryear and to understand the often crucial and ever-changing relationship between lived experience and urban morphology. Research in any field on youth and space promises to benefit from this pooling of resources.

A further aim in this chapter is to intervene in the long-standing contest between proponents who privilege an understanding of (historic) context, such as Skinner (1969), and those, such as Barthes (1977), for whom the text itself – divorced from the background of its author or social milieu – is key. Since first encountering the bald outlines of this dispute as a master's student some 15 years ago, this bifurcation has troubled me. Like the 'link boys' of eighteenth-century London – savvy youngsters who guided disoriented adults through a warren of streets by virtue of flaming torches and deep local knowledge of place – I think youthful agency can light the way in this debate. In-between the 'hard city' of policy regimes, unyielding buildings and enforcement agents, and Jonathan Raban's (1974) celebrated 'soft city' of personal imagination and will, we can for young people, and perhaps, by extension, adults, start to perceive the contours of an alternative, plasticine city: tangible and mouldable, but highly ephemeral. Such a proposition calls attention to context as well as text, but places the task of getting to grips with the changing *texture* – the feel and fabric of cities past and present – as the central purpose. Perhaps I should have committed this argument to paper sooner, for others have started to move towards the same conclusion in recent years (Creswell 2014, 6–7). No matter. For in thinking about the texture of the urban – the felt experience of life in the city, as well as its representation, social effects and the striations carved within it by different social groups trying to make their way – an exciting object for interdisciplinary study opens before us.

Becoming space

And so to colonial Melbourne. As a historical setting in the later nineteenth century, the city could be said to possess a particularly youthful spatial culture. First settled by white colonists in 1835, aspects of Melbourne's built fabric were still striking observers as inchoate towards century's end. Muddy thoroughfares and open drains troubled pedestrians (Gane 1886, 36), while English journalist Richard Twopeny, visiting in the early 1880s, condemned the 'rather higgledy-piggledy look' of the city and noted 'gaps in the line of the streets' caused by the settlement's precipitate outward expansion (Twopeny 1973, 2, 4). As freshly planted shrubs and saplings took hold in the city's new parklands (Figure 9.2), Melbourne's central streets were ripped up to accommodate the laying of tram tracks and the belated construction of sewers. The rapid unfolding – and attendant growing pains – of the city's frame can be seen to share more than a little in common with the 'adolescent urbanism' noted by town planning expert Paul Burton as characteristic of city growth on Australia's Gold Coast for a much later period (Burton 2009). Perhaps in a city rendered from its foundation as an upstart metropolis, such markers of adolescent awkwardness were inevitable. But contemporaries impatient for civic improvements were unforgiving – 'Melbourne', wrote a newspaper columnist at the close of the century, 'though still among the youngest of great cities, is now old enough to regard itself no longer as justified in excusing its deficiencies on the score of youth'.[4]

Within this nascent space, Melbourne harboured an especially youthful population. The discovery of gold in 1851 in country Victoria transformed the city as

Figure 9.2 Charles Nettleton, *View from observatory in Flagstaff Gardens (2)* (albumen silver, 1866)

Source: Reproduced by permission from State Library of Victoria

immigrants poured in to seek their fortunes. As these immigrants settled and married, they fostered a baby boom of unprecedented proportions: by 1871 fully 42.2 per cent of the white population in Victoria was aged 14 and under (by comparison with 18.6 per cent at the most recent census). Centred on the colonial capital, Melbourne, a still higher proportion of Victorians – 49.6 per cent – was 19 years old and under (Sleight 2013, 3). In turn these young people produced an 'echo boom' as they rose to maturity and produced their own offspring, pushing the city towards a population of half-a-million souls in 1901, an increase of well over 450,000 people in just 50 years. Associated discussion of the perceived merits and shortcomings of colonial sons and daughters and the place of an imagined Australian nation within the imperial household further underscored the youthfulness of the city (Sleight 2013, 32–9). Converging on Melbourne, this Young Australia debate drew upon the real-life activities of young city-dwellers coming of age within an urban landscape prone to accelerated change.

Spatial cultures of youth

What role did the incipient spaces of colonial Melbourne play in cultivating an urban youth culture? And how, in turn, were those spaces affected? To answer these questions, we need to draw upon a wide range of primary sources located not within a discrete collection (a hitherto favoured port of call for historians of childhood) but scattered across multiple archives. Repopulating the spaces of the past using sources including memoirs, police files, photographs, travelogues and the correspondence of Melbourne's Town Clerk encourages conceptions of landscapes rather than sites. As Elizabeth Gagen and others have noted, young people are especially dynamic movers through space and blind to adult mappings of territory (Gagen 2008, 407, 411). Urbanists, therefore, need to abandon any preconceptions based on adult-centric patterns of city space and instead follow youthful footprints. Such tracking exercises reveal a layer of urban life largely missed by historians. Elsewhere, I have termed this layer the 'youthscape': an overlapping, underlying and competing terrain produced by young people in common opposition to adult intentions (Sleight 2013, 51, 53).

Examples of this youthscape in colonial Melbourne are diverse. Soon after the influx of gold-seekers in the early 1850s, at least one group of girls and boys played at 'lucky diggers' outside their family home, further extending their games across the city amidst swampy ground and fallen trees soon re-imagined as ships, mountains or houses (Chomley 1944, 19). Similar renaming by young Dudley Ricketts later saw a thin strip of church ground in central Melbourne designated 'the cigar' and occupied for games of football (Ricketts, c. 1981, 2). Part of the process of 'children's symbolic work and creativity' observed for more contemporary times by educationalist Kim Rasmussen (2004, 157–9), naming an area implies a claim to its rhetorical possession. Adult demarcations of space mattered little in this schema.

Going well beyond linguistic appropriation, young Melburnians' colonization of space often gravitated towards the city's newly built fabric. Cable trams, tracks and railway cuttings proved popular from the 1880s onwards. Children attached strings of jam tins to tram cables, watching in delight as their handiwork clattered off into the distance (Collingwood History Committee 1979, 51). Other youngsters jumped on and off trams to test their balance, sometimes by virtue of handstands or through leaping backwards (Loughlin 1988, 4). The pastime was not without its hazards (Sleight 2013, 59). Similarly dangerous, young people attempted to make trams derail by bouncing on the cars (Stretton 1976, 8), placed small stones on train tracks in 1884 in an effort 'to see the train jumping' (Sleight 2013, 60), and pushed idle train bogeys down inclines (Joynt 1979, 4). Sloping ground in the city centre also attracted children after heavy downpours. Down the very same open gutters pictured in Figure 9.3, Frank Matthias and his friends would sail crude wooden boats in the 1890s (Matthias 1985, 4). Betting pennies on the outcome of such races attracted press concern;[5] likewise, the gambling habits of juvenile news-vendors engaged in street games of 'pitch and toss' (see Figure 9.4) troubled city fathers, for whom such precocity suggested moral waywardness.[6]

Adult city-dwellers often found young people's unsanctioned use of city spaces annoying. In the summer of 1873, the correspondent 'Anti-Waste' noted water gushing from four street plugs as children splashed around 'with evident enjoyment'. Unable to secure sufficient water pressure for his evening bath, the ratepayer did not look kindly upon the scene.[7] Similarly, in 1880 an infuriated clergyman found no joy in the act of children opening the rear doors of occupied water closets which backed onto laneways. Caught indisposed once too often, the police were called upon to intercede (Sleight 2013, 55). Elsewhere adults complained of being forced to 'run

Figure 9.3 Thoroughfare or urban playground? A view east along Melbourne's Little Bourke Street – note the gradient of the gutters and the vacant corner block (glass photonegative, 1875)

Source: Reproduced by permission from Mitchell Library, State Library of New South Wales

the gauntlet' of flying balls to access their commercial premises (Sleight 2013, 55), or else dodged slingshot ammunition or the mouldy fruit hurled around by rowdy adolescents at the city's Western ('Rot') Market.[8] Not content with causing disturbance in that location, across town young roller-skaters held court in 1894 amidst the dismantled stalls at Melbourne's Eastern Market.[9]

In Melbourne's parklands, modifications to urban space found perhaps their fullest expression. Here civic authorities including the police, gardeners, foremen and even the council's dog catcher waged an ongoing battle for the right to shape the city. Children were reported by the parks curator as 'running riot across the beds' and 'dancing among the tubers' (Sleight 2013, 47). Flowers were taken, hose pipes used for drinking, and birds' nests raided for eggs or else destroyed with catapults. City fathers looked to green spaces with hopes of civic improvement and popular edification, but to the city's children, Melbourne's parks were a malleable resource supplying all manner of wants. Boys and girls were intercepted climbing the young trees, and breaking branches for a variety of uses including making hockey sticks or fishing rods and for firewood. Sometimes youngsters simply damaged shrubs during

Figure 9.4 F.A. Sleap, 'Incidents in the Life of a Street Boy', *Illustrated Australian News* (Melbourne), 5 August 1885

Source: Reproduced by permission from State Library of Victoria

games of hide-and-seek or during the search for stray cricket balls (Sleight 2013, 47, 61).[10] In a telling case, a resident of north-east Melbourne, E.J. Jubbs, expressed his anger about the use of Curtain Square. In an emphatic letter to the city's Parks and Gardens Committee in June 1897, Jubbs complained:

> [On Sunday] during the whole of the afternoon the Gardens are filled with children who simply turn it into a large play ground, and their calls, screams and laughter are a perfect nuisance to residents adjacent to the square … the noise is simply unbearable. Surely when your committee granted the opening of the Gardens, it was never intended that a crowd of unruly children would run wild all over the place in the way they do.[11]

A follow-up letter sent by Jubbs the next month stated that 'the week days are now nearly as a bad as the Sunday afternoons'. But further correspondence from the local police sergeant and parks curator determined that as only very occasional damage had occurred in this location and because the children concerned seemed respectable, a clampdown was unnecessary.[12] In this instance the authorities sided with the children in determining the park's proper function; not all adults were set against the raucous activities of the young people in their midst.

The same could not be said regarding attitudes towards the raffish young people found congregating on vacant ground, the last of the city spaces I wish to discuss here. As noted earlier, in the late nineteenth century the speed of Melbourne's expansion left behind many empty parcels of land across the city centre and flanking suburbs. These 'unprogrammed' spaces – to borrow a later term from American urban planner, Kevin Lynch (1977a, 13–21) – were subject to an ill-defined legal status and consequently hard to regulate. Reports regularly appeared in the press and in correspondence to the Town Clerk concerning activities within these areas of 'larrikins': bands of working-class youths who shared a subculture based around fashion, popular theatre and showy acts of aggressive, often violent, behaviour (Sleight 2009). These male and female adolescents threw stones intended for street-making at telegraph posts or passers-by, and gathered to sing favourite music hall songs, dine on pilfered produce, swear liberally and often and show off to one another their step-dancing moves (Sleight 2013, 140–6). Adult observers noted that the larrikins were 'coeval' with the progress of Melbourne, somehow linked to its urban structure.[13] As visiting Anglo-Indian writer Henry Cornish put it in the late 1870s, 'Melbourne, like some of the gawky, weedy "larrikins" in her streets, has grown too fast in her youth, and has thus run too much to arms and legs. Her frame-work is big enough to last her for the next fifty years. What her constitution now requires is consolidation' (Cornish 1975, 93–4). Here Melbourne's growing pains are compared directly to those of its city inhabitants and urban form is elided with the attributes of city dwellers. Caught themselves in the divide between child and adult, Melbourne's larrikins were seen to flourish within the gaps of their city.

Why did young Melburnians seek out such spaces and – in an era before the construction of fenced-off playgrounds – look to turn entire swathes of the entire outdoor city into settings of their own? Historical contemporaries noted that it was the 'small and stuffy' conditions of working-class lodgings that forced youth outside[14] and that 'it is the comfortlessness of firesides which peoples the streets'.[15] Functionally poor school yards often consisting of flat and featureless earth or bitumen (Pascoe 2011, 140) also increased the attraction of nearby green spaces; lunchtime attendants were stationed in Yarra Park in the early 1880s in an attempt to supervise the expected student influx. Moreover, in an era before the tighter enforcement of compulsory universal schooling (introduced in Victoria in 1872, but barely affecting the proportion of school-age Melburnians to be found at their school desks 20 years later), young people were everywhere in the city, playing, working, spending and socializing (Sleight 2013, 51).[16] Among the 'swarms of untaught children' observed in Melbourne in 1884 were certainly to be found more boys than girls, but by the close of the century the outdoor city had also become an increasingly welcoming place for females in their teens, 'pioneers' of social space as they ventured across the city, 'walked out' with boys and pursued their own leisure interests (Sleight 2013, 51–2, 76–82). Melbourne's Chinese community, by contrast, produced proportionally less urban space, its young people by and large sticking close to Melbourne's Chinatown, located within the city's theatre district (Couchman c. 2009).

The activities outlined above often left no physical mark on the fabric of the city. Sometimes dropped marbles or orange peel, broken branches or the 'horrid sketches and obscene writings' left by factory girls in the toilets at the Eastern Market alluded to young people's presence (Sleight 2013, 53), but like the chalked street games or messages in other contexts (Pugh 1906, 266–71; Davies 2009, 15), such evidence often vanished quickly. Melbourne's historical outdoor youthscape is hence often cultural heritage at its most intangible. This layer of past urban life is not something that one can point easily towards in today's city, albeit that folklorists have highlighted the remarkable persistence of aspects of children's 'playlore' across time (Beed Davey *et al.* 2013), and notwithstanding that some of the fixed playgrounds built by adults in response to children's wide-ranging city play are still extant. It should also be stressed that even within the time period discussed here, the youthscape was an impermanent phenomenon, varying from day to day, by time of day, by season and in part according to a tangle of legislative enactments concerning the regulation of public space (Sleight 2013, 64–6). To remain viable, it required continual manipulation by its creators, based on an ability to access the vacant grounds, parks and other spaces where the youthscape was crafted and cultivated. As early as the 1880s, some observers perceived that opportunities for free play in city streets were being discouraged (Brown-May 1995, 28); by 1909 a columnist noted that the vacant allotments where children still played (and where larrikins had gathered) were now being filled in at great speed.[17] Such trends towards diminished free space in central metropolitan areas would only exacerbate with the incursion of the car into young people's street territories and the eventual 'islanding' of young people into separate – and nowadays often commercial – spaces across the city (Gillis 2008). But diminution should not mask the vitality of this plasticine city in its earlier guise, or curtail assumptions that the model can reworked once again from similar materials. Contra the judgements of Stevens (2007, 27–8), true 'ludic' creativity is not the preserve of adults. This chapter finishes with a move from the historical towards the contemporary, and a broadening out to discuss recent international initiatives to encourage the repopulation by young people of city spaces.

Pliable cities

For a period of around 15 years from the early 1960s, a cluster of international scholarship on the uses of the public space implored urban planners to show greater appreciation for the sensory requirements and physical well-being of children. Perceiving the negative impacts of trends which had commenced in the nineteenth century, such writings advocated the restoration of the 'lost educational functions' of the outdoor metropolis (Parr 1969). Instead of containing urban youth in discrete and functionally dull sites, urged Colin Ward among others, we should 'help them climb out of the sandbox and into the city' (Ward 1978, 211). An earlier oral history study by Lukashok and Lynch (1956) had noted that urban children were most contented when situated in a manipulable environment; over the following years this observation chimed with the conclusions of Jane Jacobs on the importance of 'unspecialized' spaces and 'sidewalk niches' (1962, 81, 86) and Roger Hart's concern for the maintenance of 'dirt play' in facilitating the symbolic construction of children's worlds (Hart 1979, 335–6, 343). Such entreaties were inspired in part by witnessing the re-colonization by young people of bombed-out city spaces and the attendant origins of the 'junk play' movement, as well as by the authors' recollections of more expansive childhood freedoms enjoyed earlier in the century.

More recent discussions regarding 'toxic childhoods' (Palmer 2006) and an associated 'logic of containment', as identified by Gill (2007, 14), might well be taken to suggest that this advocacy on behalf of young city-dwellers has failed. Look beneath the headlines, however, and evidence can be found both of international schemes to make urban environments more malleable for young city-dwellers and to highlight the persistence of the youthscape in the contemporary city. From 'Home Zones' (or *woonerf*, their precursor in the Netherlands) to 'Living Streets' across England; from 'the Land' project in Wrexham to the '78th Street Play Street' in Jackson Heights, New York: at local level a series of international initiatives – of which the above represents only a fraction – are attempting to improve the useability of city spaces for young people. Focused especially on encouraging outdoor play among younger children and the banning of cars, these adult-initiated schemes have garnered broad-based community support and frequent press attention. The ultimate limitations of localism, however, hold back such programmes and arguably do little to reverse the overall 'islanding' of young people into discrete zones.

The now international Child Friendly Cities initiative promises a more thorough reassessment of urban space. Originating in 1996 and hosted in Italy, Child Friendly Cities aims to promote children's participation in making cities around the world more suited to their collective needs. From a public space perspective, this means making cities healthier, safer, more playable for younger children and more welcoming to older youth. Scores of municipalities around the world have so far implemented projects inspired by the scheme (UNICEF 2014). In 2009, for instance, a University of Melbourne study by Whitzman *et al.* (2009, 8) observed that four local governments in Victoria (the Cities of Melbourne, Greater Bendigo, Brimbank and Port Phillip) had at that time formally committed to making their cities child-friendly. The report noted that the adoption of Child Friendly Cities policies has had a positive impact on planning policies for and with children, but that the potential for effecting children's independent mobility throughout the city remained 'as yet largely unrealized'. A recommendation by the authors for the Victorian State Planning Policy Framework to 'explicitly refer' to the rights of children to use and enjoy public space is yet to be implemented (Whitzman *et al.* 2009, 39), although in May 2014 an updated version of the Framework did specify that 'open space is designed to accommodate people of all abilities, ages and cultures' (Department of Transport, Planning and Local Infrastructure 2014). Still wider campaigns championing 'the right to the city' – the right of all urban-dwellers 'to shape, intervene and participate in the unfolding idea of the city' (Chatterton 2010, 235) – have also yet to yield the practical effects for which proponents aim (Sleight 2013, 221). Little wonder that many adults prefer the more immediate benefits conferred by closing a single street, or turning over an empty plot to junk play.

Conclusions

If the city of the Victorian era can offer any templates, foremost among them is the notion of the metropolis as a playground city rather than a city of playgrounds.

Figure 9.5 Painting the laneways: the production of street art in Little George Street, Fitzroy, Melbourne, 26 June 2007

Source: author

Younger Melburnians ranged far more freely across the colonial city, and at younger ages, than is the case today. For good or ill they felt the power to amend what they found, bending the city to suit their own purposes. This desire to change the environment is likely near universal: as Colin Ward observed in *The Child in the City*, his celebrated 1978 study, 'Children *will* play anywhere and with anything' (Ward 1978, 86). Perhaps urban authorities charged with policing contemporary cities should allow a freer rein where transformative play in parks or noisy occupation of empty plots or street corners is concerned. Certainly sociological evidence suggests that the sociability fostered by such activities can be good not just for young people, but also for adults (Hüttenmoser 1995). The relatively tolerant attitude of the city council towards laneway street art in Melbourne in recent years (see Figure 9.5) is an example of best practice here. Allowing an outdoor urban canvas for teens and twenty-somethings has offered freedom of expression as well as economic benefits to the city. Graffiti-strewn laneways have become magnets for tourists and locals alike, and locations in turn for new coffee shops, galleries and bars. Many would argue that the texture of the city has been enhanced as a result, with the hasty tags of the past replaced by more thoughtful artistic murals.

Leaving to one side advocacy for youth's right to the city, the cause of researching young people's lived experiences – and, by extension, those of their elders – is also served by moulding together insights from past experiences with a range of recent scholarly approaches. In suggesting the model of the 'plasticine city' to think with, it has been my intention to learn from the abundant creativity of young people and to draw attention to the myriad moments of shaping that occur within – and to – the parameters of urban space. Cities, it might be concluded, are neither hard nor fluid; it is rather their material pliability that renders them so engaging for social groups and individuals alike. The task of scholars, then, is to attempt to know what city life in past times felt like, what it feels like today, and how, perhaps, it might be experienced tomorrow.

Notes

1 My thanks to the editors for their assistance in bringing this chapter to fruition. Thanks also to Edward Carey and his publisher, Picador, for permission to use the epigraph with which this chapter opens. I wish to acknowledge my collective affiliations in the writing of this chapter, namely with the Department of History at King's College London and also (as Adjunct Research Associate) with the School of History at Monash University.
2 Deyan Sudjic, 'The broader picture; the streets are paved with plasticine', *Independent*, 9 November 1997. Accessed 13 November 2014. www.independent.co.uk/arts-entertainment/the-broader-picture-the-streets-are-paved-with-plasticine-1293151.html. David Rodgers, 'Plasticine model of future city needs new home', Building Design Online, 21 March 2011. Accessed 13 November 2014. www.bdonline.co.uk/plasticine-model-of-future-city-needs-new-home/5015298.article.
3 'Bishkek to host exhibition of miniature plasticine city of Goldenburg', *AKIpress*, 31 March 2014. Accessed 13 November 2014. www.akipress.com/news:538476/.
4 *Age*, 1 April 1899, 2.
5 *Age*, 8 April 1899.
6 *Argus*, 24 June 1897, 6.
7 *Argus*, 11 December 1873, 6.
8 *Argus*, 28 October 1872, 5.
9 Town Clerk's files. Public Record Office Victoria, VPRS 3181, Series 1, Unit 553, 'Markets', item 3876. Letter dated 8 October 1894.
10 Also: Town Clerk's files. Public Record Office Victoria, VPRS 3181, Series 1, Unit 748, 'Parks', item 3215. Report dated 22 August 1892.
11 Town Clerk's files. Public Record Office Victoria, VPRS 3181, Series 1, Unit 753, 'Parks', item 1954. Letter dated 9 June 1897.
12 Town Clerk's files. Public Record Office Victoria, VPRS 3181, Series 1, Unit 753, 'Parks', item 1954. Cover note dated 3 July 1897.
13 *Argus*, 25 October 1899, 11.
14 *Age*, 17 January 1899, 4.
15 *Argus*, 10 May 1872, 4.
16 Also: *Age*, 11 July 1900, 11.
17 *Argus*, 24 March 1909, 9.

10 What has the future of urban parks to do with their past?

Katy Layton-Jones

Introduction

The conventional narrative of the British public park is predicated upon pastoral nostalgia and a detached romance that eludes any other feature of the modern city. As these green landscapes often incorporate caves, lakes, hidden walks and dells, it is unsurprising that they have come to be championed as remnants of a lost age, within which childlike innocence still resists the utilitarian precepts of the modern world. For examples of such features see: the caves and grottos at Sefton Park, Liverpool (opened 1872), Jesmond Dene Public Park, Newcastle upon Tyne (laid out during the 1860s), Wanstead Park (purchased 1880), and the rockeries of Burslem Park, Stoke-on-Trent (built in c. 1894), and Birkenhead Park (1847) (See figures 10.1 and 10.2).

The fantastic nature of these spaces has often outlived the equally whimsical architecture of the Victorian villas and terraces that surrounded them, as the latter were lost to bombing in the early 1940s, large-scale council developments in the 1960s and 1970s, and, more recently, the higher-density schemes of private property developers. At each of these stages the imaginative character of the public park has acquired greater scarcity value. Always a diversification of the urban realm, they are today, more than ever, viewed as an exception in the wider spatial politics of urban life. Although their form and features are now familiar to us, they remain divergent from the surrounding streetscape. In this sense, they have more in common with the alternative urban 'wildscapes' sought by many of today's urbanites than the Victorian streets that they were originally designed to complement (Jorgensen and Keenan 2012). Yet, as with many exceptional environments, there is growing pressure for them to revert to the uniformity of the surrounding city, both aesthetically and governmentally. As a result, the urban public park has become a spatial and ideological battleground. Representatives from across the political spectrum continually jostle to propose their own solution to the 'park problem', but thus far there is little consensus as to the nature of that problem, let alone a cohesive strategy for its remedy. As Sjerp de Vries has recently observed, 'although few people doubt that green spaces and natural elements benefit human health, such insights are not commonly applied in policy making in relevant sectors such as nature conservation, spatial planning and public health' (de Vries 2010, 77). In some instances the challenges faced are historical and pre-date the creation of urban parks; others are the consequence of both accidental and wilful ignorance of historical precedent; and yet more are the result of an historical precedent being exploited with little attention to whether or not the model in question was successful or misguided. Acknowledging historical precedent and

Figure 10.1 Birkenhead Park, 1906
Source: author

historical attitudes to urban green space does not equate to appreciating and reflecting the current value of those sites as historicised landscapes. Nowhere are the implications of these oversights more apparent than in questions surrounding ownership.

The park is public

Although many early parks were created as components of private, speculative, housing developments, these were usually the least successful years of their existence. Parks such as Princes Park, Liverpool were gifted to or bought by local authorities as a necessity in the late nineteenth and early twentieth centuries. Those funded through private subscription, real estate sales, or charitable patronage, floundered financially between the 1840s and 1918. As hybrids between private gardens and public service providers, urban parks failed on both counts. In fact, despite widespread concerns about raising rates, as early as the 1850s local authorities demonstrated widespread preference for taxpayer-funded schemes as the only practicable means to guarantee the survival of municipal parks. As a result the fundamental principle of land that belongs simultaneously to the individual, the community, the city and the nation as a whole, with equal benefit for all, became central to the perceived value of urban parks in the late nineteenth and twentieth centuries. An account of Stanley Park in Liverpool published in 1870 boasted that 'the park is for the people – the people should use it without let or hindrance; and the only conditions imposed on those who enter should be propriety of behaviour ... The beggar and the prince should be free to use it on this condition'.[1] A similar principle

Figure 10.2 Sefton Park, Liverpool, c. 1914
Source: author

defined common lands of the medieval period and, for many urban residents in the nineteenth century, this was their closest reference for these new environments. In public hands, urban parks conferred on those residents with neither property, nor the wealth to acquire it, an important sense of collective ownership. For town councils, the benefit of that ownership was not merely a reliable funding stream, but also community wellbeing.

This vision of parks as democratic landscapes was clearly in the minds of civic leaders, designers and commentators alike, even if equality of access did not necessarily result in equal investment. In some instances this principle was made explicit, as in the nicknaming of Birkenhead Park the 'people's park', a title adopted as the formal name for Halifax's new facility ten years later. These 'Arcadias for all' were not seen merely as independent from the principles of private land ownership, but specifically as an antidote to its effects (Lasdun 1991, 167).

The correlation between the physical openness and accessibility of the public park landscape, and its status as a public asset is enduring and powerful. For many the value of the public park resides not merely in its physical qualities, but also and even more importantly, in its perceived immunity from private acquisition. This is no historical accident. In 1857 two of the conditions for the transferral of the freehold of the People's Park from the donor, Mr Frank Crossley, to the Halifax Corporation, West Yorkshire, was that 'the park be open to the public during the whole year' and that 'no charge shall ever be made for admission to the park, nor any use made of it for private or municipal pecuniary advantage'.[2] If this condition was ever broken the land was to revert back to Crossley or his heirs. The principle of accessibility and public ownership is therefore long-standing and, whether welcome or not, holds at least as much significance today as it did in the mid-nineteenth century.

At the time of their creation, the case for public parks was also predicated upon the promise of their permanence. As the town clerk for Wolverhampton promised, while debating the creation of a park in 1875, such a site should offer 'an opportunity of recreation in perpetuity'.[3] Far from diminishing over time this principle of parks as permanent *and* public spaces has intensified. While the temporary or 'pop up' park may provide a brief relief from the built environment, its transient nature frustrates and undermines the tradition of parks as the physical realisation of an enduring civic contract. The Victorian public park promised a permanent and corporal acknowledgement of a free citizen's absolute right to open space, solitude and the forms, scents and murmurs of the natural world. It is unsurprising then that communities are reluctant to compromise or erode such landscapes. The tension between public and private interests is thus not merely specific and localised, but rather has defined the very terms on which urban green space is valued and, for many, the urban park remains the physical expression of a principle of civic responsibility, collectivism and common interest.

As the wider streetscape has become increasingly removed from public control, with the creation of gated communities, 'defensible space', CCTV and malls without walls, existing public parks have, in most cases, remained beyond the grasp of private interests (Minton 2009; Newman 1972). There are exceptions, such as the sale of formerly public sports facilities within parks to private operators, the long-term leasing of tennis courts or bowling greens, and even the granting of exclusive use of park facilities to nearby schools.[4] However, the intention of many Victorian park designers to create sites of passive recreation has often made them difficult to adapt for commercial purposes. This is due in part to the fact that vistas and large, uninterrupted greenswards are fundamental qualities of many parks and these features are easily corrupted and undermined by even small-scale intrusions (see figure 10.3). Serpentine paths bisect greenswards, making them unsuitable for football pitches, and the undulating, naturalistic topography preferred by nineteenth-century landscape architects for picturesque effect has, to some extent, protected the sites from development. However, the scarcity value of historic features and the integrity of design within a park, both of which attract the greatest attention (and funding) from the heritage sector, are, for many park users, merely the physical evidence of a more fundamental principle; the tacit right of citizens to claim and reclaim collective ownership of public spaces.

In many cases, nature itself has played a part in protecting the park landscape from the encroachment of the city. The presence of natural watercourses or marshy land made various locations undesirable for house building and it is these areas that were most commonly designated as parks. Locations such as South Park, Ilford (1902), and areas of the Tooting Commons (see figure 10.4), include terrain that has deterred generations of developers.

Although this unsuitability for commercial exploitation presents challenges to park managers it has ensured a relatively high survival rate, in size if not condition, of the urban parks canon. In the absence of statutory protection, a combination of natural topography and public attachment has thus far guarded the unique qualities of the urban park. However, one consequence of this stark contrast with the increasingly commercialised surrounding streetscapes is that the park boundary has become a

Figure 10.3 The Public Park, Barrow-in-Furness, c. 1920
Source: author

Figure 10.4 Tooting Common, 1904
Source: author

symbolic front line between the collective enterprise of civic life and the private profit nexus of the modern property market. As Karen Jones and John Wills have observed, 'the park landscape has commonly functioned as a paradigm of reaction against contemporary social problems, promising a green tonic, a natural remedy, to civic alienation' (Jones and Wills 2005, 173–4). When the effects of this civic alienation are perceived to be encroaching on the park itself then residents do not merely lose a resource, but also the remedy with which to mitigate or cure that sense of loss. This is not a peculiarly British perspective. Across Europe and America, the public's pursuit of non-proprietary space has intensified over the past two decades. Of the commons in Christiania, Denmark, Maria Hellström Reimer has observed that they have 'come to represent a non-proprietary, collective form of spatial organization based on a localized relationship to land, resources and ideas' (Hellström Reimer 2012, 125). Of course, public parks were never wholly unregulated landscapes and sanctioned uses and proscribed activities were stipulated upon their creation. However, the values of the many were privileged above private advantage. These values and attachments existed even when there was little historical precedent for an urban green space. In fact, the act of reclaiming the private domain for the public good can forge an even more powerful attachment between communities and urban parks.

In the 1820s and 1830s, as urban populations grew, previously industrial spaces were increasingly adapted as formal and informal urban green spaces. In some instances they also served more pragmatic functions, such as cemeteries, but the narrative was invariably one of improvement and service to the needs of the wider community. In

Figure 10.5 St. James' Cemetery, Liverpool, c. 1835
Source: author

Liverpool in the late 1820s a former stone quarry in the area of St. James' Mount underwent just such a transformation (see figure 10.5). It was first utilised formally as a cemetery and informally as an urban walk, as the *Mirror of Literature, Amusement, and Instruction* described in 1833: 'the lower part of the cemetery is laid out as a garden, and forms a valuable addition to the public walks of the town. The north, west, and south sides slope from the upper wall, and are planted with trees'.[5] More recently, the New York High Line (2009) and the *Promenade Plantée* in Paris (1993) similarly required the reclamation of previously inaccessible urban infrastructure. The resulting 'parks' are essentially detached from the city proper, while enabling the user to be constantly reminded of the novelty of their experience within an urban context. This state of simultaneous knowledge of, and detachment from, the commercial landscape of the modern city echoes the escapist narratives that underpinned nineteenth-century attitudes to the municipal public park. Today in New York and Paris the pedestrian literally transcends the city without denying its presence, while nineteenth-century park advocates saw the close proximity of the built environment and urban green space as a means of accentuating the distinctive character of both. On the new aerial greenways, as in the sunken quarry gardens of the nineteenth century, the contrast between that which falls within and without the park boundary enhances the sense of escape and fantasy. In short, the implicit public contract that traditionally underpinned public parks remains a key principle among the most successful newly created facilities. If the city encroaches physically or ideologically upon the park, that contract is compromised, potentially irreparably. The threat of this occurring is more imminent than ever and one early sign is the growing popularity of terms such as 'green infrastructure', which serve to subtly integrate the park into the wider, increasingly privatised, townscape. The threat originates in some unexpected directions, including traditional advocates of park protection who are seeking to satisfy the new government appetite for a quantitative evaluation of urban parks, not least in relation to health and fitness.

The park is 'healthful'

Perhaps the most powerful and persistent claim made by champions of public parks, both throughout the nineteenth and twentieth centuries and today, is that they improve the health of an urban population. In the nineteenth century the greatest threats to the health of urban populations were contagious diseases and malnutrition, while today they include the 'obesity epidemic' and the complications that accompany old age. These challenges place heavy budgetary burdens on national and local government and so, unsurprisingly, politicians and health workers are keen to explore the potential of public parks to mitigate their impact. At the same time, and while under pressure to deliver new residential developments alongside budget cuts, many councils have adopted the nineteenth-century public health narrative to attract funding and protection of their parks. Yet, although understandable, there are reasons to challenge the reduction of the defence of urban parks to an invocation of the historical public health argument, both in terms of the success of original park schemes in achieving this objective, and the applicability of the comparison to the modern fitness agenda.

Throughout the nineteenth century the rapid growth of Britain's urban population resulted in unregulated development and increases in the frequency and death tolls of epidemics. Liverpool's role as a seaport made it particularly susceptible to outbreaks of contagious diseases, which could arrive in the town via the numerous ships that docked at the quayside every day. The threat to public health posed by the city's mercantile character was noted by Dr Trench, the Medical Officer for Liverpool in his 1863 *Report* on the city's health in which he argued that 'the migratory habits of our poor, and the vast stream of people flocking to the banks of the Mersey for pleasure, business, or in transit to other lands, by providing a succession of fresh victims, maintain the constant activity of contagion' (Trench 1863, 9). Campaigners like Edwin Chadwick, who published his famous tract *The Sanitary Condition of the Working Population of Great Britain* in 1842, also identified the physical environment as a crucial factor in the rapid spread of contagious diseases (Chadwick 1842). In the densely built courts and infamous cellar-dwellings of Liverpool diseases such as typhus and cholera could decimate an entire district. In 1832 and 1846 the city, like much of Britain, was hit by two particularly savage cholera epidemics. Westminster responded to the latter epidemic with the Public Health Act of 1848. The main intention of the Act was to improve housing, the street infrastructure and sewerage, but it also encouraged environmental improvements such as green space provision. Still, it is worth reconsidering the extent to which parks could have assisted in combating diseases such as cholera and typhus.

In nineteenth-century newspapers and reports on urban health, park creation was particularly associated with improvements to air quality. As early as 1829 the landscape architect John Claudius Loudon framed his proposals for improving green landscapes as 'Hints on Breathing Places for the Metropolis, and for Country Towns and Villages' (Loudon 1829, 686–90). Parks were thus championed for providing pockets of clean air, devoid of the heavy black smog that hung over residential and industrial districts. Residents and observers were instructed to view parks as an antidote to the physical symptoms of urban life. However, the extent to which this was ever a reality prior to the Clean Air Act of 1956 is debatable. The working population tended to live in the most polluted regions, close to employment opportunities and usually some distance from these urban 'lungs'. Being located on the outskirts of the towns as they were, even large parks such as the 'ring of parks' of Newsham, Stanley and Sefton in Liverpool (1868–72) could have only limited impact on the health of slum dwellers. In addition, cholera is waterborne, not airborne. The disease certainly had a greater impact in densely populated areas, but this was because a large number of people could potentially share a polluted water source. Similarly, typhus, which is carried by lice, would spread quickly where residents lived in close quarters, but the only way to combat the threat was to improve the residential districts themselves. In many instances, the provision of public parks on the outskirts of towns made little impact upon the squalid conditions of the slum districts themselves.

For those recovering from these illnesses, the opportunity to enjoy public green spaces would surely have proved beneficial. However, as those who were most in need of clean air were likely to live in the most polluted districts, their nearest park was also likely to be one most heavily compromised by air pollution. In an account of the opening of Stanley Park in 1870 the *Illustrated London News* noted that the view from the terrace there was 'commonly obscured by the smoke of the factory districts'.[6] Clearly, from the time of its creation, that park promised little improvement in air quality in comparison to the old city-centre gardens. The old Mount Gardens on the site of the modern Anglican cathedral was criticised in 1861 by the local newspaper *The Porcupine* for being a 'cheerful scrap of blackened grass plot, with all the smoke of the whole line of town and docks agreeably uprising on one side, and the vapours of a charnel-house … steaming upwards on the other'.[7] If the newly laid out Stanley Park suffered from the same problems of air pollution as its town-centre predecessor, then we really should re-evaluate our assumptions about the direct atmospheric benefits of park provision in the nineteenth century.

One element of the 'health' debate that remains almost unchanged from the nineteenth century is that of 'moral health'. Although today, the problem is more likely to have the nomenclature of 'anti-social behaviour', the concerns remain strikingly consistent. In the nineteenth century, as today, it was when problems such as prostitution and alcoholism were noticed in parks in affluent neighbourhoods that a city's ruling elite expressed concern. In Liverpool, this occurred in Sefton Park, as the local newspaper *The Liverpool Citizen* lamented in 1883:

> We have heard a great deal recently about social purity, and the rest of it. What do the Parks' Committee say to the condition of Sefton Park after dark? The few keepers are useless. The plain truth had better be said publicly. The condition of the park after dark is a scandal and a disgrace to the city and Corporation of Liverpool … At present it is a hot-bed of iniquity instead of a healthy breathing place for the community.[8]

As the reputation of Sefton as a polite, even exclusive, destination was essential to attracting respectable visitors, efforts were made to control access and deter the disreputable, even when that included the very sectors of the population most in need of the facility; a response not dissimilar from the defensive architecture and zero-tolerance policing advocated by many commentators today.

It is reasonable to question the extent to which public parks really improved general public health in the nineteenth century. Parks were neither hospitals nor gymnasia. The parks movement in the 1860s was not about creating sports facilities. Indeed many sports were banned in parks, with only the 'polite' sports of archery and cricket permitted. Football was prohibited in Sefton Park, and although it was permitted in Stanley and Newsham parks, it was only to take place in designated areas which did not attract any funding from the Improvement Committee. A similar policy was common across the country (see figure 10.6), with Crossley stipulating that 'no games, such as cricket, bowls, of hockey &c., be allowed, nor bathing' in his newly gifted park.[9]

In fact, the sports fields of many sites retain their pre-park appearance and quality to the present day; they are literally fields. Rather than investing in sports facilities the original park designers preferred to mimic the features of private estates. Stanley Park (in a district not well known for its horse-owning elite) had its very own 'Rotton Row' for horse riding from the late 1860s until the 1920s. Only during the social crisis that followed the First World War did the city's leaders finally concede defeat and rename it a 'cycle path'. This adaptation is more comparable to the model currently championed. However, whereas earlier park providers were mindful primarily of public health, today the focus is the fitness of the private individual and active recreation. This shift in emphasis is about more than mere semantics. The greenswards that comprise large portions of the historical park landscape served originally as surrogate countryside and so were valued aesthetically as well as practically. In the future, it is likely that they will have to deliver quantifiable impacts on the fitness of individuals. One risk is that this will, over time, result in a de facto commodification of the park

Figure 10.6 Muirhead Gardens, West Derby, c. 1925
Sources: author

landscape, just as the private health centre or athletics track is a *productive* space, the value of which is appreciated exclusively in terms of standardised and quantifiable fitness outcomes. The threat is not that such outcomes could occur in addition to the values of alternative, public non-proprietorial space, but that the two systems of values are fundamentally incompatible.

In a battle between the utilitarian benefits of the commoditised park landscape and the less tangible but enduring principle of the public space there seems little doubt that in the current fiscal regime the former will triumph. In so doing it will potentially erode progress made over the past century in the lifestyle quality of urban populations. Such a reorganisation of the civic contract would necessitate a categorical change in the spatial politics of the public park. The sense of ownership and therefore custodianship of public space could be replaced with unguaranteed use of prescribed services. If, as Jones and Wills suggest, 'from the very beginning, the park has been part of our creative voice … Put simply, the park is our invention', then to a great extent the appeal of the park landscape is that its presence acknowledges that human needs extend beyond mere utility (Jones and Wills 2005, 1). To reduce the status of the park to that of a mere pragmatic solution to urbanisation would be to diminish its value permanently.

Conclusions

If the provision of public green space in the nineteenth and twentieth centuries was motivated by the need to compensate for the condition of growing cities, then this compensation was just as much ideological as it was practical. The *idea* of an alternative to the slum or tenement was as important, if not more so, than the health benefits of green environments. This should not be exploited as a reason to undermine the very real value of urban parks to individual health and fitness. However, it does highlight the limitations of the current emphasis on quantifying the impact of parks. What distinguishes historical models from those proposed for the future, is the current lack of potential for the coexistence of the practical, utilitarian model and the principle of accessible, publicly owned, space. Previously, citizens determined how the space functioned for them, enabling their personal engagement with, and use of, the land to evolve over the course of their lives. As the functions of green spaces are increasingly and restrictively prescribed, this high level of engagement may be denied to future park users. While the future of public urban parks in Britain is far from certain, the myriad initiatives, reports and policy revisions of recent decades make one fact abundantly clear: one way or another, the future of urban parks will remain inseparably wedded to their past.

Notes

1 *The Porcupine*, 21 May 1870, 77.
2 *Leeds Mercury*, 15 August 1857, 8.
3 *Birmingham Daily Post*, 14 May 1875, 5.
4 For recent examples, see the tennis courts at Ravenscourt Park, Hammersmith and Fulham, and the battle over the Old Football Pitches in Hyde Park.
5 *Mirror of Literature, Amusement, and Instruction*, 14 December 1833, 639.
6 *Illustrated London News*, 28 May 1870, 545.
7 *The Porcupine*, 29 June 1861, 146.
8 *The Liverpool Citizen*, 11 July 1888, 3.
9 *Leeds Mercury*, 15 August 1857, 8.

Part III

Historical cities, contemporary spatial cultures

11 Between urban and digital spaces

Vinicius M. Netto

Distance is a form of the attainable.[1]

(J.L. Borges)

Introduction

The idea of digital spaces – an invisible network of interactions and information overlapping the historically concrete and durable space of cities, until recently the sensory condition of experience – has captured contemporary imagination. These terms evoke a somewhat unusual encounter between apparently separate phenomena, distinct materialities. Scholars from different areas have been attracted by this new materiality, which in itself contains the potential for connection between events, people and places, taking the form of networks of hypertexts and possibilities for interaction on unprecedented levels – new, paradoxical forms of the attainable, to recall Borges' words.

Indeed this intriguing fabric of materiality and its ubiquity even begins to mean endless connectivity. Digital networks are considered as a new space-time experience in which the properties of *multiplicity* and *simultaneity* of world events become increasingly clear – and attainable as never before. Understanding the unveiling of this world based on new material foundations and a culture experiencing rapid technological acceleration means understanding how the coexistence of urban spaces and digital networks arises and affects our lives, our ways of acting, experiencing and socialising in material conditions that are also partially new. It is interesting to note that until recently the relationship between the concrete nature of urban space and the elusiveness of the digital even seemed to be a contradiction – an impossible synthesis, as if the digital put a check on the concrete, in an electronic overpowering of the old space, making it obsolete. A destructive and inexorable dualism was suggested.[2]

As we shall see, this dualism has been overcome. We know that the rift between concrete spaces and digital networks has not occurred. But if there has been no rift, can we say that there has been an interweaving instead? Confirmation that there is no contradiction does not suspend the fundamental problem that digital networks and urban structures have entirely different natures: the volatility, elusiveness, invisibility of one; the tangibility, rigidity and constant presence of the other. Where are the connections, the points of convergence between the 'digital' and the 'concrete'? How do these planes of different materiality actually involve our practices (linguistic and nonliguistic, semantic and corporeal)?

This chapter aims to explore these questions beginning with an investigation of the nature of urban and digital networks and their roles in the emergence of social practice. We shall then move on to see how practice *bifurcates* between them, i.e. how practice unfolds into two distinct realms, as webs of acts simultaneously and sequentially interconnected and continuous. The aim is to understand (1) how two seemingly different material layers that constitute the fabric of the social relate to and intersect each other; and (2) how social practice emerges through these realms, i.e. how our individual acts relate to the acts of other people in different places and times in the process of *association*. Put simply, the idea is to understand how practice becomes *social* practice once this process begins to emerge through different material channels and technological media. This chapter addresses the coexistence of these flows and materialities, indeed resolved and naturalised in practice. It is theory that now needs to capture the conditions and modes of this coexistence. Three points of connection and return of practice from the digital to the concrete (and back) are suggested: *meaning, the body and the centrality of the subject acting in the place*. Finally, the chapter shall propose that, in the unfolding of practice in diverse material networks, urban space is becoming increasingly removed from its original role as a material environment that is highly central in social reproduction for one among many environments. On the other hand, and rather counterintuitively, such process would trigger the incorporation of a new ontological role, precisely as a *fundamental counterpoint to the elusiveness and partiality of digital communication networks*. This chapter will pose the paradox of a diminishing role for urban space that is also potentially reasserted and renewed by the growing complexity of the social world.

In different spaces

Some questions seem central to the understanding of the way in which the two supposed forms of 'space' relate to each other, where their contact touches on human practice and how it affects our associations. A first question is how to grasp the complexity of the framework of actions unfolding in two types of networks. In other words we have to understand how our actions unfold in interactions that spread partly through urban space and partly in electronic networks and mobile communications. We also need to understand the material, practical and cognitive connections that make them possible and give them substance.

Understanding these problems means finding new descriptions. A first description relates to how we act immersed in spatial and temporal dynamics that increasingly break down the profound association between distance and time inherent to the historical constitution of practice and demand constant updating and changes in our understanding of the world that is presented. A second description would touch on the conditions that ensure integration of these frameworks and their potentially endless ramifications.

Let us start with urban space. An unusual question might help us to see more of the place of space in such ramifications. What spatialities other than the urban could have the effect of stimulating interaction? Someone might point out the possibility of interaction without needing to be co-present – precisely through distance-communication technologies. But could those technologies be attained without the density of co-presence and communication historically provided by the densities and structures of the city in the first place? The city means the possibility of performing acts not through a shapeless world of fragmented or labyrinthine spatialities, which would render the emergence of communication mediated by the body and co-presence highly problematic. Spatial economics has shown us that the production of urban structures increases the potential of space in putting actors in conditions of exchange. But the effects of these spatial structures echo through all the spheres of practice mediated by the body, articulating every form of interaction. Urban spaces have effects on the relational nature of our acts and can intensify the passage from individual to social acts.

So the potential for the emergence of communication mediated by the body is deeply embedded in the actual materiality of the urban. We may see space as a material condition for stimulating (or controlling, depending on the society, group and urban area in question) the possibilities of communication in the form of: supporting the randomness and contingency of interactions in the social system, based on a kaleidoscope of encounters and access to places of activity; increasing the potential for complementary actions finding continuity or for desired interactions to occur – say, in urban centralities where complementary activities seek to locate themselves close to one another; and asserting conditions for what Giddens (1984, 73–78) addressed as *seriality of encounters*, thus ensuring the continuity of interactions and their passage to a wide range of actual relationships, which in turn will perform the role of establishing continuities in the symbolic and material reproduction of groups and social systems.

But how does social practice emerge through concrete space? How can space be part of the communication networks that constitute association? Although largely ignoring the problem of materiality, Niklas Luhmann (1995) offers interesting ideas in this sense. He explores a Husserlian view of meaning as *reference*: '"[M]eaning – in communicative speech – is always bound up with such an indicative relation' (Husserl, 2001, 183). Luhmann pushes Husserl's concept of indicative meaning into a concept of self-referentiality: every meaning brings within itself references to other meanings in a Lifeworld. Social systems are immersed in informational production: meaning becomes their environment. Such an approach emphasises communication and discusses the place of meaning in association and in 'ontological relationality', the integration of social and material reality.

Could space and digital networks take any part in such chains of references? According to Luhmann, an act is not an isolated phenomenon without connections to its surroundings. Luhmann ignores the materialities at play in this environment of communication, but we may recognise its importance through notions like the *contextuality* of situated practices asserted by social theorists like Goffman (2010), Garfinkel (1963) and Giddens (1984) and human geographers like Hägerstrand (1970) and Thrift (1983). The intrinsic connections of acts in the world immediately relate our acts to their outputs, their context (acts arise through spaces and places) and other acts, their outputs and contexts. Neither can urban space be experienced in isolation, nor have effects only upon itself. While our actions continue to be related (that is to say produced through references to other actions, their contexts and their effects), their *spaces* will take part in those referential connections. Luhmann's theory allows us to see that these connections are shared meanings produced in actions[3] and conveyed in urban spaces. In other words, a substantial part of the inherent relationality[4] that makes up social and material reality is achieved through references between our actions and space. Examples are frequent – say, looking for a particular service in a certain street. Association requires actors to generate relations as they act, a construction that includes physical presence and communicative exchange; connections that include a dimension that is as informational as it is material; connections that incorporate the spatial contexts in which they act.

However, this description of action and space as linked together is still too general. How do such connections materialise in everyday life? Urban space seems to have different presences in association.

1. *Space is appropriated as a source of information* for social practices.[5] Urban space is a means to know the social world and the way in which it is works, a way of the social world presenting itself to itself. We are participating in a social situation if we know that it possibly occurs in a particular place. We are guided to that place by its meaning, that is to say by understanding its social content as a context of certain acts that interest us. As we know the city and its material and semantic structures we can anticipate the possibilities for interaction. The city provides a reference system for our actions.
2. Then there is *appropriation of space at the moment of interaction*. The spatiality of places makes way for interaction, but considerably beyond being a mere physical support and visual setting. Space can be seen as an ambivalent environment – both material and informational – that supports communication. Such space laden with meanings, a *semanticised* space, is the result of our practice (Wittgenstein, 2001). Moreover, space is only fully semanticised when it is enacted *jointly and jointly recognised as such*, i.e. when another actor perceives it as such by interpreting practices that take place in it. That space of socially recognisable meanings is the result of the actual association that it supports and expresses. A semanticised space is the sign that space has been appropriated and recognised as a reference for acts.
3. Finally, this semanticised space becomes part of the *informational connections between acts* in progress or to be produced at other times and places. This is the formation of communication networks beyond the temporal and spatial boundaries of the event and its place, when actors interact with absent actors through distance communication (through the exchange of artefacts, accessing texts, hypertexts or with mobile devices). This third moment leads to the *formation of complexes of interaction* and includes space as mediation in the production of a highly ephemeral form of social structure, produced as networks of communicative exchange in the 'here and now'.[6] We relate to spatial contexts so that the effects of our actions can relate to each other through them.

That is the central idea of *association through a 'referential space'* – and the place occupied by space in the production of references that connect our acts and their effects to their contexts of realisation (Netto 2016). Let us see now how digital communication takes part in these referential networks.

Practice and the invisible world of the bit

Notions of 'cyberspace' and 'digital spaces' have acquired considerable popularity, but what exactly do they involve? Key features of the change we are experiencing include the growing use of computers in carrying out actions, our progressive immersion in signs and content in the continuous production of information allowed by new technologies. From a functional point of view these technologies have led to increased speed and precision of actions and the control of their results. Actions previously carried out in the physical world of presential interaction and materialised in the analogical register of printed signs are now realised in the technical interface with electronic media. Products of action lose tangibility and are re-codified in the apparent immateriality of the electron, represented as machine code; *re-materialised* in the invisible and impalpable world of the *bit*. The use of computers involves and absorbs our actions – making the 'digital world of the bit' inside computers the projection of actions, the means for the materialisation of their effects. It launches our actions into a kind of *third nature* – beyond the ecosystem and the physical urban system.

The sense of partial invisibility of the operations and products of practice mediated by digital technologies is further expanded by another particular aspect. Texts and images acquire the possibility of moving and replicating themselves infinitely, instantaneously re-materialising in any other place that is digitally connected. The effects of our acts are able to travel without the material support of the physical artefact and can affect people and contexts we have never seen and never will see.

This invisible environment is formed by the apparent omnipresence of the bit and takes the historical form of an *inexorability of the immersion and conversion of practice to the digital*; an immersion in the 'unbearable lightness of the bit', to ironically paraphrase the fiction writer Milan Kundera, exponentiated by the possibility of emitting the effects of our acts, now converted into the imponderable flow of electronic bits to anywhere with a device capable of decoding the flow and re-materialising it on a machine screen. This invisible environment does find physical form – and certainly a physical *connotation* in our imagination, since we can no longer imagine all its ramifications, or where its nodes really come together: the *electronic web* that connects computers and everyone in conditions to use them. The web becomes what Mitchell (1995, 44) might call 'another place'. Socialities seem available at the click of a mouse. Practices of interaction across distances begin to form in the paths of this endless maze.

The relation between space and the status of human practice in this new ontological condition is still tensioned by an *acceleration of the historical rupture with the dependence of tangibility and presence*. Such ruptures require the subject's redefinition (consciously or otherwise) of her possibilities in the face of the increasing means of conversion of practices and their effects into such virtual, invisible maze of connectivity.

The encounter between practice and the technologies and flows of the bit seem to appear to the human subject, in their extent and consequences, as a *growing virtualisation of the world*. Practice does not become virtual in the sense of being less real but instead in the sense of becoming partially invisible, imponderable, since a large part of the 'world within my reach' (Schütz and Luckmann 1973, 36–40, 51) is substantially produced and experienced in a world whose frameworks are becoming increasingly digitised. This new status of practice's spatiality new, endless connectivity seems part of the difficulty in understanding the changes being experienced.

The communication network that became popular with the introduction of the internet is now associated with another: *telematic webs* of mobile communication are popular and evolving in new technologies, exponentiating communication with the mobility of transmitting devices. A new period of communicative explosion came into being with the passage from the static interface of computers and fixed digital-web nodes to the mobile interfaces of cell phones and other portable devices with internet access, the recent convergence of two kinds of communication network through mobile technologies. Smartphones are practically everywhere and connected to the internet, wi-fi and Bluetooth short-range networks, along with other portable devices like tablets and laptops. These hybrid networks form the new media of 'post-mass function' with increasing bilateral exchange of information. Cyberspace now relates intimately to people on the streets, enveloping actors in a 'total environment' of connections.

Mobility and its relationship with other means of transpatial association, breaking with the extent and contiguity of space, and the de-materialisation and re-materialisation of the effects of practice transmissible and replicable, form impressive ontological conditions. We are experiencing a materialisation of Henri Bergson's concepts of 'simultaneity of events' and 'simultaneity of flows', the apparent possibility of events entering within a single, instantaneous perception (Bergson 1999) in a connection with places, socialities and subjects on a global scale.

We also experience the impression of the ubiquity of another property, *connectivity beyond contiguity*. The ontological place occupied by spatial contiguity in our readings of social and material reality is somehow destabilised by the assertion of a topology beyond concreteness, possible only through a break with the absolute continuity of space and able to align globally distributed events into a single temporality in our cognition. Channelling practice in multiple shortcuts of the concrete materiality of urban and geographic spaces leads to the establishment of what we might call a *topological understanding of the world*, in which socialities, places and subjects suddenly seem attainable. The ontology of a social and material world becomes penetrable and intelligible in its opaque totality through the concept of 'network', popularised and turned into a paradigm, as the relational graph representing the whole social world becomes flatter and flatter. This is an ontology in which historicity itself seems to either collapse or become invisible through the impression of an *eternal present*. It is introduced by access to the simultaneity of agencies and situations appearing globally, all the time – visible and recordable as never before by cameras and communication devices carried on the body.

Precisely because of this, we are affected by the exponentiation of an overwhelming property: *complexity* – in terms of information in volumes that are impossible to be processed (pages and hypertexts that will never be known; actors suddenly becoming closer and closer through digital webs which cannot be sufficiently explored). Contemporary complexity is even felt in the problem of *choice* in virtually infinite options of information and counterparts in communication – an aspect visible in Luhmann's concept of complexity.

These are some reflections on the nature of urban spaces and digital networks of transpatial communication as reference systems for the emergence of social practice. Let us discuss now how practice unfolds in these two entirely different materialities.

Unfolding practice into urban and digital networks

What is the connection between the actions we perform in urban and digital networks? Both networks become part of the connections between actions and are fundamental means of communication. We naturalise these connections and take them for granted. Given that human practice has become materially more complex as it has bifurcated into distinct, albeit interconnected, planes, one problem is to know how these interconnections are produced and kept together in practice. Thinking again counterfactually, we can also understand what *prevents* such a system becoming so complex as to be partially disconnected, through gaps in its frameworks or loss or difficulty in the search and selection of information and cooperating actors. I would like to suggest five instances of connectivity between these different material realms.

1. In cognitive terms, the interweaving involves the form in which circulating meanings connect with the concrete. I would suggest that the passage between acts materialised in urban and digital networks occurs according to *mutual references* through which connections between things, acts, information and places are established. Acting while plugged in to digital networks means bifurcating our acts and instantly merging their meanings and effects with those circulating in other places, which connect us to actors who are not present. But how do those references occur in practice?
2. In practical terms, the interweaving involves forms in which we convert references shared between actors into sequences of acts performed on different planes of materiality. Practice converts signs and images using mouse, keyboard or screen into codes, invisible bits, emitted and infinitely replicable. The products of our acts converted into bit flow and transmitted via digital and mobile networks are then reconverted into words and images in other places where, their meanings recognised, they can intervene in and affect the actions of others. Once related to new actors and places and transformed by their acts, they can once again be electronically conveyed to previous and other places and actors. The power of the references contained in the meaning structures opens out the connections latent in these exchanges and allows them to circulate coherently in multiple planes, places and agencies.
3. In spatial terms, there is the role of space in asserting social and material connectivity. New information technologies have the effect of generating an increasing *complexity of the social world*: information in volumes that are impossible to process (Castells' [1996] definition of complexity) and the problem of selecting from an unlimited choice of signs and meanings (Luhmann 1995). We are forced to filter and discriminate between an enormous number of options every time. Commerce offers a notable example. In the traditional retail system of clients purchasing products in physical stores, the level of choice is restricted by the amount of space on the available shelves. Electronic-commerce websites, however, offer selections that can never be fully known. Search tools and websites are one way of converting virtually limitless choice into a manageable set of recommendations. In this context, choice limited to a local spatial scale makes the decision-making process easier.[7]

   These observations suggest that, paradoxically, the role of urban space is potentially renewed by the exponentiation of informational complexity. The city is a form of projection of actions, which stimulates interactions and therefore the production of meanings (in objects, speech, texts and hypertexts), thus increasing social complexity.[8] At the same time, urban space is a mode of social information that enables actors' capacity for knowing and engaging in acts that make up the social world. The concrete materialisation of actions, in the form of production and occupation of spaces, arises as patterns of distribution of recognisable and reachable activities. The production of space as the changing location of actions and agencies consists of an arrangement and 'pre-selection' of situations of socialising action. As they find recognisable structures, cities form a powerful form of reduction of social complexity. Ultimately, cities become a fundamental form of projecting, increasing and reducing social complexity – at the same time, consistently with Luhmann's (1995) concept of *autopoiesis*. This cycle takes on a self-referential form that defines the city as an active material counterpart in social reproduction – as space becomes a way of making the relationship between acts, actors and their production possible, rendering social reproduction sufficiently unproblematic.

4. Hugely complex, substantially invisible, never entirely knowable, the flows of referential connections between acts and their effects – even digitalised – can be entered and referenced through urban space. Connections of actions are still substantially produced and structured through meanings latent in places – including the activities of production of digital networks themselves, spatially concentrated (see Sassen, 2001).
5. We come finally to the role of new electronic spaces. William Gibson, the author of *Neuromancer*, said, 'I was never interested in Facebook or MySpace because the environment seemed too top-down mediated. They feel like malls to me. But Twitter actually feels like the street. You can bump into anybody on Twitter' (Gibson 2010, 1). The parallel between urban spaces and digital networks in Gibson's observation illustrates both the paradigmatic position of public space and the limitations of electronic networks. The street is the space where distinct identities can be co-present in a non-programmed way. The chances of effective communication are slight, but such a fortuitous and volatile encounter is where the possibility of recognising the Other is constructed. Structurally social systems need both what Hillier and Hanson (1984) call *randomness of contact*, and what Giddens (1984) calls the *seriality of encounters* in the generation of social relationships. Electronic communication networks seem to offer less space for the randomness of encounter. By tending to be used in the reproduction of ties between members of pre-identified and selected groups of individuals, they act as forces for affirming social fields rather than transversal integration of fields. These risks of dissociation need to be compensated for – given that their elimination is impossible. And the city is a place of unforeseeable exchanges, which cannot easily be replicated in electronic spaces of distance communication.

Cities as convergence of materialities: concluding notes

Let us take this exploratory reflection to the starting points of the passages between actions performed in situations of co-presence and those performed across digital networks. Even the frameworks of actions that perform through elusive materialities have their origins in the concrete spaces of the city and return to it. This return is assured by the ontological primacy of three nodalities through which such flows must pass:

The referentiality of meaning. We have seen that the cognitive connections that make up human interaction take the form of meanings, in different media and materialities. Communication is a volatile fabric of meanings, transmitted, interpreted and transformed, in perpetual circulation. These meanings have a property of establishing endless connectivity between things, even of different materialities. This concept allows us to find the material traces of the relations between our everyday acts – traces active at the precise moment of association and which form the possibility of any association.

The return to the body. Far from reifications of bodies transforming into cyborgs or extended interconnection systems, the ultimate materiality of the experienced world is related to the body and to presence. In a time when different material systems and spaces overlap and interpenetrate, Santaella (2010) suggests that it is the presence of the body with its driving, perceptive and cognitive force that rules. Even when immersed in cognitive or digital spaces, we still inhabit the indivisible body.

The inescapable centrality of the subject and her place. We can see that digital networks offer new conditions for social practice and its connectivity. So we return to one of the initial questions: how do digital networks affect urban space? Some approaches claim that digital networks would redefine and reconfigure urban spaces. Nevertheless, meanings produced in 'informational territories' (Lemos 2010) are not *inscribed* in space. These re-significations are either *volatile*, active in the place during the limited time of our practices and access to information circulating in digital networks and may remain while emission persists, or exist as a somewhat *enclosed system*, accessible only by those with the digital technologies or who are part of the electronic networks that share them.

These re-significations still depend on an ontological primacy of subjects. However, seemingly sharing a phenomenal symmetry, subjects have ontological precedence over *created* tools and technologies. In fact, subjects have a creational role. There is no symmetry when we consider the role of creativity. So despite Heidegger's (1962, 2000) assertion of the ontological place of tools and current assertions of the importance of technologies as social *actors* (Latour 2004; Harman 2010; cf. Thrift 2008), we cannot ignore their ontological condition as non-autonomous tools.

These introductory descriptions refer to the fundamental issue of what it is that maintains the integration of a social world – namely, in the form of frameworks of material and informational connections. These frameworks include language, space and the structuring means of social reproduction, such as technologies. These different subsystems provide the ontological infrastructure that connects acts; the infrastructure through which the products of actions are transmitted and disseminated to other actors and places and absorbed and synthesised into new actions. Digital networks and their progressive convergence are considerable additions to the possibilities of integration of social systems – as shortcuts in the extension of concrete space.

When the material means of integration multiplies, the historically privileged role of urban space in practice is reduced. Considering the material condition of digital networks, a counterpoint to their extraordinary properties is needed: a counterpoint able to (1) compensate for the pervasive but largely invisible presence of information flows; (2) constitute a material and informational resource in managing the growing complexity of the social; (3) constitute a means of introducing both randomness and recursivity in the generation of encounters and social relationships. Urban spatialities therefore retain their centrality as the locus of communication mediated by the body, the situated production of meanings and information – a way of providing support to practice in times of growing social complexity and de-materialisation. There is in short a shift in the ontological role of urbanised space as a material means with central task of integrating the social system into one among other means (such as information and communication networks), but with the incorporation of a new role: as a fundamental counterpoint to the elusiveness of digital flows. Digital networks are certainly going to extend and penetrate more and more into the concrete, with the multiplication of types of networks, technologies and media; but it seems that they will do so in constant dialogue with urban space, in its unique position in the ontological structure of reality.

Notes

1 'La lejanía es una forma de lo alcanzable' (Borges 1956, 1).
2 Virilio (1991), Castells (1996) and Cairncross (2001).
3 The concept of meaning as an effect of practice derives from Wittgenstein (2001).
4 See Latour (1999, 2004).
5 Cf. Haken and Portugali (2003).
6 The concept of social structure as communicative achievement derives from Luhmann (1995).
7 See Florida (2011).
8 I derive (and spatialise) the notion of increasing social complexity from Luhmann (1995).

12 Artefact and rhythm

Reflections on the introduction of Location-Based Services in the city

Frederik Weissenborn

Introduction

With the rise of digitisation – a phenomenon which includes the continued growth in computational power, the spread of ubiquitous computing, and the ever more refined capacity for capturing, processing and analysing data in real-time – human geography is experiencing a moment of significant fluctuation. The recent introduction of cheap environmental sensors has allowed scientists to capture and analyse the torrential amounts of data churned out by 'smart objects' – such as turnstiles, bicycles and thermometers (see for instance Martinez *et al.* 2004) – thereby creating new avenues of research. At the same time, the advent and popularisation of Geographic Information Systems (GIS) has enabled geographers to visualise and treat data in ways that would have been impossible just a few years ago (see Maliene *et al.* 2011); as well opening up new interpretative possibilities. In light of these developments, it has been suggested that human geography is undergoing a paradigmatic shift similar in impact to the so-called 'spatial turn', with some referring to this new paradigm as 'the digital turn' (Rogers 2013). Such claims arguably overstate the influence of digitisation on the discipline, but they also point to the increasing imbrication of the so-called "urban question" with digital systems. To the extent that digitisation may have a transformational impact on the problems faced by human geographers, it seems important to keep at least two things in mind. On the one hand, one must take seriously the potential changes brought about by digitisation both in terms of adjusting research methodologies and in re-evaluating existing theories in light of the data now available. On the other, one must be careful not to discard the potentially valuable knowledge already present in the field, investigating new empirical findings against the established theoretical architecture of human geography.

This chapter explores an important socio-spatial phenomenon – 'quotidian urban rhythm' – in the context of urban digitisation. It is divided into three parts. First, the notion of rhythm and its deployment in the theory of social space is introduced. I here devote specific attention to the work of architectural theoretician Bill Hillier, in whom a particular (but latent) kind of urban rhythmicity is defined. Second, the link between artefact and rhythm suggested in Hillier's writings is developed in the context of André Leroi-Gourhan's contention that sociogenetic rhythms may be externalised into the artefactual environment as a whole. This makes it possible to conceptualise the rhythms of urban space alongside those caused by other technologies. Third, the effects of new digital technologies and their deployment *in* and influence *on* the rhythms of urban life are examined. Drawing on recent research into so-called 'Location-Based Services' (LBS) the rhythmic potential inherent in such applications and their relation to deep urban rhythms is considered. The conclusion of the chapter reflects on some of the possible repercussions of a change in quotidian urban rhythmicity brought on by LBS and how these may be articulated.

A material approach to the question of urban rhythmicity

Human geography tends to divide rhythm into two archetypical categories. On the one hand: the mundane rhythm of *quotidian* life. On the other hand: the rhythm (or rhythms) of more exceptional events, typically of a *ceremonial* kind (e.g. processions, coronations, state funerals). Whilst such definitions provide a useful starting point for addressing the problem of urban rhythm they are, nevertheless, not the clearly demarcated categories that they sometimes are presented as being. With stylised behaviour finding its way into even the most mundane of acts and vice versa, they rather are fuzzy instances at either end of a continuous spectrum and for this reason it is imperative that one employ them with caution. Arguably the most famous account of *quotidian rhythmanalysis* is that developed in the work of Lefebvre (1992) in whom the problematic is exalted to great lyrical – if not necessarily analytical – heights. But in fact a great number of thinkers touch on the subject before him. These include human geographers (Hägerstrand 1970); social scientists (Giddens 1984); historians (Braudel 1979); philosophers (Bachelard 2000); even poets (Baudelaire 1999). Whilst these authors provide different perspectives on the problem of quotidian urban rhythmicity they also share a profound concern for the mundane *materiality* of quotidian life. In the absence of 'great narratives' – arguably the main rhythmic principle at work in ceremonial socio-spatial activities – such enquiries let the seemingly insignificant series of daily events take centre stage.

The particular phenomenon of *urban* rhythmicity may still designate many things. What in fact makes a rhythm 'urban' in any sense of the word? In answering this question, one may take as a point of departure the daily movements of the individual, the global movement flow of the urban population, even the buildings that encapsulate these flows (architecture, as Goethe would have it, is but frozen music). Yet, whilst these *kinds* of urban rhythms have been explored independently of each other, few have pursued in a rigorous way the link between them. Here, the work of Bill Hillier stands out (e.g. Hillier and Hanson 1984; Hillier 2007). In his writings, one finds a (latent) rhythmic component unfolding between the materiality of the urban artefact and the social behaviour of the people that populate it. This is a complex proposition; one which Hillier arrives at through several steps. First, he demonstrates a statistically significant regularity in the global flow of people passing through the city. Known in the literature as 'natural movement', this regularity is explained as a network effect arising from the urban morphology itself, with more central or 'integrated' streets carrying more movement than segregated ones, all things being equal (see Hillier *et al.* 1993 for an elaboration). Second, the rhythm of this natural movement is conjectured to feed back into the distribution of social activities in the city, thereby instantiating a different order of rhythmic behaviour which manifests itself in emergent patterns of socio-economic activities. The principle responsible for this ordered distribution of activities arises from the regularity of natural movement, with social actors exploiting the movement potential afforded by the material city. (Hillier for instance shows how retailers tend to capitalise on augmented levels of footfall by setting up shop on busy thoroughfares, whereas various forms of anti-social behaviour – e.g. vandalism, threatening behaviour, etc. – typically happens on less busy street segments.) This, effectively, constitutes what he calls the 'social logic' of space.

Hillier's sociogenetic theory of urban space, however, only accounts for one half of his engagement with the material city. Crucially, he also explores the processes involved in the genesis of the urban artefact itself, uncovering an emergent morphogenetic principle in operation. This has profound implications for interpreting his theory of the city. Rather than being the *representation* of an anterior social fact, the city to Hillier constitutes a material fact of its own, the logic of which is effectively irreducible to external sociological pressures.[1] More than just a reification of a social fact, the order of the urban artefact thus arises in and from its proper morphogenetic process. This process may best be conceived of in terms of an initial problem: how can an urban system maintain both *visibility* (essentially long lines of sight) and *integration* as it evolves? It is the progressive resolution of this *material* problem which leads to the organisation of urban space: accordingly, it is from this resolution that the precondition for an ordered kind of global movement – and the emergent patterns of urban social existence that Hillier argues this movement potential gives rise to – is created. By theorising, in this way, the urban artefact as the resolution of an immanent material tension, and by endowing it, in turn, with a sociogenetic efficacy, Hillier presents a relationship between urban *artefact* and urban *social fact* which has important sociological repercussions. No longer the docile carrier of analytically anterior sociological forms, the urban artefact 'takes on a life of its own' and acts back on society thereby making the relationship between the two more complex than is conventionally held. This makes his space syntax theory a useful if difficult starting point for the engagement with the rhythms of urban space.

Having made this point, it should also be noted that Hillier's conception of the phenomenon of urban rhythmicity is a somewhat impoverished one. As a concept, it remains largely implicit in his *oeuvre* (it is mentioned neither in the discussion of natural movement, nor in the discussion of the genesis of the urban artefact and therefore must be teased out from the latter. The problem furthermore is never explicated with respect to other artefacts – whether analogue or digital. There might be many reasons for this – the most significant of which is probably the already considerable level of complexity that Hillier's proposition involves – but its one-dimensionality means it has not been very receptive to analyses of digitally mediated behaviours. Whilst Hillier establishes a highly original relationship between material urban form and socio-spatial activity, his theory can only form a starting point for approximating the question of rhythmicity in the digital city.

Rhythm, artefact and societal memory

One finds a much more sustained enquiry into the notions of artefact and rhythm in the work of French palaeontologist, André Leroi-Gourhan. Bringing together the phenomena of *rhythmicity*, *artefacticity* and *sociality* in an original conceptual architecture, Leroi-Gourhan sketches out the blueprint for the artefactually-minded form of materialism that this chapter will bring to bear on the problem of urban rhythm and digitisation. A prolific writer, Leroi-Gourhan introduces a wealth of concepts in his oeuvre, two of which are particularly relevant to this enquiry. One is his notion of 'the technical lineage'; a concept which designates a group of artefacts that are structurally homologous and that therefore share certain morphological *and* morphogenetic principles (if they show similarity in *form*, then this similarity pertains to a shared ontogenetic process of *formation*). The other is that of 'societal memory exteriorization'; a concept which designates the process whereby human societies embed (or 'exteriorize') socially significant information in the artefactual or technological environments that they inhabit. It is effectively in the open-ended dynamics between these two phenomena that an original form of artefactual materialism becomes possible.

Leroi-Gourhan's technical lineages are autopoietic phenomena. As such, they follow developmental paths that are informed by tensions arising from within the individual lineage itself and their characteristics may therefore not be reduced to human ingenuity. Each technical lineage perpetuates a structural order that it may be said to *involve* or *imply*, and this order itself is constituted at the intersection of a multiplicity of forces including material forces (what are the limitations of the material(s) utilised in the production of the artefact, and how do they guide the forming process?), functional forces and aesthetic forces. If the technical tendency is irreducible to the social sphere it may nevertheless have certain functional potentials folded into it. As Deleuze and Guattari (1987) show, the form of the sabre is, for instance, optimised for attacks that take an oblique angle – thereby lending itself well to attacks from horseback as this was favoured by the Tatars and Mamluks that popularised it – whereas the two-handed sword found in various European cultures is optimised for forward-facing attacks executed from a standing position. The two artefacts thus carry altogether different functional potentials; a finding that may be extended to most artefacts.

The relationship between the technical tendency *involved* in a given artefact and the functional affordance that the artefact may be said to *express* has some important sociological implications. In time, social groups may come to rely on artefactual potentials for their own perpetuation,[2] thereby setting up an interesting but complex dialectic between *social fact* and *artefact* not unlike the one explored by Hillier.[3] In this way, exploiting a functional potential inherent in a particular artefactual lineage enables the group to store bits of the sociogenetic code involved in its proper reproduction in the artificial world, thereby perpetuating a certain 'ethnic tonality' particular to it. Leroi-Gourhan refers to this process as social 'memory exteriorisation', but it is important to note that memory exteriorisation here does not necessarily involve a cognitive process in any recognised sense of the word. It should rather be taken to denote the complex process whereby a given social group places in its artefactual environment a function that in some way or another assists in the reproduction of features that are essential to it. If this process is 'complex', it is because the evolution of the social group itself cannot be detached from the evolution of the technical milieu that it inhabits – a point that Stiegler (1998) has also made – and so the social fact and the artefact envelop one another even if their essences remain irreducible.

Crucially, Leroi-Gourhan explores the link between technical lineage and societal memory by way of *rhythm*. Accordingly, the morphogenetic principle at the heart of a technology may be said to *involve* a particular rhythm – 'technical rhythm', he writes, 'materially transforms untamed nature into instruments of humanization' (Leroi-Gourhan 1993, 272) – just as it *expresses* a particular rhythmic potential by affording certain rhythmical figures rather than others. Devoting an entire book to this problem,[4] Leroi-Gourhan demonstrates how different artefacts concurrently involve and express different rhythmic potentials and how the analysis of those rhythms may in turn provide a powerful starting point for investigating a given social group and its underlying social solidarity.[5] He writes:

> *[The] analysis of the rhythms and the organization of human society in space and time [helps] define more clearly the close connection between social behaviour and the technoeconomic apparatus* and to place both in a dialectic comparable to that which governs the relationship between the body and the mind as sustained by the nervous system.
>
> (Leroi-Gourhan 1993, 148; emphasis added)

Although the analogy with the relationship between the body and the central nervous system seems tenuous, it is clear that Leroi-Gourhan's rhythmic 'dialectic' – suspended between the many rhythms of the technoeconomic apparatus and those of a given form of social solidarity – is both a highly operational and bold concept. It attempts to define a way of theorising the technological impact on society without reducing the one to the other, thereby eschewing overly anthropocentric models of explanation as well as models steeped in technological reductionism. It furthermore provides an interesting platform from which the impact of digital technologies on urban forms of social behaviour may be investigated. Leroi-Gourhan's dialectics of rhythmic involution and rhythmic expression furthermore demonstrate a strong compatibility with the material problematic of the city as this is investigated by Hillier. The urban artefact may thus be seen to involve a technical tendency which does not reduce to a social fact, just as it may be said to express a rhythmic potential which feeds into a particular sociological order that is inherently urban. Extending Leroi-Gourhan's rhythmic dialectic to the theorisation of the city, a series of artefacts – among them Hillier's urban artefact – appear to suffuse the urban social fact like so many sociogenetic rhythms. Together they constitute a complex network of affordance navigated by the social group.

Rhythmanalysis and the digital city: group perpetuation in the context of situated feedback

Traditionally, the city has been regarded as a place of social and economic transaction. Some have described it as 'a mechanism for generating contact' (Hillier 1997, 303). Others have conjectured how this mechanism may have constituted a central precondition for deeper socio-cultural shifts such as the transition from a *kinship-based* social system to a system based on *residence* (Childe 1951) or the emergence of a specialised *division of labour* (Durkheim 1984). During the formative period of digital cities scholarship some twenty years ago, the expectation was that this functional aspect of the city would be lost. The 'bricks-and-mortar' city, so ran the conjecture, would give way to a digitised world where virtual space would play an increasingly important role at the direct expense of physical encounters, thereby progressively making one of the most significant properties of urbanism redundant. While elements of this prophecy can seem prescient – see for example the debate on the decline of the suburban high street as popular shopping destination (for example, Carmona *et al.* 2010)[6] – it is nevertheless doubtful whether its realisation has been as comprehensive as some voices suggested it would be at the time (Cairncross 2001). Indeed, the effect of digitisation on urban social life has proved to be more ambiguous and context-specific than that. Virtual or *transpatial* transactions may be replacing spatial transactions in certain areas of the social realm – e.g. suburban retail – but this does not seem to be the case in others. Take for instance social network perpetuating interaction. As Eliott and Urry show, it is still uncommon for people to completely forgo face-to-face interaction when this is geographically feasible, even if social relations could in theory be maintained purely by way of telecommunication technologies (Elliott and Urry 2010). A wholesale reconfiguration of the functionality of urban space at the hands of digital technology therefore seems to overstate the potential change urban social existence is undergoing.

To add further complexity to this problematic, the integration of digital technologies does not necessarily entail the negation of space that it has conventionally been taken to signify. While certain technologies are self-evidently about overcoming space – as, for instance, a Skype call to a geographically remote locale – others draw on space as a resource that they can exploit. A case in point is the recent introduction of so-called Location-Based Services (LBS), now available with commonplace 3G/4G mobile technologies. LBS such as Foursquare, Facebook Places and Google Latitude enable users to broadcast to their social network their geographical position in real-time, just as they may draw on digitally available information as they move about in space. A LBS user may thus share her location with her social network by 'checking in' to a particular locale, just as she may consult her LBS to see what her friends are doing and where. Such behaviour may have some interesting sociological repercussions; a point not lost on human geographers. It has for instance been speculated that LBS might boost group-perpetuating encounters by providing members of a social network with real-time information of that network and its spatial distribution. In theory, this would provide the group with a means for augmenting unplanned encounters, thereby reinforcing the relations of its members. As Bava-Cavia (2011) notes, 'tracking the location of friends in real-time can have social benefits by encouraging social encounter'.

This hypothesis appears to be substantiated by recent research carried out by scientists at Cambridge Computer Lab. In a study based on data 'mined' from Foursquare, a team of researchers have shown how certain subgroups found within the Foursquare network appear to strategically exploit the social potential provided by the LBS to congregate in space, thereby creating a phenomenon which they call 'placefriends' (Brown *et al.* 2012). For such placefriends, LBS may be said to promote or encourage group encounters by providing members of the group with a tool for monitoring the movement of friends and thereby for meeting up face-to-face. Rather than negating space, this form of digital technology thus promotes spatial encounters by harnessing the power of real-time digital feedback and the awareness that it provides. In the terminology of Leroi-Gourhan, the LBS here becomes a means to externalise the sociogenetic DNA of the group to the environment with real-time data becoming part of an artefactual memory that contains both a spatial *and* a digital component. This process of externalisation clearly involves a particular rhythmic logic that may influence group encounters and thereby possibly accelerate the self-reinforcing tendencies of the clique. Yet, one must be careful not to approach this rhythm purely by way of digital data lest one risks disregarding the deeper-lying programme (or rhythm) of the city identified by Hillier. Unfortunately, this spatial aspect of the impact of LBS is only treated rudimentarily, if at all, in the literature, and where it is discussed – as for example in Brown's paper – the problematic is grounded in a purely social terminology (does the encounter occur in a restaurant? a health centre? or a transport venue?) without any real regard for the spatial logic of the wider urban system.

In such analyses, the rhythm of the city and that of the LBS remain separate. They thereby perpetuate a logic whereby a social phenomenon is either social or spatial but never both at once. However, nothing in fact suggests that a given event may not incorporate several concurrent logics or *rhythms*. Indeed, this is in many ways the main gist of Leroi-Gourhan's notion of the technoeconomic structure: that the production of a particular rhythm (the socio-spatial event) prolongs itself over a series of other rhythms which it momentarily subsumes and presides over. This is an important perspective that should not be lost sight of in all the banal excitement that usually follows the introduction of new technologies. With LBS becoming increasingly popular, there is little doubt that the real-time feedback which those services offer will influence the constitution of a variety of urban phenomena. But affirming this does not entail that other, more ancient rhythmic technologies must necessarily be superceded. If, as Leroi-Gourhan holds, '[t]he integration of individuals in the urbanized organism is accomplished by *rhythms* that control the conditioning of the group' (Leroi-Gourhan 1993, 330), then these rhythms – in all their *plurality;* in all their multiplex *temporality* – must form the point of departure for any meaningful analysis of that 'organism'. The technoeconomic structure particular to the contemporary city may thus be said to instantiate a multitude of different rhythms. The disentanglement and analysis of these represents a major challenge for future urban research.

Conclusions

This chapter has explored the notion of urban rhythmicity and the influence certain digital technologies has had on this. It has suggested that the subject of digital urbanism be approached by taking a so-called 'materialist' approach, which it defined by drawing on the work of Hillier and Leroi-Gourhan. According to this approach, a great deal of sociogenetic developments sustained by a particular social group occur outside of the direct control of the group. This implies that technological innovations – the logics of which are themselves complex – may well cause nonlinear shifts in the realm of social practices and that social fact and artefact must ultimately be theorised alongside each other even if their essences are irreducible. This means that patterns of social interaction or 'rhythms' arise from an artefactual ground and that these rhythms may, in turn, be instrumental in perpetuating particular forms of social solidarity. This proposition helps to formulate more clearly the question raised by digital technologies: will their introduction into the urban realm in any way interfere with the current programme of the city, and – if this is found to be a possibility – what would such a shift signify?

Since Brown and colleagues carried out their study, LBS have become increasingly commonplace just as the services that they offer are now more differentiated. Whilst LBS at the time of the study were predominantly related to already established online social networks, they now involve a host of other activities, including location-based *promotion of goods* (a service provided by companies such as Groupon, Tippr and LivingSocial) and location-based *dating*, such as it may be found in the popular Tindr and Grindr apps. With such new 'offshoots' to the technological lineage of LBS, the question is *if* and *how* existing patterns of social behaviour will be altered by the fine-scale modulation of social interaction that the former make possible. Will targeted LBS advertising, for instance, impact on the creation of consumption patterns in any measurable way, either intensifying or disseminating entrenched *patterns of taste* as these are discussed by Bourdieu (1984)? Or will dating apps – the rationale of which is to promote social encounters in space between people who do not necessarily know each other beforehand – instigate forms of social co-existence currently unknown? This remains to be seen. Yet it is clear that the question of LBS and the real-time feedback that it makes possible contains a strong spatial component and that this component in turn relates to the formation of various kinds of social patterns, some of which are spatially determined.

Taking a broader, more sociological perspective, one cannot help but be reminded of Deleuze's description of Foucault's disciplinary society: 'to concentrate; to distribute in space; to order in time' (Deleuze 1992, 2). Yet it is clear that the modulation that LBS make possible is of different intensity to those of the disciplinary society. Whilst disciplinary society relied on the relative obduracy of enclosed spaces – the school, the prison and the factory all being technologies of a certain duration – the modulation of behaviour that LBS strive for is of a more finely scaled and open-ended character. As such, their logic is arguably much closer to what Deleuze refers to as 'the society of control', characterised by 'ultrarapid forms of free-floating control that replace the old disciplines operating in the time frame of a closed system' (1992, 5). As open-ended systems of control, LBS do not seek to predetermine behaviour – predetermination being the constitutive aspect of the disciplinary logic – as much as they aspire to concurrently modulate *and* learn from situated behaviour. As geographers such as Anderson have shown, this is a logic that seeks to anticipate and influence discontinuous futures as these are played out in space, not unlike the online search engine attempting to anticipate the search query before it is made (Anderson 2010). As more aspects of urban life are digitised, one may foresee a whole new research agenda unfolding from this question of control; one which would necessitate several investigations, including:

- the identification and taxonomisation of the various stages and components involved in group perpetuating dynamics as these are mediated by different kinds of LBS;
- the analysis of check-in patterns relative to urban space and social networks;
- the analysis of consumption patterns relative to space and social network.

Crucially, such enquiries would have everywhere to exhume the intricate relationship between technology, space and rhythm and to uncover – on a deeper level – their complicity in the reconstruction or modulation of the social group and its underlying solidarity. In this manner the new fluid 'dispositifs' of power may be mapped and understood.

Notes

1 At least, this holds for the subgroup of cities that Hillier calls 'organic cities', all of which converge towards a particular form of global structural stability. He also describes another group of cities – the ideal cities – which in their essence are representational and which for this reason will not be dealt with here. For an elaboration of this discussion see Hillier (2007, 262–86).

2 Lemonnier (1993) has described this propensity to invest in certain technologically induced practices at the expense of others, finding examples of it in both ancient and modern societies. 'Societies', he writes, 'seize, adopt or develop certain technical features (principles of action, artefacts, gestures) and dismiss others' (Lemonnier 1993, 6). It is from this open-ended process of adoption and dismissal that new techno-social relations are forged whilst others disappear.

3 Indeed, the conceptual architectures of the two theoreticians may be seen as highly complementary even if this complementarity is never fully explicated by Hillier. One may note that Leroi-Gourhan provides the epigraph to Hillier and Hanson's influential *Social Logic of Space*, yet this is the only reference to Leroi-Gourhan in Hillier's entire oeuvre. Instead, Hillier builds his central proposition regarding society and space around the work of Émile Durkheim whose structuralist notion of 'the social fact' is – in the eyes of this author – to a large extent incompatible with Hillier's much more complex and astute conceptual architecture. The quotation, rendered in the French by Hillier and Hanson, reads: 'The foremost human trait is perhaps less the creation of tools than the domestication of time and space, which is to say, the creation of a human time and a human space' (Hillier and Hanson 1984; my translation).

4 Book 3 of his seminal *Gesture and Speech* is called 'Memory and Rhythm'.

5 The term 'social solidarity' may appear to be a pleonasm or a conceptual redundancy. It is however a widely used term in sociology and it is in the sociological sense that this article deploys it. Durkheim, for instance, discusses what he calls 'the different species of social solidarity' (Durkheim 2014, 52). In his theoretical framework, social solidarities are a kind of entities or 'species' that emerge from a process of speciation operating according to an anterior social logic (the so-called 'social facts'). This chapter may not agree with Durkheim on the particular logic of speciation – it ascribes sociogenetic efficacy to artefacts where Durkheim is only interested in abstract 'social facts' – but the concern for sociogenetic processes and the different kinds of social solidarity that these produce is the same.

6 See also Griffiths (2015b) on the historical materiality of traditional high streets.

13 Mediated spatial cultures

Place-making in London neighbourhoods with the aid of networked interactive urban screens

Wallis Motta and Ava Fatah gen. Schieck

Introduction: theoretical perspective

In what follows we consider *space* as a socio-material and cultural construction; a site of flows and becoming, constantly produced and re-produced through myriad everyday practices (Thrift 2008, 20). Thrift, a proponent of non-representational theory conceptualises space as an emergent property of practice that occurs prior to conscious definitions. Hence, space is a site of open-ended possibilities, a non-discursive affordance to life in a given locale. Conversely, *place* is a more conscious and meaningful construction that people use to understand the space they live in. Although it is a complex construction in its own right, 'a constellation of processes rather than a thing' (Massey 2005, 141), place is distinguished by including social narratives and meanings. The concept of place, therefore, extends beyond non-reflexive practices and material flows, incorporating instead purposeful social action in the form of political activities, values, norms, memories, stories and histories. This interplay between spatial materialities, socialities and culture, which transforms a generic *space* into a culturally charged *place*, is conventionally known as *place-making*. Place-making activities occur when we make efforts to define, understand, experience, navigate, utilise, modify or regulate what is happening in and with space. Therefore, place-making activities bridge between space and place. They draw on the mundane practices that serve to extend or maintain local social networks through face-to-face interaction, in addition to more purposive efforts to build a sense of place and belonging. Place-making then, is about the *production of locality*: the set of processes that communities used to build a sense of place out of space (Appadurai 1996; Feld and Baso 1996). It involves turning a seemingly generic, ambiguous, unruly or 'fuzzy' site of spatial possibility into a more meaningful and social entity, where we recognise ourselves in relation to others.

When the everyday activities of place-making become patterned, prominent features of a locale they denote a distinctive *spatial culture*: a constellation of related materialities and meanings that individuals recognize as familiar. This is not to say places have a single, unified and unproblematic spatial culture. Anthropologists have long been aware of the provisional nature of places, which are understood as fragile achievements, inherently mutable and always under construction. Places are sites where cultural identities and differences are produced relationally, in contact with diverse movements of people and objects through translocal conversation, while also being part of a local/spatial product (Appadurai 1996). This chapter focuses on the process of place-making rather than the 'end product' of place. We contend that examining place-making practices can help elucidate how the material possibilities of a spatial culture unfold and become meaningful in particular social contexts. In this sense, place-making offers a bridging concept between space and place that can help unveil different cultural *genres of practice* that people use to produce a sense of place. A cultural genre of practice is a set of socially learnt activities that enable people to construct a coherent biographical worldview (Kapchan 1995). Cultural genres of practices are not exclusive to place-making. Place-making practices are a particular subset of cultural genres of practice that help simultaneously to characterise both people and places.

Place-making is constrained by the material reality that people contend with in their everyday lives. The notion of spatial cultures necessarily implies a sensitivity to the material objects that enable sociality in public spaces. In this chapter the focus is on the materiality of digital media technologies rather than public space per se – in the construction of a locality's spatial culture. Amin (2008, 8) argues that 'formative sites of urban public culture – collective forms of being human through shared practices – need not be restricted to those with a purely human/inter-human character, but should also include other inputs such as space, technological intermediaries, objects, nature, and so on'. In this spirit we enquire how both digital media and the built environment enable place-making practices in urban communities. To approach this question we develop the notion of *mediated spatial cultures*, defined as cultures that use media/objects to support the performances and practices that assert, enact or contest the communal possibilities of place. The notion of *mediation* suggests that there is 'something' in between people's activities helping to achieve social relationships amongst them (Livingstone 2009, see also Silverstone 2005). Many anthropological studies recognise that 'social relationships are the basis of ... space, yet these relationships necessitate materiality ... to work as a medium' of social interaction that help to build a sense of place (Low 2009, 18; see also Low and Lawrence-Zúñiga 2003). It is necessary to be attentive to the relation of all of these material elements to achieve a mediated culture approach, understanding both how place is constructed and how people perform being-in-place.

Methodological considerations

To explore these theoretical propositions this chapter presents anthropological research into a particular digital mediating object – interactive urban screens – and examines how these can be incorporated into people's daily practices of community place-making. According to Gell (1998), an anthropological approach is *biographical*, since it attempts to replicate the process of meaning creation from the perspective of the subject's life journey. For this research project, focusing on the lived experience and cultural dimensions of the human subject, a biographical approach was deemed most appropriate. Ethnography was used as the main data gathering methodology. Ethnography entails 'sharing a process of sensory apprenticeship in order to appreciate ... the way of seeing of the ethnographic subject' (Pink 2007, 244). Our informants were drawn from participants in community workshops, run in parallel with the installation of the digital media screens in the two community buildings (details of the workshops are given in the following section). Consistent with the ethnographic approach we joined our informants at various life stages in their daily activities to understand more about their social and cultural practices of place-making. We not only talked with our informants during our workshops, but also observed their daily lives on numerous occasions. The ethnographic methodology proved a good way to explore cultural genres of practice in local communities. It helped to reveal the diverse processes through which different people, things, public spaces, media, sensory embodied experiences and behaviours become entangled to form a sense of place.

In this ethnographic context, we found in mediated spatial cultures a useful conceptual approach to exploring how new public interactive networked screens become implicated in the production of locality. The study of mediated spatial culture is of particular interest in locales where there is a large heterogeneity in local populations. Our fieldwork site was located in London, which is home to people from more than 179 countries and speaking more than 300 languages (Vertovec 2006). The borough of Waltham Forest where the screens were situated is one of the most ethnically diverse in Britain,[1] and can be considered *super-diverse* (Vertovec 2010). Waltham Forest is characterised by a multiplicity of demographic differences in countries of origin, languages, migration histories, religions, educational and economic backgrounds, both among long-term residents and newcomers. In these cases, a sense of place does not arise implicitly as an incidental product of long-held traditions and shared cultural understanding. Instead, the local inhabitants of demographically diverse areas must make a greater effort to explicitly create a sense of locality through the purposive endeavours of place-making (Appadurai 1996). This process can, in turn, foreground the genres of everyday practice that people invoke to build a sense of place.

Screen nodes, mediated spatial culture and place-making

Our research project involved implementing a network of four interactive touch screens, designed to encourage public participation and to explore how screens located in public spaces can augment the urban experience, supporting local communities (Fatah gen. Schieck *et al.* 2014). The two main screen node applications discussed in this chapter are *Slideshow* and *ScreenGram*. *Slideshow* loops through a set of pre-loaded images in the screen node; *ScreenGram* enables local users to generate content by uploading images to the screen nodes via Twitter and Instagram, making use of hashtags. For further description of the applications see North *et al.* (2013). In this chapter we describe the social interactions that occurred at the two London based screen sites.

The first screen was located in the Mill in Walthamstow, a building run by a non-profit organisation as a community hub; the second screen was located in Leytonstone Public Library that fulfils a similar community role to the Mill in a municipal building. Both the Mill and Leytonstone Library are popular public buildings where people organise and enact initiatives that actively encourage a sense of local identity, in their immediate neighbourhoods and of the borough as a whole. Both sites bring together a diverse cross-section of local residents to pursue varied social activities. In order to engage with the range of social groups we carried out two workshops in Leytonstone and four in Walthamstow that introduced the community to the digital media screens. A total of 70 people from the locality participated in the workshops. More information on the workshops and our urban screen software can be found in North *et al.* (2013), Fatah gen. Schieck *et al.* (2014), Behrens *et al.* (2013) or Motta *et al.* (2013). In addition to the workshops, we also carried out ethnography for over a year, and conducted 64 semi-structured interviews with locals belonging to different social groups. We divided our informants into three approximate age-related groups: children/teenagers, adults/young adults, and seniors. In the following section we discuss how our informants integrated screen nodes within these pre-existing place-making efforts.

Playing in the present

One of the ways in which children and teenagers generate a sense of belonging, community and place is by playing in public spaces such as parks, markets, malls, playgrounds or sites like the Mill and Leytonstone Library. Through play, children and teenagers assert a certain degree of control over their spaces and their social experience within them, eventually claiming these places as an extension of their own selves. To a great extent, this is done *praxeologically*, that is, through a highly embodied and physical engagement (Warnier 2001). A park or public space, then, only becomes theirs *in corporeal action*: through play and interaction both *in* the space and *with* the space. For most informants this play was accompanied with interactions with friends, family and strangers. The material environment of public spaces, such as parks and play areas, opens up the possibility of mixing with others, getting to know people living in the area in different ways, forging both friendships and a sense of place.

Overall, we observed that children and teenagers have a heightened *embodied* relation to public spaces when compared with other age groups because they place a greater emphasis on physical interaction. For instance, teenage informants mentioned that one of their favourite activities in parks is running from one place to another through different routes, and thus learning how long it will take them to get from *A* to *B* in different conditions. This mastering of the physical environment, in turn, enables them to engage in social games with other teenagers (for example, through spontaneous races, surprising them by following unexpected paths and having the upper hand when chasing them or hiding from them). Since this intimate embodied knowledge of open spaces has a social dimension, it also helps them to relate to others, and to build a sense of place and social belonging.

Children also interact with public spaces (and other children their age) in a more embodied way than adults. For instance, children love to play in playgrounds with swings, experimenting with the force they need to exert to achieve certain height in their motion. No two swings are exactly the same. They have their own different histories of use and maintenance. They were perhaps purchased at different times, or are made of different materials that have withstood the weather differently. Each swing might pose slightly different physical demands to children, who experiment to find the appropriate amount of force to achieve the *right* height – that which they, their friends and their parents feel comfortable with. When they play with a swing, children generate both corporeal feelings and emotions in themselves, but also in friends and adults. By mastering the swing, and hence, the feelings and emotions that it triggers in themselves and others, children also learn how to play with other children and relate with their parents, negotiating simultaneously physical and social constraints or possibilities.

Children and young teenagers used the same physical, action-based approach when interacting with the interactive digital screen node. They banged the screen or touched it softly to ascertain if this made any difference in the response of the

Figure 13.1 Children simultaneously using the screen through heightened embodiment
Source: authors

machine. They used hands, feet or even elbows to test the machine response in relation to their bodies. They observed the reactions of their peers and parents in relation to their physical experimentation, and use these reactions as a means to engage in social exchanges. They liked to explore the opportunities provided by the technology: the sensitivity of the touch screen foil, the fluidity in application response, and their ability to exert control over the system. Whether it is by simply moving the pointer or by systematically exploring the visual feedback to their touch, their approach to the screens was dominated by the *actions that they can perform*, rather than by the explicit interactions that the developer envisaged for the application running in the node. In these ways, children and young teens used the node in novel ways, inventing games that were simultaneously competitive and collaborative, and which relied heavily on physical interaction (for example, free pointer motion). Furthermore, their interactions with the screen software were framed by their interaction with family and friends, and the screen capabilities were explored in conjunction with the emotions and responses of others. Children and young teens experience the screen node spatially just as a park or a swing, and social behaviour is enacted accordingly (see Figure 13.1).

Another way in which interaction with the screen resembles the way in which children and teens interact with open spaces is that they challenge individualistic views of ownership, in favour of a social use of screens that is simultaneously collective, competitive and communitarian. Both the Mill and Leytonstone Public Library are places where children attend playgroups and learn to share the space and resources with other children. They run, play and perform for others in these shared spaces. In terms of material culture, they enjoy books and toys that are not theirs, but which they borrow to play. These are not seen as alien because they are communal, since 'public property may become viewed as private possessions and thereby potentially contribute to a sense of self (and of place)' (Belk 1988, 150). In a way, children explore the possibility of something being theirs whilst simultaneously belonging to everyone else, as well as competing for use of public resources and the mastery over them.

It is interesting to note that place-making practices also impinge on the media consumption of children and teenagers, particularly in the case of interactive media. When they play video games (such as the extremely successful *Angry Birds*),[2] children and teenagers go beyond those actions provided in-game and develop physical interactions with friends that enrich their game play. These interactions include pushing, prodding, joking and teasing, and can be directly attributed to social consumption of video game play in public spaces. Conventionally, most of our children and teenager informants consumed *Angry Birds* on their parents' smartphones, while waiting for the adults to complete chores and other tasks in public spaces, when they could share the game experience with friends.

The experiences of these informants with *Angry Birds* provided them with a referent to imagine future uses of the screen node. A six-year-old child stated he liked the screen node '[because] you might be able to play *Angry Birds* on it too! [Wide eyes of excitement]'. In this case, he imagined not only a technically improved experience (e.g. a bigger screen) but also the opportunity to more easily share this improved experience with friends in a different and more embodied way. Hence, when considering the screen node, children imagine its consumption in similar terms to that of smartphones, but offering enhanced physical and communitarian opportunities that allow them to assert themselves in their personal and social identities. In this sense, children and teens envision the screen node as an extension of their current place-making practices. Overall, children and teens fully incorporated the screen node to their usual playing practices in public spaces. In addition, the children and teens imagined potential uses for the screen that enabled them to engage physically, not only with the device itself, but also with other people next to them.

Remembering the past

One important way in which seniors generate a sense of place is through remembrance narratives involving the past of the neighbourhood and their lives within it. This activity is one of *anchoring* (Giddens 1984), in which subjects bring the past to inform the present, and ensure that valuable historical lessons remain with us. Although this is particularly visible when they interact with each other during community group meetings, these narratives are also important in other contexts. Whether sharing these stories with children in local schools or historical societies, seniors enjoy talking about the past in public spaces.

Given this predilection for reminiscence, our senior participants enjoyed the node applications that enabled them to preserve and share memories. Hence, they quickly identified the *ScreenGram* and *Slideshow* applications as a potential collective photo-album, which could help preserve and rework memories, acting as a vehicle for their remembrance narratives. In addition, since the screen node was located within one of the social hubs in which our senior participants congregate, the node could also serve as a focal point to talk about these memories while looking at relevant pictures. For these informants, the visual elements of the screen facilitate the verbal storytelling, which they traditionally employ to build a sense of belonging.

Seniors enjoyed the social opportunities afforded by using the node to spend time watching and discussing images that held personal importance for their own lives, in relation to those of others in the neighbourhood. They stressed that they would have liked a pause button to dwell on images and comment them with others at their own pace, instead of having a timed loop as featured in both applications. Capturing a moment in a photo, and then re-capturing it again by pausing the image display in the screen node would better enable them to build intimate connections with places and people. This process is about reclaiming the experiences of others as their own, and enabling their own experiences to be reclaimed in the same way by others. For instance, some of our senior informants were very enthusiastic to discuss and display an image of their childhood that showed a local school dinner, which they celebrated with other neighbourhood children the end of the Second World War (see Figure 13.2).

For many of our senior informants, this intimate relationship with images as memory aids extended well beyond the local screen node. Several have a digital photo camera as one of their cherished possessions, which they carry with them at all times. In the storage cards of the device they safeguard snapshots they have taken of old printed photographs, whose negatives have been lost, as well as images of recent events in their lives and neighbourhood, documenting local history. I was shown pictures of social events in the neighbourhood, paintings they have created inspired by classic artists, and vignettes of the flora and fauna in their gardens. They are often ready to take out their cameras and show its contents in conversations with others. Repeatedly we found ourselves chatting to seniors in public meeting spaces and

Figure 13.2 Biographical history gets placed through Instagram on the screen
Source: authors

getting to know their families, their friends, and their cherished memories of recent and past events, aided by the small viewers of these cameras.

Seniors regularly used the images in these cameras as an aid when interacting socially with others. Both their understanding of the screen node and the value that they see in it is an extension of the way they mobilise objects, images and conversations to reflect on events and forge memories of a local history. Seniors were not able to fully incorporate the screen node into their remembrance practices, since they lacked digital skills or tools to upload their images. However, they imagined the screen could be a local history living repository, just as their cameras (or even themselves) are. This new device could help extend the labour of public consciousness of the past they are engaged with, as to constantly consider the past in relation to the creation of new senses of place.

Bringing about the future

When constructing a sense of belonging and place, adults emphasise less the physicality of the environment or its local history. Instead, they take a more instrumental, idealistic and future-looking view. They are interested in using the screen node to create what they consider a *positive* local environment. They seek to improve educational opportunities, increase awareness of future or imminent problems, and even foster better communication flows between different groups within their neighbourhoods. This place-making perspective is about enabling future ideals to happen in the present. It has resulted in the formation of many neighbourhood initiatives, from the transformation of an abandoned library into a community centre, to the foundation of local development economic districts.

Our adult participants see place-making is an opportunity for the purposeful pursuit of change in the locality; it is a *scaffold* (Greenfield 1984), an instrument to mediate between their imagined view of their locality and the reality in which it stands. Hence, our adult participants see their place-making activities as a springboard to generate local culture and bridge the gap between what their locale *is* and what it *could become*; a stage for the display of the unrealised potential development of the locale. Adults and young adults hence showed a proclivity towards place-making as activism.

We found many examples of this proclivity for activism as place-making that become relevant, both because of the community-centred discourse that accompanies them and because of their physical and social effects in the locale. One specific instance of this involves a local blogger, who in 2012 decided to produce and sell 'Walthamstow Dad' T-shirts. He wrote in his blog:

> Proud Dads of Walthamstow, I salute you … Last year we sold 60 Walthamstow Dad T-shirts and raised £250 pounds for the Maternity unit at Whipps X. I just wanted to say a big thank you to all the tweeters for the gossip … Stella Creasy (the local MP) for championing yet another local cause, StowScene and its amazing website.[3]

Another instance of local activism that brings the ideals of the future into the present are the two neighbourhood *Art Trails*, in which people showcase their artwork by opening their houses to the public. These events take place every year, during one week over the summer. During these events, people from different walks of life open up their homes to the public and showcase around 300 different art exhibits they create themselves. These events foment contact with people with different cultural values, backgrounds, worldviews and life choices. Furthermore, these events generate a *third-space*, where everything comes together: 'the abstract and the concrete, the real and the imagined, the knowable and the unimaginable, the repetitive and the differential … mind and body, consciousness and unconscious … everyday life and unending history' (Soja 1996, 56–57). *Art Trails* help generate an alternative public space, where residents enact the vernacular cosmopolitanism (Georgiou 2013) and mixing ethos (Wessendorf 2013), as social norms and values. Thus, residents come together to celebrate art, participating in public life through the respect, recognition and the defence of difference. The *Art Trails* offer a glimpse of a possible heterogeneous but harmonious community. They constitute an intellectual, symbolic and material practice that not only enables place-making, but also promotes culture as 'the human capacity to expand worlds towards other potentially distant horizons and more complex outcomes of life' (Miller 2011, 215).

Although not all adults or young adults participate to the same extent in local activism or the *Art Trails*, many like to at least keep abreast of local issues. This happens both online and offline, usually through public notice boards, local newspapers, personal conversations in public spaces and direct communication from the local government. All these become part of place-making endeavours because they constitute a readily available system of references that is shared by adults in the locale, and which helps to create a sense of personal and community identity. Similar shared references arise from the shared experiences of immigration online through blogs. A 27-year-old local blogger thus explains the challenges of living and belonging in diversity:

> My time in the UK has made me appreciate my heritage as a Black American Woman to a greater extent, all while learning to appreciate other cultures around me. I have met people from all over the world … I feel alone at times, get … frustrated, feel like … moving back home, but then I remember a quote that I learned growing up: 'A test of a man is not where he stands in times of comfort and convenience, but where he stands in times of challenge and controversy' … I'm learning that despite the differences people have, they tend to be more alike than they are different. We might have different backgrounds, stories, and life experiences, but we all laugh, cry, bleed and love the same … Walthamstow, London is my home, and I'm looking forward to being a part of its community and way of life [*sic*].[4]

With regard to adults' use of the screen node, those explicitly engaged in transformative activities (such as local activism) saw it as a means to extend their efforts, informing and educating the local population about relevant projects. Members of a knitting group, for instance, urged researchers to use *ScreenGram* to advertise knitted characters they were trying to sell to raise funds for their local community centre. Local artists used the node to extend the available physical exhibition spaces by imagining it as a window to display digital art, and proposed to integrate into their *Art Trail* projects. The managers of the Mill community centre imagined the node as a virtual notice board that extended their physical meeting space, and enabled users to connect with each other even if the building itself was closed. For those adults or young adults not involved in specific community-building projects, the screen node was approached as an extension of their online place-making activities and became an outlet for useful local information, locality-specific news, and a means to connect with other locals in conversation about future ideals.

As shown above, most adults and young adults in our workshops and interviews reported that they expected the screen node to embody a higher purpose. They approached the screen as an extension of their pre-existing place-making endeavours, uploading photos and short messages relevant to their communities. For instance, Figure 13.3 shows photos adults commonly uploaded to the screen. These photos aimed to get others involved in either local political protests or the imagining of alternative neighbourhood landscapes.

For many, the mere physical presence of the screen node legitimised previous efforts for place-making, showing these to have been efficacious (as new resources for the improvement of the community had been produced). Activism-inclined informants presented the node in local events with pride, as a symbol of their achievements to bring a multi-cultural community closer to its unrealised potential. Those with less political engagement saw it as an arena of constant small-talk and debate. However, in both cases locals saw themselves simultaneously as unique individuals with interesting ideas to provide, and as members of a vibrantly diverse greater community.

Closing remarks

Material objects, including the built environment, digital cameras and screen nodes, provide ways in which residents construct and convey their biographical worldviews. They are means of ordering social relationships and connecting with wider collectives. In other words, 'they are a medium through which people construct their world (their neighbourhood) and reflect on it' (Tilley 2006, 62). When interacting with the screen nodes, participants (irrespective of their age) enact *genres of practice*: socially learned and appropriate patterns of behaviour 'whose repetitions situate actors in time and space, structuring both individual and group identities', enabling a sense of place (Kapchan 1995, 479). In the Waltham Forest case studies the genres of practice were: playing in the present, remembering the past and bringing about the future. It was through these patterns that residents constructed a mediated spatial culture: self-consistent units of behaviour that are supported by media and objects, and which exist only as they are performed.

When our urban screen nodes were installed, there were no prior referents to guide the neighbourhood residents in their interactions with them. This allowed people to think about how the screens resonated with their biographies, values, desires and habitual strategies to connect with others in public spaces. By inserting this foreign object in the locale, the various genres of practice and biographical dynamics governing the place-making of different groups became more explicit. We observed that 'when a new medium arrives on the scene, the first and most important impact tends not to be anything which is itself radically new. Rather the tendency is to seize upon it in order to finally realize some desire that was already present' (Miller 2010, 115). This observation is evident across all the cases presented, since residents approached the screen as an extension of their regular place-making practices. In turn, the screen enabled residents to fulfil their pre-existing desires to connect with others, in accordance with their life stages.

Figure 13.3 Local activism and alternative landscapes displayed through the screens
Source: authors

In this chapter we have shown that interactive public screens acquire particular meanings and uses through various biographical dynamics and place-making strategies. We emphasised the continuities between the already existing place-making practices and those practices enacted when interacting with the screen nodes. For children and teens we observed that this new medium was much more exciting than it was for seniors or adults. This was because they had previous interaction desires that had been frustrated so far. Public interactive screens enabled them to play embodied collective computer games in a highly unpredictable, public, setting. This stands in stark contrast with current game consoles, which can only be played in private even when they allow collective use. This is a very predictable, and thus boring, environment for play. The new medium and its setting is attractive because it allows for unexpected and highly embodied interactions, more akin to what parks and playgrounds can offer. In contrast, adults and seniors already used digital media (blogs, images and websites) to connect with other residents.

Since place-making practices rest between space and place, oscillating between the discursive and non-discursive registers, there are aspects of them that cannot be easily verbalised. Hence, an ethnographic approach becomes necessary to flesh out the genres of practice operating in a locale, which is itself necessary to making sense of the ways in which apparently aspatial technological infrastructures become significant and embedded in the production of locality. In this sense, a digitally mediated spatial culture approach that spans its analysis across material culture, media, place-making practices and the biographies of people, can be a useful addition to conceptualising the impact that the deployment of a new technology can have in a given urban community locale.

Notes

1 Council, Waltham Forest. 'Data and Statistics: Key Facts and Figures'. Last Modified 18 January 2015. www.walthamforest.gov.uk/Pages/Services/statistics-economic-information-and-analysis.aspx.

2 Wikipedia. 'Angry Birds'. Accessed 29 January 2014. http://en.wikipedia.org/wiki/Angry_Birds.

3 Walthamstow Dad Blog. 'Proud Dad's of Walthamstow I Salute you'. Accessed 25 May 2012. http://walthamstowdad.blogspot.co.uk/2012/05/proud-dads-of-walthamstow-i-salute-you.html.

4 Mary, comment on Walthamstow Wildlife Blog. 'Living in Walthamstow'. Accessed 10 February 2010. http://walthamstowwildlife.blogspot.co.uk/.

14 Commuting with others

Infrastructures and interactions in public transport space

Anna Plyushteva

Introduction

Throughout its history transport geography has paid close attention to the commute. However, geographical perspectives on commuting have undergone important changes since the beginning of the twenty-first century. In this chapter, the commute refers to the familiar, routine, urban and suburban trips which many people doing paid or unpaid work have to undertake several times during a typical week. Specifically, the discussion which follows focuses on public transport: the cultures and practices of urban dwellers travelling together on shared conveyances along pre-determined routes.

Historically, geography's interest in the commute had often focused on passenger flows, the quality of infrastructure provision, and the interdependencies between the journey to work and land-use patterns (e.g. Levinson 1998). While such scholarship was often able to produce in-depth and quantitatively sophisticated representations of transport use, many geographers felt the approach failed to engage with social and technological developments and went too far by reducing human experience to the rational weighing of costs and benefits. Despite various criticism, transport geography's framing of commuting changed little between the 1950s and the 1990s. The practice of applying the rules of economics to analysing travel behaviour remained the dominant paradigm (cf. Shaw and Hesse 2010). A major shake-up of the geographical study of transport began in the early 2000s. Originating in sociology rather than geography, the 'mobilities turn' which is associated with the work of John Urry, Mimi Sheller and others, challenged the purely utilitarian understanding of human movement (Sheller and Urry 2006). In mobilities research, transport users are not regarded as purely rational, atomised agents moving between A and B. Rather, the focus is on the experiences, meanings and interactions with human and non-human agents involved in journeys. While retaining a focus on the physical movement itself, mobilities scholars have often drawn attention to the affective, cultural, social and political circulations which permeate travels. These considerations could be applied in equal measure to the 'big' journeys of international migration and tourism, and the mundane and familiar trips such as the commute. This chapter stems from semi-longitudinal research on the changing metro commuting practices in Sofia, Bulgaria. The project was designed to coincide with the 2012 opening of the second line of the Sofia metro and aimed to explore the experiences of the 'new Line 2 commuters': those who either lived, worked or studied near a newly opened metro station and who changed all or part of their everyday journeys to make use of the new metro line.

These changes came at a time when many of Sofia's inhabitants felt that the city's mobility system was at breaking point. Since the 1990s, rapid population growth, an antiquated bus, trolley and tram system, and explosive gains in car ownership rates meant that Bulgaria's capital became increasingly congested, polluted, unequally accessible, as well as stressful to the people going about their daily lives (Petrova 2012). In this context many decision-makers, civil society representatives and citizens placed high expectations on the metro. The metro project had been in planning since the 1960s, but its first line was not completed until 2009 (see Figure 14.1). This chapter is based on a study which aimed to explore, from a passenger perspective and using qualitative methods, the meanings of success or failure in the context of the metro expansion and the ways in which the new infrastructure became part of, shaped, and was in turn shaped by the everyday lives of its users. Data were collected between 2012 and 2015 using repeated go-along interviews with metro users, observation and auto-ethnography. The go-along interviews were based on the methods developed by Wood and Latham (2015) in their work with bicycle commuters in London, and combined visual and audio data. Aiming to contribute to discussions of sustainable urban mobility, the research focused particularly on those commuters who switched from being a driver or car passenger to travelling by metro. However, those who switched from other modes of public transport to the metro were also interviewed and both groups are represented in the interview excerpts quoted in this chapter.[1]

This research builds on mobilities literature in conceptualising the embodied subjectivities of public transport users. At the same time, a greater emphasis was placed on the changing and unstable nature of everyday travel habits than in most mobilities research to date. Drawing on a case study of newly launched infrastructure, this research was well-positioned to question the dominant assumption of a commuter's 'typical' journey. It emphasised the always-changing shape of urban transport infrastructures, technologies and individual routes and routines. Apart from refining the theoretical understandings in mobilities studies, this dynamic conceptualisation of travel behaviour also has important implications for research into transitions to sustainable urban transport. Some of these implications are the focus of the discussion section at the end of the chapter.

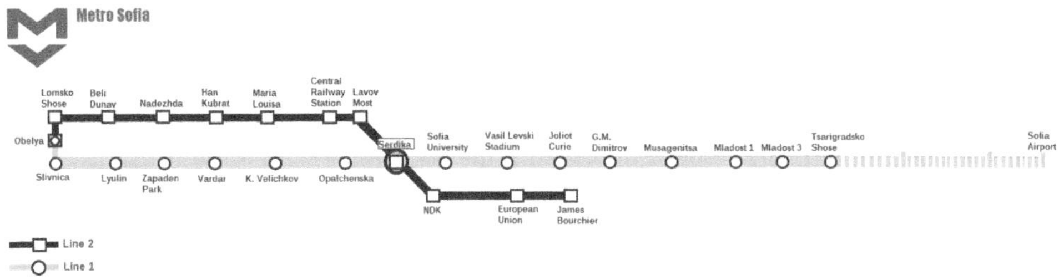

Figure 14.1 Map of the Sofia metro showing operating lines and planned extensions as of 2014

Source: author

Practices of commuting

In planning and carrying out research on metro commuting in Sofia, two theoretical considerations underpinned the data collection and analysis. The first, as described above, was attending to the fact that travel habits, unlike the word 'habit' might suggest, are in fact fluid and unstable. A commuter's habitual journey is likely to undergo frequent changes, at both major and more subtle levels. From switching transport modes to occasionally meeting a co-worker on the train, commutes change more often than is recognised in most scholarship on the subject (Lyons and Chatterjee 2008). The second consideration referred to the social nature of public transport commuting. While commuters can often be thought of as individual agents moving between origin and destination, their experience is shaped to a great extent by interactions with humans and objects along the way. It is these socio-technical collectivities that this chapter focuses on in particular. The notion of practices is particularly useful in thinking about commuting as both dynamic and collective. It locates the socio-technical activities related to commuting in the space (physical or virtual) which individuals and objects share, rather than within discrete individuals. This is important in empirically engaging with the mechanisms through which practices travel, the connections they create among human and non-human agents, and the resulting adoption, discarding and modification of particular ways of performing everyday activities. The doings and sayings which often define what come to be thought of as the spatial cultures of particular locales are thus not seen as always emerging from the intentions or decisions of individuals, but also as distributed, embodied and pre-cognitive. In addition, this conceptual framework offers a much more nuanced view of commuting habits than the dominant psychological-economic view of their development as linear, starting off as a rational decision (e.g. 'I'll walk this way to the station as it is quicker'), before gradually solidifying into autopiloted mundanity ('My legs just carry me to the station every morning'). The term practice as used here originates in sociological theories of practice (Schatzki 2012). Contemporary theories of social practice such as those developed by Elizabeth Shove *et al.* (2012) provide a helpful approach to exploring the meanings, skills and materials involved in performing everyday activities.

Circulating practices: learning commuting skills

The spatial cultures of particular urban environments are often shaped by the specific behaviours, the small doings and sayings which actors perform. These shared practices can often be particularly visible in the relatively regulated spaces of public transport, and can include things like standing on the right on escalators, or avoiding eye contact with other passengers on a metro train. In the case of Sofia, it was of particular interest to explore how such practices were deliberately learnt or unintentionally spread across the new transport infrastructure. Doing research before, during and after the opening of Line 2 in 2012 created an opportunity to discuss and observe socio-technical change directly. This data offered insights into the processes through which individual activities were gradually translated into both enduring cultural practices and fleeting situations, i.e. into the totality of shared moods and ambiences which David Bissell (2010) has referred to as public transport's affective atmospheres. An example of a practice which was discussed by many respondents was the habit of placing one's entire wallet on the card reader on entry, instead of taking the travelcard out in order to validate it. This gesture, entirely familiar and taken-for-granted in many other metros around the world, was suddenly made visible and dynamic in Sofia through its newness and ambiguity. Some research participants had overheard co-workers talking about it, while others had observed other commuters doing it and had decided to try it for themselves. The agency of the cards and card readers was also significant in this process of learning and spreading a practice: 'They have similar validators on the trams as well', observed one commuter, 'but those are very weak, and you couldn't have your travelcard read through your wallet: it just displays an error message. I've tried'. In the metro, however, the practice spread rapidly. Interviews, observations and the researcher's own experience of 'becoming a commuter' in Sofia suggested that this was achieved through the interweaving of three processes: the technological capacity of the metro validators; the time-saving and convenience of the gesture; and not least through the performative aspects of the practice, which clearly convey both the haste and the expertise involved in being a busy commuter.

Numerous other examples demonstrated the role of public transport socialities in how commuters both acquired and modified their embodied knowledge of the new infrastructure. One such example concerned a one-way pedestrian tunnel in Serdika station which was extremely busy during rush hour and generated a lot of comments from research participants. Some felt the tunnel's designation as one-way was illogical, and made a point of using it in both directions. Others abided by the regulation and explained that through their behaviour they hoped to demonstrate the general importance of observing the rules – not their approval of the specific layout. As the metro authorities quickly gave up on having a member of staff regulate the flow of passengers, the number of 'rule-breaking' commuters gradually increased, until a two-way use of the tunnel (which, arguably, did work better in practice) eventually became the norm.

Thus, the spread of some commuting practices occurs through deliberate learning based on discussing, observing and copying. Other practices circulate through purely affective, embodied and pre-cognitive channels. This is illustrated in the following excerpt from the author's auto-ethnographic diary on becoming a metro commuter in Sofia:

> I am transferring from Line 1 to Line 2 at Serdika. I walk through a tunnel to reach the vast underground concourse at Line 2 [Figure 14.2]. In the tunnel, both lines are too far for any sound to reach us, and since it's early in the morning, it is very quiet. However, many commuters speed up into a run as soon as they enter the tunnel. And a strange thing keeps happening – I start running with them. At first I possibly suspect they know the Line 2 timetable, so I find myself following them. Gradually, I realise that at least half of the time there is not a train in sight when we reach the platform. As months pass, I begin to realise that it is the not-knowing, rather than the knowing, that spurs people forward into this run. That quiet corridor, the anticipation, the possibility that the train might be slipping out of the station just as we are, oblivious and silent, walking the last metres towards it; our bodies choose not to take the risk.

Figure 14.2 Serdika station
Source: author

Navigating collective commuter timespaces

As discussed by Laurier and Lorimer (2012), the skills and resources of wayfinding, e.g. the ability to retain just the right amount of attention on the question of 'Where am I?', are a key aspect of moving in general and commuting in particular. Although knowing one's way can be a complex task requiring significant mental effort, human bodies are also equipped with a range of non-cognitive, habitual capacities for wayfinding which are both effective and efficient. These capacities make the most of a body's sensory and affective presence in particular familiar spaces. The presence and proximity of others on public transport, the multiple opportunities to interact with and observe them, shape commuters' approach to navigation. A new metro commuter might be unsure which way the exit is when they step off a train, yet they often can just follow the crowd of others heading in the same direction. While clocks and signage provide the 'objective', official orientation tools in commuting timespaces, the relevance of these can often fade in habitual journeys, making room for navigation skills which are embodied, unthought, and intimately linked to one's own routines and priorities. Thus, one commuter described not really noticing the actual time on a particular clock in a metro station, 'I just subconsciously register whether it is later than 8.25, which would mean that I'm running late'. Another described the development of an 'internal clock' during the first few months of commuting along the new metro line. While initially she was more likely to pay attention to announcements and displays, gradually she discovered she was able to 'tell' a form of personal time which was specific to her journey: 'Not how many minutes have passed, but roughly how many stations are left. It's like being a teacher – you "sense" when the 50 minutes of the lesson are coming to an end, you don't need a watch.' In other cases, people and objects were 'mobilised' as devices for navigation, such as getting off at the 'first station where doors open on the left'. Or in this example: 'Serdika is where most people get off; when the carriage empties I know that I need to get off at the next one, without even looking up from my book.' Some of these highly situated skills amounted to sophisticated navigation mechanisms, as described by this commuter:

> I don't know the timetable of the trains, but I do have a little trick to know whether to speed up or not as I approach the metro station. Basically, if the crowd of office workers exiting the station has reached as far as the market, it means the train arrived 3–4 minutes ago; so I need to hurry up. If the commuters are only just emerging from the exit, it means the train has only just arrived at the station, and I can take my time.

At the same time, the refinement of practical knowledge on how, when and where to position oneself is not a linear process. Some metro commuters developed detailed knowledge of where is best to wait on the platform, only to gradually stop using this ability a few months later. Reasons for this were sometimes rationalised ('Because it will be near the exit, everyone gets on the train there. So I'd rather not be, and go to a less crowded carriage') and sometimes evoked more felt, embodied processes ('I don't care that much anymore; one minute doesn't make a difference'). These dynamics underlined the fluid and unstable nature of practices, as well as the affects, meanings and skills which came together in reconfiguring them, retaining them or replacing them with other practices.

Enjoying and enduring commuting with others

Both lay and academic discussions of public transport commuting often focus on the stresses and annoyances of collective transport, contrasting it with the private, domestic-by-extension space of the car (see, for example, the comparative research of Wener and Evans (2011)). Thus, commuter journeys shared with others have sometimes been discussed in terms of the degree of privacy (O'Dell 2009) and the kinds of proximities with strangers they configure (Wilson 2011). However, just as commuting has been demonstrated by mobilities scholars to be more than a derived demand, the lived experience of public transport journeys amounts to more than an inferior alternative to the car. Publicness and privacy, and enjoyment and stress can be intertwined in the commute in ways which are both complex and temporary. Commuters' accounts often provided nuance to the assumption that there is a straightforward correlation between privacy and enjoyment. In studying the changing practices of everyday mobility in Sofia, it was often evident that a single commuter's experiences complicated the dichotomies of pleasant/unpleasant, and private/public repeatedly, often several times over the course of a single journey. These categories were made and revised in practices such as mobile phone use:

> I find it very frustrating that now everyone is glued to their smartphone, like a zombie. People don't look up at all. But I suppose it also means there is less scope for rudeness, no one is up for a good argument over a backpack or a shove or whatever, because everyone is completely immersed in their phone [*laughs*].

The non-linear changes in the practices of making privacy and publicness are once again underlined in the following excerpt:

> In the first month, I was really paying attention to everything and everyone. It's interesting to look at people, you know, make up little stories about them. Then, there was a period when I was reading, and also listening to music on my headphones. I wanted to completely isolate myself and have this time to myself. But after a while, my headphones broke, and in the time it took to replace them, I lost the habit of listening to music during the commute … I guess [lately] I prefer to hear city life around me.

This account of the participant's changing practices once more draws attention to the active role of objects and infrastructures in the making of the socio-technical space of public transport commuting. While commuters are often themselves responsible for shaping and spreading mobility practices, the role played by objects in such circulations should not be underestimated. Above, the failing of the headphones started a process which reconfigured the sensory experience of making time for oneself. On another occasion, a respondent described her approach to choosing a handbag for the day: the small one, if the book she was reading fits into it; otherwise the big one. Another respondent discussed specifically buying small paperback books which he can read on the metro and which would neatly fit into the back pocket of his jeans. The agency of objects and technologies on public transport should be of particular interest to those concerned with designing and managing it. In foreseen or unexpected, sometimes subtle, and occasionally dramatic ways, non-human actors like luggage, ticket barriers and voice announcements can shape experiences of comfort and conviviality on public transport. At the same time, neither the infrastructure of the metro, nor the multitude of objects which move in and around it, should be seen as determining the public transport experience, or the behaviour of its users, in straightforward ways. Instead, the above account has sought to highlight the way in which different agencies and intentionalities are always woven together to continuously reshape the ways in which people travel together. Rather than predictable and habitual, the commute can be seen as a domain in which to explore the connections between the rational, embodied and environmental drivers of behaviour.

Discussion

Design and planning decisions, political choices, dominant models of development, relations of power and economic forces play out in daily mobilities and the interactions they draw commuters into. At the same time, commuting on public transport is not simply another aspect of everyday urban life, another arena where structural forces play out. It is different to playing in the park, to shopping for groceries, to seeking medical treatment, to browsing the Internet. To understand it as a particular fragment of urban culture, an empirically-grounded engagement with its situated interactions, infrastructures, meanings and technologies is needed. One possible application for qualitative studies of the emergence, modification and transmission of micro-practices of mobility is in the field of sustainable urban mobility. Transport places great demands on cities in terms of scarce resources such as energy and space. Sustainable mobility is in turn critical to social justice and the creation of inclusive and healthy cities. As argued by Banister (2008), public transport provision is fundamental to the urban sustainability agenda, not least because of the way in which it enables the most sustainable transport modes – walking and cycling. There is growing recognition among policy-makers that both changes in behaviour and changes in transport infrastructure are needed. This involves a gradual, yet speedy reconfiguring of the component parts of urban mobility systems away from privately owned motorised vehicles and towards more environmentally and socially sustainable ways of moving. Introducing policies which cause significant shifts in these complex socio-technical arrangements is usually both expensive and fraught with political peril. In this context, behaviour change has become an oft-cited concept. However, it can frequently be framed in terms of a switch to some loosely defined 'desirable' way of acting, without giving in-depth consideration of existing practices.[2] In the context of everyday life, behaviour change is a complex process. It should be understood as occurring within situated, lived and dynamic bundles of practices. It involves the intricate interplay of individual decisions and collective sways and cannot be reduced to either. In the case of the Sofia metro, a qualitative examination of the take-up of the new infrastructure demonstrated that the successes and failures of the project played out in the small everyday interactions of the new users. This meant that they were subjective and provisional, but also that the infrastructure itself could be thought of as less permanent and more malleable than often assumed. With the small doings, sayings and things involved in the everyday use of its spaces, the metro was continuously modified. These findings mirrored those discussed by Furlong (2011) in the context of Canada's water infrastructure, where small devices often transform the uses and meanings of large water systems. The notion that through small changes in the practices the overall utility, conviviality and ultimately, success, of the metro as urban infrastructure could be enhanced, can have important implications for the design, management and appraisal of such large projects. At the same time, such considerations cannot replace the adequate provision of infrastructure for sustainable transport. Instead, this type of knowledge can support efforts to uncouple urban cultures from ubiquitous automobility through the provision of infrastructure which is more open-ended, flexible and more-than-rational, i.e. transport infrastructure which is more like mobility practices themselves.

Amid the rapid development of smart transport technology, a growing supply of 'big data' and their expanding role in planning urban transport, it seems particularly important to highlight again and again the passenger perspective on everyday mobility. This involves producing scholarship which foregrounds the differences between the discrete moving dots of transport flow visualisations and the lived interactions of embodied mobile subjects. As transport planning's capacity for data processing continues to grow, aggregation should not make the minute disappear from view.

Notes

1 Unless otherwise indicated, text quoted from research data is a direct translation of excerpts from anonymised interviews with Sofia-based metro commuters, who participated in this research project in 2012–2014.

2 See Schwanen *et al.* (2012) for a critique.

15 Grindr Guys #7

Difference, traces and spatial practices

Regner Ramos

Introduction

Cerebro is a super-device used by the mutant Dr Charles Xavier, which first appeared in Marvel Comics' *X-Men* #7. In a context where mutants were to be feared, marginalised and ostracised from society, it was uncommon for them to reveal themselves as what they were, unless they were forced to do so because of physical mutations that gave them away. With many mutants changing their birth names and creating aliases that represented their particular mutations, privacy and anonymity were their allies. Created by Charles and Magneto, Cerebro's function was – to state it in a few words – to track down mutants.

Throughout the past decades, science-fiction has been a successful medium through which people have been able to project their understanding, as well as their fears, concerning the embedding of technology within bodies and society. An example of this lies in the cyborg theories of the 1980s, though mentioning the term 'cyborg' might make people think more about cinema and comic culture than of the theoretical figure used in feminist discourses and in science. The term 'cyborg' was coined in the 1960s by Manfred Clynes and Nathan S. Kline in an attempt to solve man's problems and limitations in science, specifically in space travel, but it was not until the 1980s, when feminist Donna Haraway revived the term in her exemplary 'A Cyborg Manifesto', in which she defines the cyborg as a hybrid machine and organism and as a creature of reality as well as a creature of fiction. The cyborg is a figure worth perusing, especially after it has been so vacuously portrayed and misinterpreted in mainstream culture and science-fiction; Haraway's vision of the cyborg, intended to be a groundbreaking, political figure, currently struggles with notorious, social misconceptions that strip it from its agency, rendering it laughable and nearly foolish.

The cyborg is Haraway's metaphor for the subversion of gender codes in the 1980s, when the arrival of new technologies were becoming increasingly prevalent and were thus redefining/reconceptualising social relations as well as social power structures. Marxism, psychoanalysis and feminism failed to address the rapid changes that took place in the mid 1980s, and Haraway's cyborg attempted to break down divisions and blur boundaries, not only between race, class and gender, but also between body and machine, advocating a different theoretical discourse fuelled by science and technology. The cyborg represents a breaking down of subjective singularities by 'embracing the skilful task of reconstructing the boundaries of daily life, in partial connection with others, in communication with all of our parts' (Haraway 1991, 27). Through this she argues that 'cyborg imagery can suggest a way out of the maze of dualisms in which we have explained our bodies and our tools to ourselves' (Haraway 1991, 27). However, Cyberfeminist theory is also about presenting new tools to reconstruct the ideals that it tore down in new ways, particularly by imagining a world without gender, which would be one of the first steps in fracturing one's embodiment and identity. Gender and body become ambiguous in light of late twentieth-century machines and the lines between what is artificial and what is not are blurred. But in 'A Cyborg Manifesto', as a counterpoint to the formation of identities rooted on gender, Donna Haraway asks a seemingly simple question that can be translated into our current technological and spatial context: 'What about other embodiments?' She continues, 'Why should our bodies end at the skin, or include at best other beings encapsulated by skin?' (Haraway 1991, 25). In light of the twenty-first century's technological advances these embodiments acquire new meaning, no longer encapsulated by skin, but rather extended into pixels and data which interconnect users with each other and their environments.

In her text *Cyberfeminism with a Difference*, Rosi Braidotti approaches the postmodern, technological debate by viewing postmodernity as a threshold of new re-locations for cultural practice and, like Haraway, proposes a shift in perspective, in which technology is seen as co-extensive with and intermingled with the human, thus putting technology in a position in which it is spoken about as a material apparatus (Braidotti 1996). The intermingled relation of technology in day-to-day life is a pervasive phenomenon when we think about mobile phones. The relationship between users and their mobile phones and apps is being designed to become increasingly seamless, a to-and-fro of interactions between the person, their location and the technology. It is then that Haraway's ideas become particularly relevant to the current condition of the twenty-first century GPS-app user. She states:

> Communications technologies and biotechnologies are the crucial tools recrafting our bodies ... Technologies and scientific discourses can be partially understood as formalizations, i.e., as frozen moments, of the fluid social interactions constituting them, but they should also be viewed as instruments for enforcing meanings.
>
> (Haraway 1991, 13)

Though Haraway's postulations emerge as a tool for the female gender, it would be contradictory to limit the political agency of the cyborg to being applicable to one gender: cyberfeminism is not a theoretical framework that sustains one gender; it is a discourse that celebrates the construction and the performance of subjective, gender, identity and/or embodied difference. In *My Mother Was a Computer*, N. Katherine Hayles writes that:

> Human action and agency are understood as embodied processes sharing important characteristics with the processes taking place within computational media, including possibilities for evolution and emergence. In my view, an essential component of coming to terms with the ethical implications of intelligent machines is recognizing the mutuality of our interactions with them, the complex dynamics through which they create us even as we create them.
>
> (Hayles 2005, 243)

For this reason, studying the affective properties of GPS-based mobile apps is key to understanding the impact they have in society, culture, identity and space. For millions, downloading and tapping through apps is a habitual act so commonplace, very little thought – if any – is given to how these technologies, their performance and design are slowly creating new subjects, just like previous technologies did. This is as true for the arrival of the car as it is for the television. The presence of these machines within our daily lives changes our socio-spatial realities, altering the way we perform in space. Hayles claims that the person who emerges after engaging with technologies is changed; much like reading a book, even if it is slightly, the person who engages with technology is not the same person after the interaction ends. Thus, through their interface and performance, mobile technologies and the apps which have become so important to us create a sense of othering. Through the process of reading the app's interface – of being disassembled and reassembled when the user crosses the threshold from physical to digital and vice versa – difference, in a variety of forms, emerges.

Grindr

Released in the United States in March 2009, Grindr is advertised as the world's largest and most popular all-male location-based social network. Although Grindr is used for a vast number of reasons, it is most notoriously known to be used as the digital substitute for gay cruising – the act of walking or driving in particular areas looking for a sex partner. The one thing each app user has in common is that no matter what they use it for, they want to find Others. Since its launch, the app has had over ten million downloads in 192 countries. With 263,700 users, London ranks as the city with most active users per month, making the app an appealing subject of study, particularly because of its correlation to urban space and histories of gay culture in London.

Grindr is the modern-day, real-life Cerebro, which instead of tracing mutants, discloses gay men. Where previous Internet technologies were completely stationary, tying users to a particular location such as a desk, mobile apps like Grindr possess cyborgean qualities that go beyond their mere prosthetic condition. The figure of the cyborg has commonly been used to describe the prosthetic relationship between organic bodies and machinic attachments; and while the coupling of Grindr user and their mobile phone can be interpreted as a cyborg-like relationship, the term encompasses a greater scope. The overlap between human and machine enables the user to extend their senses in the McLuhian sense of the word, allowing them to view digital Others that surround them as if they were seeing through walls. Grindr situates each individual user in a particular place in the city and furthermore positions them in relation to each other. Using the app makes Grindr Guys aware of their location while altering the way they understand their immediate space and the people who surround them. Even in the design of its interface the app is uncomplicated, so the users are not distracted into dedicating time to the software in itself, but rather into making use of the technology and taking advantage of its performance dealing with interpersonal relationships. These social interactions raise questions regarding subjectivity and othering, as each user on the Grindr grid is a virtual character, a self-constructed portrayal of themselves.

Like with Cerebro, Grindr users are able to see an intimate and personal aspect of their neighbouring bodies, which at times is an area of speculation: their sexual orientation. Differing from other social networks, where users visit profiles of other who do not have to be close or online, Grindr offers a kind of interaction which deals with being synchronised in space and time, reverting to modes of interaction where synchronous communication was indispensable; users have to be nearby and simultaneously online (or relatively close to each other's online periods) to interact. Before the arrival of the telephone if a person wanted to interact with someone else, they would have to go out into the city and socialise with other people who were *there* as well. There used to be a need for people to be synchronised in *time and space* in order to carry out social functions. Even with the arrival of the telephone, there was still a need to be synchronised, as both people needed be fixed within a space that hosted the telephone number being called. With modern man's newfound ability to break spatiotemporal restrictions, human interaction within the built environment is increasingly being reinvented and re-represented.

As technology and society evolve, it becomes important to redefine concepts or misconceptions that become outdated. The fact that apps like Grindr are synchronised with space and time does not weaken the figure of the cyborg; in the twenty-first century, the cyborg does not wish to live a virtual life but rather one that is wholly connected, one that they can fluidly navigate from digital to physical as seamlessly as possible. The strength of these GPS technologies lies in their need for bodies to negotiate the boundaries of space and time in order to carry out actions both physically and digitally. Adriana de Souza e Silva's research on 'hybrid spaces' has led her to conclude that these technologies have the capacity to *strengthen* users' relationship to the built environment, as opposed to other studies which suggest 'that cell phones withdraw users from the physical space in which they are' (de Souza e Silva 2006). For architect William J. Mitchell, in his book *City of Bits*, cybernetic social spaces are performing in new, vivid ways 'the traditional urban function of creating opportunities for chance encounters between strangers' (Mitchell 1997, 113), which we could relate to Henri Lefebvre's notions on the production of urban spaces. The relationship between these theories and Grindr stems from the nature of the app; being somewhere while using it is important, because *where* the user is instantly affects *who* he's able to socialise with, as this 'Cerebro' will only track the nearest 100 other Grindr users within the specific radius.

Placelessness

Grindr addresses issues related to gay identity and the built environment, when one considers histories of gay subculture and its correlation with spaces ... or rather, lack of. Therefore, before beginning to describe Grindr as a tool that reconfigures, and arguably enhances, a users's relation to the built environment, it becomes necessary to contextualise the relationship between identity and built environment. Because there has been a historical and social placelessness within the built environment for citizens who identify as being gay, the condition of being spaceless has fostered a cultural move into cyber-spatial grounds. Sharif Mowlabocus's book, *Gaydar Culture*, based on the popular UK dating/cruising site 'Gaydar' gives insights to the historical and social displacement of gay men and helps explain why the Internet has played such an important role in the formation of gay identities. Mowlabocus states:

> Prior to the decriminalisation of homosexuality in 1967, the homosexual was not considered to be a member of society; he had no rights and he was not a citizen. As such, there were no 'legitimate pleasures' and there were no legitimate

> spaces in which such pleasures could be openly validated. The gay man's only option was to seek pleasure within the sex-zones of the city. These zones included the parks and lay-bys as identified by Califia, but also the homosocial spaces of locker rooms, toilets, prisons and dormitories. Homosexual sex was public sex (Mowlabocus 2010, 138).

Throughout the years gay men have been the only sexual minority to eventually set out and claim specific zones within certain cities and turn them into queer spaces. This is the case of London's Soho and Vauxhall, New York City's Chelsea and Greenwich, Sydney's Oxford Street and Madrid's Chueca, and while other cities may be added to the list, there is still a great number of countries in which homosexuality remains perceived as an illegitimate expression of normal sexuality, meaning that the public manifestation of gay identity is discouraged and thus placeless.

Being cast aside by the law, expressions of homosexuality were left to be manifested in private, and thus spaces like the public bathroom or toilet, according to Mowlabocus, became:

> A symbol of oppression and contained within its walls is a history of queer desire, covert negotiation, fear and entrapment. The cracked tiles and filthy porcelain, scuffed concrete and scribbled walls all operate as signifiers of pre-liberation homosexuality, a time of agent provocateurs and watch queens, of silent fumblings in lunch hours and guilty pleasures stolen on the way to and from work (Mowlabocus 2010, 139).

In the cases of urban spaces for gay men to congregate, such as gay bars, their physical locations would sometimes be difficult to find unless the visitor knew exactly where to look. Finding another man to have a sexual encounter with was often an act contained and limited to these spaces through the sexual practice of 'cottaging', the British term for having casual gay sex in public toilets. By cottaging, some gay men would partake mostly in sporadic, intimate encounters with different gay men in the area, through an act of surveillance known as 'cruising'. It then not only becomes evident that the gay man was given an abject body by society, because heteronormativity is challenged by homosexuality, but also that gay expression is given an abject connotation by the way that it was carried out as well as the spaces that served as stages for its encounters. Even though there is still something abjectifying and objectifying about it, Grindr has changed the way that gay cruising works by making surveillance a digital, rather than a purely physical, act. This in turn has had vast socio-spatial reverberations that extend beyond the sexual discourse that is so heavily associated with the app. These reconfigurations in the way users relate to, use, perceive and perform in public spaces are what the chapter seeks to highlight.

Visual importance

According to the Grindr website as of 1 January 2015:

> To get started with Grindr, simply grab your mobile device and download and open the app. Then, give yourself a profile name (if you want), upload a photo (also optional, but we encourage it), and answer a couple of questions about yourself (your choice, too). Then, get out there and start chatting, trading

> photos, and meeting up with the many men in your area – wherever you may be! Whether you're traveling, new in town, or just hanging out at home, Grindr's the fastest, easiest, and most fun way to connect with the biggest network of nearby guys. And it's available on not just the iPhone but also the iPad, iPod touch, Android, and BlackBerry.
>
> (Grindr 2015)

Once the app is opened, the main screen shows a grid of pictures, most of them with user names (some people leave them blank). At the top left, your own profile is shown, and the guys who are displayed in the screen are displayed in order of proximity, with the first on your right being the closest one; the further away they get from your picture, the further they are from your actual location radius. Similarly, the profiles displayed are guys who are either currently online or were online less than an hour ago, so at any given time of the day, the grid changes, having users appear and disappear, or slide left and right, depending on whether they have gotten closer or farther from you. Just as if they were out for a stroll in a public plaza or park, Grindr Guys find themselves looking around, trying to spot someone that sparks their attention to interact with them.

The visual within gay culture holds vast importance; being seen in certain places or wearing a certain article of clothing, for instance, can act as codes of disclosure, and they are ways to perform one's identity in public settings, such as gay zones in the city. Even though these zones are an important part in the formation of gay identity for citizens in cities, they are no longer the sole spaces where gay identities can be formed. They are also being constructed online, in digital spaces. We have to acknowledge, then, that surveillance plays a dual role and is not unidirectional; it is not merely about gay men acquiring visibility in the city or even so much as gay men being able to see other gay men. Visibility in digital spaces has also vastly to do with self-surveillance and introspection, in order to create a virtual identity to be displayed on the digital forum. The self, thus, becomes fragmented and reconstructed, making each user the architect of his virtual persona.

Creating a profile in any social network, as easy as it is, requires an unspoken act of self-reflection and of auto-evaluation. As Mowlabocus (2010, 92) argues: 'To look at a profile is to see the subject *from his own position* – imagined or otherwise'. If Donna Haraway was right upon stating that the development of writing and the visual organisation of life gave way and made possible the discovery of individualism and introspection, then digital social networks make vital the fragmentation of the physical self through a process of introspection and digital-identity-reconstruction in online worlds. In the majority of cases, the profile picture says more about how the Grindr user wants to be seen/read/understood/interpreted than it does about what the Grindr Guy actually looks like. Whether the picture is representative of the person or not will become evident when and if there is a meet-up, but in either case, it is a self-assessed representation of what the Grindr user wants to show on the grid, usually guided by a general understanding of what others deem desirable.

Issues of self-surveillance as well as external surveillance play a crucial role in the way that humans construct a body image of themselves. The experience of opening up Grindr and immediately seeing scores of men at any given time and at any given location recalls the experience of physically walking through a public space and watching people. Whether you are sitting down at a park and seeing faces walk past you or whether you are walking on the pavement passing people yourself, Grindr has a similar effect because of the way it functions, providing a constant flow of everything from chiselled faces to suggestive torsos. Through Grindr the gay male body is exposed in grid of dozens and dozens of men, each contesting in a sexual survival-of-the-fittest. The primary and most important factor in the Grindr grid is a display picture, and it is precisely this condition – dictated by the software and interface – that highlights the extreme importance of a body, or shall we say, of a picture of the product that is to be consumed. By dictating these conditions, Grindr is, perhaps, bringing back Freudian notions of the libidinal content of the spectator's look: 'visual impressions remain the most frequent pathway along which libidinal excitation is aroused' (Mowlabocus 2010, 98).

When the physical body becomes a 'consumable' product, objectified and displayed on the Grindr grid, the user enters a state of self-critique. In their paper, 'The Surveillant Assemblage', Kevin D. Haggerty and Richard V. Ericson write:

> A great deal of surveillance is directed toward the human body. The observed body is of a distinctly hybrid composition. First it is broken down by being abstracted from its territorial setting. It is then reassembled in different settings through a series of data flows. The result is a decorporealized body, a 'data double' of pure virility.
>
> (Haggerty and Ericson 2007, 605–622)

Spatial relevance and cultural variations

The spatiality of Grindr presents cultural variations as well, in terms of the way it is used in particular locations. Though the interface, design and functions are exactly the same regardless of the city, the interactions that take place on it have subtle variations. The rules of etiquette and the way interaction is executed varies, and these are often notable by the way that men initiate contact. For instance, Grindr in Puerto Rico's capital city, San Juan, performs differently than in larger, dense cities like London, in terms of the app's GPS radius's size. With a lower number of Grindr users and with Puerto Rico's urban and suburban sprawl, the nearest guy might be thousands of feet away; whereas in London the nearest guy might be next door. Thus, within the hundred guys that a user could be able to see and contact on the Grindr grid, there are users who are not even located in San Juan but rather are located in other cities, miles away from the capital. This affects the ability for men to reach the ideal status of being 'zero feet away' because the distance is too great, and driving becomes a necessity. Consequentially, it is unsurprising that the typical opening line for a Grindr user in Puerto Rico is commonly 'Hey. De dónde?' Meaning, 'Hey. Where are you based?' Thus, the initial interaction responds to the spatial condition of the app.

Grindr seems to be able to, in a sonar/radar-like manner, allow guys to see through walls and buildings, finding other men who they could talk to. This issue of 'finding' people is of particular curiosity, because they are not referring to finding the gay men who are 'hidden in the closet'; they are referring to finding *any* man, hidden, exposed, closeted or out. Though guys on Grindr are visible and physically present in the city, whether they are recognisably gay or not, the new subjectivities of today often find it difficult to approach each other in person. Grindr seems to be providing a virtual space where neighbours can access and be rid of the confines of physical boundaries. This is to say, the built environment – as a physical space – is removed to reveal these digital embodiments who have reservations about initially interacting and meeting in urban spaces. Paradoxically, Grindr seems to also be able to make a large city like London quite compact; people who you have seen on the grid become familiar faces that stand out in the crowd when spotted at public locations. At times, these faces become recurrent sightings, spotted in different parts of the city, making it seem smaller, friendlier and more humane than one would normally expect from seeing a map. Where walls obstructed views, the digital screen acts as the x-ray or peep-hole that gives a glimpse into the lives of Grindr Guys. So when the guys go out into the built environment they are recognisable by other users. The two-way gaze is exponentially magnified by the Grindr grid on the digital screen.

Digital imprints

Although the app is meant to show users who are nearby, bringing people together through spatio-temporal synchronisation, there is a condition in the software itself that provides a time-lag: when a Grindr user closes the application, his profile remains visible in the grid for a certain amount of time, which is usually equal to one hour – though this may depend on the volume of users online in any given area in the city. This means that even though the particular user may have already gone somewhere else, Grindr still registers them as being present in the last location where they were online; thus, whenever a user opens and closes Grindr, they leave a digital, transient imprint on a particular space, detectable by other users' digital devices via the Grindr app. This invisible imprint makes a user *polypresent* by having their digital embodiment shown as present in a given space in the city, while their physical body is present elsewhere. Similarly, it is this trail which allows other users to be able to come in contact with someone who is supposed to be nearby but is no longer there, increasing the number of potential interactions for the user.

Naturally, train stations, bus routes, railways, docklands and airports are rich zones for trails and digital imprints, because users frequently use Grindr in such places to 'look around'. Once the volume of online Grindr users increases in the particular zone or the 60 minutes expire, the Grindr user disappears from the grid. A Grindr profile can disappear off the grid in a particular area, before the 60 minutes are over, when and if a user opens the app in a new location; then Grindr will update their location and will automatically move the user from the previous location to the new one, thus showing that the act of Grindring is as physical as it is virtual.

Rather than accurately documenting and transmitting information as it occurs, the technology disturbs normal space-time configurations, opening up a new territory for theories related to space and performance. Real-time data, which is one of Grindr's key characteristics is then dubious, and through these cracks in the system one can begin to see that there is a lag in the synchronisation between people's physical embodiments and their digital ones. In *Alone Together*: Why We Expect mroe from Technology and Less from Each Other Sherry Turkle (2011, 160) states: 'Now, with a mobile device as portal, one moves into the virtual with fluidity and on the go. This makes it easier to use our lives as avatars to manage the tensions of everyday existence'. Invisible to the eye, this latter embodiment can only be traced/viewed via the app's interface. Through the severance of spatiotemporal presence, new embodiments are formed. While the Renaissance idea of humanism approaches the subject as a unified self that emerges from a universal, unified state, Posthumanism recognises the disunited nature of the self in addition to upholding the belief that humans are able to fluidly manifest themselves through different identities.

Socio-spatial transformations

Slowly, but surely, by bringing different gay men from different backgrounds, cultures and lifestyles together within centimetres of each other, Grindr is changing the way men socialise. Even the issue of the occasional 'closeted' guy on Grindr leaves an underlying tension between the 'out' users and how these two subjectivities interact. The boundaries between digital and physical have been inarguably blurred and sociability finds itself in a state of transformation. This transformation is largely due to the very nature of the technology and the ease with which digital interaction can take place, but it also has to do with the fact that even though Grindr was conceived by the designers and creators as an app to find and meet up with men who are nearby – even if it is for casual sex – this same piece of technology has been appropriated in various ways by each user, granting a plethora of ways in which it becomes a tool for connection with others. The digital behaviour influences the physical behaviour just as much as the physical behaviour influences the digital behaviour, leaving 'acceptable social interaction' up for interpretation. Through it, users have found a way to evade intense feelings of awkwardness, discomfort and rejection, because even though these feelings are still agents within the app, their effects are diluted, mitigated by the decorporealised interaction that takes place behind a glass screen.

Dependent on synchronisation in space and time, Grindr replicates the experience of meeting in a public space and engaging in conversation with complete strangers. On the other hand, this also means that there is a decrease in physical interaction, because users are no longer completely dependent on having to go to certain places to meet people. Similarly, it also could mean that because Grindr has changed *where* gay men meet, it also changes *how* they meet, in terms of how they interact with one another. Although Grindr's intention is to bring men face-to-face, it is often the case where a meet-up is never the outcome or even the intention of the Grindr experience.

Having interviewed a number of Grindr users to uncover findings for this app has revealed that often Grindr is used as a means to 'look around' or to kill time; it is used as a form of asserting and practising one's identity in a digital sphere, and this practice, of which the physical body is the centre, has material outcomes. This is why Katherine Hayles calls for a disembodied approach to the subject, meaning that the technology/information cannot be thought of as something that can replace the body but rather something that can enhance it. But it could be argued that technologies such as Grindr are in several ways embodied, though not in the sense Hayles refers to. As noted in the introduction to this chapter, Hayles does not view technology as a replacement for the body. Aware that the Internet and technological innovation have become ubiquitous and vast, she acknowledges the need to rethink original concepts in cyberferminist theories. Contrasting dramatically from other academics such as Turkle, who goes as far as to fight for the nostalgic preservation of the semantics of 'community' in a refusal to acknowledge the socio-spatial changes produced by technologies, Hayles states:

> As new and more sophisticated versions of the posthuman have evolved, this stark contrast between embodiment and disembodiment has fractured into more complex and varied formations. As a result, a binary view that juxtaposes disembodied information with an embodied human lifeworld is no longer sufficient to account for these complexities. Although I have not abandoned my commitment to the importance of embodiment, it seems to me that contemporary conditions call increasingly for understandings that go beyond a binary view to more nuanced analyses.
>
> (Hayles 2005, 2)

By using digital interaction – which is typically asynchronous – tailored to a synchronous model, it redefines the figure of the *flâneur*, morphing it into a larger-than-life, spatiotemporally unrestricted cyborg-like citizen, who not only defies the laws of physics, but is also sheltered from certain human emotions: fear of rejection, of awkwardness, of the unknown, of being seen as different. It is important to discuss the kind of identities Grindr and other mobile technologies are producing. Is face-to-face, spontaneous interaction something that will increasingly continue to be perceived as atypical, making the modern-day cyborg more dependent on digital devices to fulfil its social approaches? Similarly, one must wonder if Grindr's ability to create a social space within the vast territory of cyberspace still fosters the suppression of that which it seems to want to liberate: the expression of homosexuality. If gay interaction fails to take place within the visible built environment, how will homosexuality ever be able to break away from being perceived as normative and homosexuals as different?

The answer to this question may not be black and white, and one might argue that this might be a too nostalgic way of thinking, where antiquated methodologies fail to look forward to the success and efficiencies granted by technology. Even though Grindr claims to want to foster face-to-face meetings, the initial approach for this interaction takes place in an invisible space, hosted by servers; like Charles Xavier using Cerebro to find his mutants, the Grindr machines are used by gay men to find Others. In this manner, through this transactional order, gay identity is performed in cyberspace to then be performed in physical space. Grindr is providing new configurations of social order and spatial culture, and there is an ambiguity concerning this dual relationship between Grindr and social space, between which influences which; is Grindr public space's sidekick, or is it the other way around?

Part IV

Perspectives and methods for spatial cultures research

16 So long, and thanks for the GIS

Digital spatial history

Alexander von Lünen

> Superman is, after all, an alien lifeform. He's simply the acceptable face of invading realities.
>
> (Clive Barker)

Introduction

History, as an academic discipline, has had quite a few alien invasions in recent decades. Sociology, linguistics or anthropology have added to historical methods while leaving its core intact. They were thus the 'acceptable face of invading realities' Barker spoke of.

Not so much quantitative methods. For a time in the 1950s and 1960s they seemed to be the holy grail of historical inquiry. The now almost legendary school of the French *Annales* indulged in it as much as the British dominated Social History school at the time.[1] Geography also played a big role in these studies and French historian Emmanuel Le Roy Ladurie produced thematic maps together with cartographer Jacques Bertin that don't look too different to today's GIS visualizations in principle (cf. Aron *et al.* 1972). Given that the recent 'spatial turn' in the humanities, and spatial history in particular, frequently refers to the *Annales*, one therefore wonders why GIS seems so alien to most spatial turn historians. Much of what Historical GIS (HGIS) and its advocates have to offer isn't so different from the philosophy of the *Annales*, yet it appears there is no general anticipation by (spatial) historians to use it. As a matter of fact, while spatial history is by and large about the 'de-territorialization' of space – meaning it is about abstract and idealized spaces – HGIS epitomizes a 're-territorialization' of history; it deals with concrete, physical space, i.e. location.[2]

This chapter will look into this triad of history colliding with both digital and spatial cultures as they are represented by HGIS. It will be argued that HGIS for most historians represents an 'unacceptable face of invading realities', to stay in Barker's dictum, based on an excursion into philosophy of technology and history of mathematics to shed some light on how technology – such as GIS software – might aid in investigations into history.

Enframings

It is common knowledge that the tools we use determine – at least partially – the way we work. Historians and philosophers of technology have used this vernacular to detail the nature of human technology and how it frames social interaction and scientific work. Some more pessimistic writers have regarded this pervasion of everyday life by technology as something very limiting – if not dehumanizing – such as Henry David Thoreau in *Walden, or Living in the Woods* (1854): 'Men have become the tools of their tools.' With this verdict, Thoreau – like so many others in the nineteenth and twentieth centuries – expresses a general critique of modernity and its rationalized, technical working environment. This critique was also the primary drive for German philosopher Martin Heidegger (1889–1976) and his inquiry into the nature of technology – although he would later deny it (cf. Zimmerman 1990).

The digital – shorthand for digital methods such as data modelling, data mining or visualization – would thus be for digital history what was coined as *Gestell* (Engl: enframing) by Heidegger. The Gestell 'is the way in which the real reveals itself as standing-reserve' (Heidegger 1977, 23). 'Standing-reserve' then is defined as this: 'Everywhere everything is ordered to stand by, to be immediately at hand, indeed to stand there just so that it may be on call for a further ordering' (Heidegger 1977, 17). However, attentive readers of Heidegger will notice that he writes this enframing as 'Ge-stell' in German, referring to the verb 'stellen' with this placing of the hyphen, which translates as 'providing' or 'setting up'. This is what Heidegger means by 'enframing': technology makes knowledge about the world accessible, while at the same framing and thus restricting the way and the amount of knowledge one can access. Every tool has its limits and it determines what and how one can see things, and the computer is in no way different in this regard.

Technology thus enables a quick access to the world, making resources effortlessly available with the press of a button.[3] Heidegger points out that – contrary to common belief – modern science was not the precursor to modern technology, but the other way round (Wisser 1970, 72). Modern science is characterized by its equipment: laboratories, microscopes, DNA sequencing, etc. These technologies enabled new forms of science, or rather: new levels. Note that Heidegger only refers to the transformation of knowledge with this concept. Technology consequently aids a transformation of being: knowledge is transformed from the inaccessible to the accessible. This, however, does not mean that the knowledge has been digested already. Technology only transforms it into the *standing reserve* waiting to be picked up. The *Ge-stell* is thus a 'mode of revealing', but not the revelation itself (Heidegger 1977, 20–21).

Hence, computing and computer science constitute for digital history what Michel Foucault termed the 'discursive framework'.[4] A 'frame' in discourse theory refers to the boundaries of a discourse that act as an advance organizer or scaffolding for the discourse in general and the participants in that discourse in particular (Tomlin *et al.* 1997, 90). Computer science is this framework for digital history/humanities, because not only does the current computing technology proscribe what is possible, but also do the fundamental concepts of computer science – i.e. formalization – dictate what is *principally* possible. As British mathematician Alan Turing (1912–1954) and others have shown, this can never cover the whole spectrum of human knowledge, hence the reductionism immanent in digital history. Under the current framework, historians have to adapt/reduce their research to what computing allows them to do. This is precisely what Heidegger elucidated: technology reveals, but this revelation is automatically reductionist in character.

The work that catapulted Heidegger into intellectual stardom was his 1927 publication *Being and Time*. In another thought-provoking hyphenation of a well-established German word, 'Dasein' as 'Da-Sein', Heidegger draws attention to the temporal nature and condition of spatial being. 'Dasein' means 'existence' or 'being', but 'Da-Sein' – while not a word found in a dictionary – would be best translated as 'being there' or 'being in the world'.

Heidegger, famously, argued in *Being and Time* that human existence is predominantly temporal. With this he opposed Immanuel Kant's *Critique of Pure Reason* (1781), in which Kant argued that both time and space are a priori for human thinking; they are a given and without them thinking would be impossible. Heidegger argued that while space is important, time is more important. If we had infinite time we could travel infinitely far, but not vice versa. It is only through *time* that *space* has meaning.[5] Existence – and thus reasoning – is consequently framed primarily by time, and only secondarily by space, according to Heidegger. *Being and Time* 'interprets the "there" in explicitly temporal terms' (Malpas 2012, 17).

This angle has influenced historians as well, and partly explains why they have focused on time more than space. Many historians have perceived the actual location of an event as usually insufficient to explain why it happened. In the 1970s and beyond historians would look to sociology to learn how societies and individuals interact. Scholars like Anthony Giddens or Pierre Bourdieu would stress the formative character of spaces for human behaviour, but especially the latter would point out that spaces are 'constructed' over time. Space is thus not an a priori as Kant had elaborated, but is created by society over time by way of ordering and structuring (Schroer 2006, 88ff.), i.e. it is time that shapes space.

GISpaces

One can see this tension in HGIS. Probably the most heard complaint about GIS from historians and others is that GIS is poor – if not recalcitrant – at dealing with temporal data. While it is difficult to model even modern spatio-temporal data, the often inaccurate and anachronistic historical data renders this almost impossible. Newer versions of ESRI's ArcGIS suite offer a facility to time-code data and use it to display movement or change as animation. This data has to be discrete and precise, however. While this might be achievable with modern sensor technology to, say, track bird migration, the nature of historical data will evade this feature.

Is it is for this reason that various authors have referred to GIS as a 'positivistic tool', for example Bodenhamer (2013, 27) for HGIS and Pavlovskaya (2006, 2003) for general GIS. What does this mean exactly? How can a tool be positivistic, rather than the intellectual framework it is embedded in? Once more Heidegger's *Ge-stell* rears its ugly head. It is the tool (GIS), here probably more than anywhere else that dictates our intellectual framework. In *Being and Time* Heidegger had already introduced the term 'Zuhandensein' (Engl: 'ready-to-hand'), by which he characterized the nature of the 'presence' of things. For Heidegger, this presence is not transcended by the physical existence of a thing, but is being interpreted as tool/equipment in a system of signification and intent. The use of a tool becomes second nature without the tool-user being aware of the system he or she is succumbing to. There is thus no conscious approval of the framework mandated by the tool by its user; one accepts it without spending much thought about alternative uses – until the tool doesn't do what is expected of it.[6]

GIS – and the computer more generally speaking – is a positivistic tool because its very conception was born out of mathematical positivism. First ideas of computing machinery sprang up in seventeenth-century mathematics, such as Leibniz's verdict that computation is 'unworthy of excellent men' (from a 1685 manuscript; cf. Davis 2000, 3ff.).[7] This was a first glimpse of the notion later proclaimed by German mathematician Karl Weierstrass (1815–1897) that all mathematics can be reduced to arithmetic, i.e. computation (Everdell 1997, 37).[8]

Another person to become influential in this idea of positivistic mathematics – and positivism more generally speaking – was René Descartes (1596–1650). As a matter of fact, critics of GIS often point out that this software only allows mapping into Cartesian space. Which is slightly inaccurate. 'Cartesian space' is often a short-hand for 'Euclidean space with Cartesian coordinates', but even this is not correct. 'Euclidean space' in mathematics refers to the space described by Euclidean geometry, which is a set of axioms and theorems describing the relationships between sets of points – e.g. two points describing a line, four points describing a plane, etc. (Ivanov 2011). For a time it was customary to use geometry as a way to describe modern mathematics such as calculus. This is now reversed thanks to developments in algebra, and part of this process was Descartes' *Discourse on the Method* from 1637, in which he gave an overview of his general philosophical method. One of the three appendices in this book was titled *Geometry* (cf. Boyer and Merzbach 1991, 376). In this appendix, Descartes would apply 'algebra to geometry and geometry to algebra' (Boyer and Merzbach 1991, 383). He would try to reduce geometry to arithmetic, 'i.e., arithmetic operations, such as multiplication or square root, can be solved geometrically' (Gaukroger 1992, 91).

Descartes would not, however, develop a full-fledged coordinate system as surveyors or geographers might use it (Boyer and Merzbach 1991, 385). This was done later by translators and interpreters as a means to clarify Descartes' ideas. As a matter of fact, as Boyer and Merzbach (1991, 386) point out, the system attributed to Descartes could be found to a large extent in Nicolas Oresme's work in the fourteenth century already. But Descartes' interpreters and commentators would attribute it to him anyway, hence the term 'Cartesian coordinates'.

Descartes' more significant contribution, however, was to set a train of thought into motion to see geometry as dimensionless. In his time, geometry was always attached to magnitudes, i.e. a line had to be given with its physical length (e.g. '1 meter'). Descartes, on the other hand, introduces abstract denominations, such as letters, to name lines, etc. (Gaukroger 1992, 102). Descartes thus *maps* concrete physical entities to abstract denominations. This, however, leads imminently to a philosophical problem: given all these abstractions, how can one relate this to the natural world consistently? (Gaukroger 1992, 108). Contemporaries, Descartes included, were not yet willing to completely abandon the direct relation between abstract algebra and concrete geometry – or any other, similar kind of mapping.

Rather, as Descartes elucidates in his 1644 *Principles of Philosophy*, imagination is required to make this relation between the symbolic and the concrete. He compares his work on geometry to other areas, such as physiology: 'what we see does not resemble the physical reality; we see no secondary qualities, but only spatial extended bodies' (Gaukroger 1992, 111). From this Descartes (1985, 227) infers that 'body' and 'space' are equivalent: 'It is easy for us to recognize that the extension constituting the nature of a body is exactly the same as that constituting the nature of space.' It is this concept that actually demarcates the 'Cartesian space', i.e. physical extensions as the perceivable natural phenomena. For Descartes (1985, 229) this meant that 'space' and 'place' are not anathema to one another, but are both physical qualities of bodies: 'The difference between the terms "place" and "space" is that the former designates more explicitly the position, as opposed to the size or shape, while it is size and shape that we are concentrating on when we talk of space.' With this Descartes follows Plato who had already alluded to the equivalence of space and matter (cf. Caygill 1995, 367–373).

Why is this discussion relevant for (H)GIS? Because it becomes quite obvious that GIS indeed allows one only to work with Cartesian space.[9] While the storage of coordinates and attached statistical values can be done without reference to dimensions, the visualization of them needs an extension (usually as pixels on a screen). Also, and perhaps more importantly, historical processes and events have to be pinpointed, i.e. they need to be discrete. One would need to be able to accurately define and delineate historic events, when yet it is in their very nature to be fuzzy.

Indeed, Heidegger in the preparation of his big work *Being and Time* had argued along similar lines, for instance in *The Concept of Time in the Study of History* (1916) he noted that:

> the historian cannot regard time, as the natural scientist does, as purely quantitatively and uniform. Historical time involves qualitatively distinct periods, such as the Victorian era, whose significance depends on more than their length as measured in years.
>
> (Inwood 2000, 98)

As just mentioned, to make this qualitative assessment of history work with the standard, quantitative GIS software one would have to spend a good amount of time and sophistication to find an acceptable mapping from the continuous, qualitative space into a discrete, quantitative one. This data modelling is similar to what Heidegger then referred to as 'homogenization' in *The Concept of Time* (1924):[10]

> *Homogenization* is an assimilation of time to space, to Presence pure and simple; it is the tendency to expel all time from itself into a present. Time becomes fully mathematized, becomes the coordinate *t* along-side the spatial coordinates *x*, *y*, *z*.
>
> (Heidegger 1992, 18)

This homogenization, quite obviously, applies to spatial as well as temporal aspects of history. The historic events and its actors have to be mapped into the materialistic Cartesian space. However, most scholars in modern history are interested in the dynamics of processes rather than merely the places they involve. An often referred to definition of 'process' was given by sociologist David Berlo (1960, 24):

> When we label something as a process, we also mean that it does not have *a* beginning, *an* end, a fixed sequence of events. It is not static, at rest. It is moving. The ingredients within a process interact; each affects all of the others.

This complexity of (historical) processes has to be distilled into single distinguishable entities in order to be computable – this is after all what 'digital' means: to be enumerable, to delineate one element in an algebraic set from another.

Spatial history

With GIS we are thus encountering another instance of a *Ge-stell*: while geographical knowledge is made accessible, it is done so in a reductionist fashion. GIS is only a subset of geography; namely it lets you deal only with Euclidean/Cartesian space. As a matter of fact, GIS is in this respect not much different from cartography, which is itself only a subset of geography: the reduction of geography to a *map*. The map is hence a visual construction of the world through a cartographic logic (Farinelli 2013, 257); as Ludwig Wittgenstein (1961, 7) evinces: 'The facts in logical space are the world.'

GIS is thus the geography that is mappable, for non-Euclidean geometries are 'possible to construct ... logically, but it is impossible to visualize them' (Reichenbach 1958, 43). In other words: GIS offers only the kind of geography that is logically and mathematically describable, and visualizable – which is what I elaborated on above when talking about Descartes. The *language* of GIS is therefore akin to what philosopher Edmund Husserl (1970, 51) referred to as the 'garb of ideas':

> Mathematics and mathematical science, as a garb of ideas, or the garb of symbols of the symbolic mathematical theories, encompasses everything which, for scientists and the educated generally, *represents* the life-world, *dresses it up* as 'objectively actual and true' nature. It is through the garb of ideas that we take for *true being* what is actually a *method*.

GIS is such a garb Husserl spoke about, as it dresses geography up as scientific subject when historians and human geographers are interested in something else: space as condition of living or 'spaces of identity, which are often articulated through spatial images such as mobility, location, borderlands, exile and home' (Blunt 2003, 75–76). This might be hard to accomplish with the established language within the GIS community, though it is not impossible. However, to make any headway in this area, one would need to look at the way historians are talking about space and what their methodology is. Or as Nicolaus Sombart (1992) has put it: we need to arrive at a 'topographic hermeneutics' in history.

As mentioned above, a lot of the current debates in spatial history are about rather abstract spaces, which is detrimental to the use of GIS. This was criticized by authors like Ethington (2007), who maintained that we look at the past spatially and not temporally (similar to the philosophy of Henri Bergson). That time is mentally and metaphorically fathomed as space has long been discussed by philosophers.[11] Ethington (2007, 465), on the other hand, postulates that '[t]he past is the set of places made by human action. History is a map of these places'.

It is for this attitude of material presence of the past that historical sub-disciplines such as Ancient History or Archaeology have no issue with using GIS, for their field of study deals a lot with actual material objects. Given the lack of written sources, scholars from these fields embrace technology and empiricism, for it is often the only meaningful way to get narratives from the eras they are interested in (cf. Lünen and Moschek 2011).[12] Modern historians, on the other hand, engage in some more theoretical debates about space (e.g. social construction of space). In this regard, spatial historians discuss concepts that human geographers have discussed some twenty years ago, such as in Peter Jackson's (1989) seminal book *Maps of Meaning*.

Geographers like Denis Cosgrove (2005, 28) have picked up Jackson's ideas to point out that the kind of mapping he spoke about 'include[s] all graphic representations of knowledge'.

Wouldn't this point to novel forms of GIS, one that offers different forms of visualizations, such as network visualizations? Not necessarily. I have been involved in implementing non-standard GIS for historical data, such as the Great Britain Historical GIS, which operates as historical placename gazetteer and is based on relational space (cf. Aucott *et al.* 2009). These approaches can have great value only when employed in the proper scholarly context – which they rarely are, for we are still lacking Sombart's 'topographic hermeneutic'. Perhaps it might be an idea to go back and ask the Heideggerian question: What *is* Historical GIS? Rather than talking about what it does.

Historical GIS

It should now be clear that both spatial history and digital history each for themselves represent a *Ge-stell* for history. They make knowledge about historical processes accessible that had hitherto been inaccessible for historians, while at the same time foist reductions and formalisms onto historical scholarship that at most times seem to be too hard to swallow. HGIS, on the other hand, multiplies this restriction. Not only does it impose methods of the digital and the geographical realm individually, but it combines the two. GIS does not represent the entire body of knowledge of geography, but only a subset of it; as much as its data analysis capacities only represent a fraction of what computer science *et al.* have to offer.

GIS and Remote Sensing have transformed the subject of geography to such an extent that large areas of the field have now turned into some kind of Geo-Informatics. With this, these digital tools have also established – or at least consolidated – the rift between the more qualitative oriented human geography and the data-driven disciplines such as physical geography or cartography. Over a decade ago (human) geographers started a 'Critical GIS' (CGIS) school which looks at the geographic knowledge production introduced with GIS; geographers and others critically examine how GIS and mapping are reflecting power relations. Interestingly, spatial historians are concerned with these kinds of questions as well, yet the two schools – Critical GIS and Spatial History – don't seem to mix. Neither have CGIS and HGIS, but given the topical proximity of CGIS and Spatial History, one wonders whether there shouldn't be something like a 'Critical Historical GIS' movement that marries Spatial History with Critical GIS into 'Digital Spatial History'.

Critical GIS started in the 1990s when it became clear what level of impact GIS technology would have on both the academic subject of geography and in a wider social context, sought to challenge the kind of knowledge produced by the technology in terms of its social and political background. GIS was regarded as a deeply positivist tool, created to assist in – if not drive – governance issues, such as social planning, military agendas or environmental policies. It was also asked who the stakeholders in the creation and use of GIS were. Feminist geographers, for example, began to look into the way GIS reflected and manifested a patriarchal society.[13]

There is an interesting parallel to the debate on the predominantly quantitative methods in social history and geography in the 1970s. The general discomfort with quantitative methods was that it turns statistical correlations into policies.[14]

The school of 'micro history' had a point of departure similar to the debate on quantitative geography in those years. Historians like Carlo Ginzburg challenged the presumption made by social historians that the history of the 'lower classes' could only be researched through numbers.[15] Academically it called into question the notion that quantification would be the only possible way to formalization, i.e. it would only be by anonymous numbers that social issues – such as social class structures – could be put into a formal model (Levi 1991, 109; see also Greenstein 1997, 352).

Critical GIS, on the other hand, does not dismiss GIS *in toto*. Rather, the demand goes, GIS needs to cater for qualitative methods to counteract the positivist, quantitative approaches (e.g. Pavlovskaya 2009). As cheerful as this idea of creating a 'better', qualitative, GIS may sound – putting aside that we have yet to see a Qualitative GIS in market-ready form – one aspect that becomes obvious is the enframing issue mentioned above; the Critical GIS discourse is not one that comes from within the (spatial) history community, but from the geographical and GIS community. Historians are mere onlookers in this process, as they are not 'natives' in this discourse. The methods proposed by CGIS proponents, namely a 'Qualitative GIS' approach (cf. Cope and Elwood 2009), are quite prototypical and can only come from people at home with GIS technology. Historians, by and large, are mere users and feel alienated by this technical debate. Qualitative GIS reverberates with an IT/engineering discursive framework, as the critical assessment of the traditionally quantitative GIS technology by the CGIS community usually ends up in proposing a different kind of GIS, i.e. designing/developing software. Is this a debate historians can participate in? Maybe, but yet again we have alien invaders who have yet to produce an acceptable face.

Conclusion

In this chapter I have tried to elaborate on the pitfalls of using GIS in historical scholarship from a more general viewpoint by delving into the philosophy of technology in connection with the philosophy of history. It seems all too apparent that the issue rests with what Martin Heidegger called 'ready-to-hand'. As mentioned, Heidegger uses this term to describe the unconscious approval of a tool: using a tool without even thinking about it. As he elucidates, only when the tool is defective does its user become aware of it. It is this notion that is expressed by Orlandi (1992, 36): 'there is (and has always been) both a formalisable part and a non-formalisable part of the humanities. Humanities computing is precisely the formalisable part of the Humanities, and in this sense is not a new subject born with computers.' Those scholars who have always worked quantitatively and formalistic have no issue with GIS because this tool is ready-to-hand for them. Proponents of Historical GIS have trouble even getting to grips with the criticism uttered by other scholars, as the merits of GIS for history seem all too self-evident to them. For them GIS is not broken. For those not comfortable with quantitative methods etc., GIS is broken, and they are painfully aware of its defects.

One of these defects, as I discussed, is the difficulty to model spatial data – any data – over time. Putting aside the technical issues, it is again something that goes to the heart of historical scholarship: the act of *presencing* historical knowledge. All historical knowledge is transposed to the present by the historian. This is something that

St Augustine already realised in the Middle Ages, where he stated that 'there be three times; a present of things past, a present of things present, and a present of things future' (Augustine 1853, 239).[16]

The *presencing* of the past through digital means is not a mere technical obstacle, but part of a bigger historiographical issue, namely the analytical and empiricist notion of digital scholarship. It is thus pointless to ask for 'better' software (however defined), or to demand that historians learn more about geography to master Historical GIS (with a capital 'h'), for – as I discussed in this chapter – many geographers aren't happy with GIS either. It is, as I have pointed to in a previous publication (Lünen 2013) a matter of *expropriating* rather than *appropriating* GIS for historical scholarship – to drop 'Historical GIS' in favour of 'digital spatial history' and to use GIS in a mainstream history framework rather than in the quantitative, analytical one.

Notes

1 For the *Annales* and its methods see Lünen and Le Roy Ladurie (2013).

2 As a number of authors have attested, there are very diverse definitions of, and discussions around, the terms 'space', 'place' and 'location'. The discussions of these, while interesting and insightful, are beyond the scope of this chapter. See Jerram (2013) for a discussion of the terms in the context of spatial history, and his discussion of 'dematerialization', from which I derive my term of 'de-territorialization'.

3 Heidegger mainly referred to physical resources with this. However, he would also make references to the intellectual world by looking at cybernetics (Heidegger 2008, 313) – although he did not know enough about it to make a decisive statement here.

4 Cf. Hall (2001, 73); see also Foucault (1972, 32). As Agamben (2009, 12) notes, Foucault's 'dispositif' (Engl: apparatus) is akin to Heidegger's 'Ge-stell'. As a matter of fact, Heidegger is a major influence for Foucault more generally speaking (Elden 2001, 152).

5 For a discussion of Heidgger's critique of Kant see Mulhall (1996, 160); for a more general discussion of space in *Being and Time*, see Malpas (2012).

6 See Mulhall (1996, 46–59) for a discussion of this aspect of *Being and Time*, and Heidegger (1996, 62–71) for the respective paragraphs in which he discusses *Zuhandensein*.

7 The manuscript from 1685 was titled 'Machina arithmetica in qua non aditio tantum et substractio sed et multiplicatio nullo, divisio vero paene nullo anivni labore peragantur', which translates as 'An arithmetic machine which can be used to carry out not only addition and subtraction but also multiplication with no, and division with really almost no, intellectual exertion'.

8 This, obviously, is not the case, as late nineteenth-century mathematicians had to acknowledge – but only to move on to try to reduce all mathematics to logics, as Gottlob Frege and Bertrand Russell (and others) tried; only to be refuted by Kurt Gödel in 1930.

9 There is another angle here to 'Cartesian' space: Descartes is also regarded as one of the strongest advocates of scientific objectivity, and thus the scientific method and empiricism. Many proponents in the HGIS area are arguing exactly this point: that GIS would introduce empiricism into historical scholarship (cf. Knowles 2014, 209).

10 With this comment Heidegger is laying the seeds of later works, such as *The Question Concerning Technology*. Says Malpas (2012, 35): 'Technological modernity is thus understood not only as the culmination of the metaphysical tendency toward nihilism, but also in terms of a specific modification of time and space that reduces the thing to mere resource and place to simple location.'

11 Eminent philosophers such as Heidegger and Henri Bergson have participated and driven this debate. As mentioned above, Heidegger does not dismiss the importance of space outright, but that the spatiality of humans' existence is subordinate to its temporality. In

response to Immanuel Kant's *Critique of Pure Reason* (par. A 48–49, B 66), Heidegger argued that time is a representation of both the inner and outer world of the mind, whereas space is only a representation of the outer one (cf. Mulhall 1996, 160).

12 Hopkin and Denley (1991, 65) make this point more eloquently, referring to a certain envy modern historians may feel towards medievalists: 'In this respect the modern historian envies the medievalist the relative containment and stability of his source material, which in turn enables the scholar to engage in the close study of a defined and limited range of texts. This is an enterprise in which few historians of modern politics or society can afford to indulge.' The 'containment' and 'stability' is in some way what Heidegger referred to with 'homogenization'.

13 There is now a good deal of literature on CGIS out there; the journal *Cartographica*, for instance, often features contributions to the topic and had special issues on it in the past (cf. Harvey *et al.* 2005).

14 One of those actors using quantitative methods in geography for social planning was Gunnar Olsson, and his departure from quantitative geography is exemplary for many geographers and historians in the 1970s. In my interview with him from 2011 he gives some insights into his motivation for this (cf. Lünen and Olsson 2013).

15 As Ginzburg (1976) stated in the foreword to his famous monograph, *The Cheese and the Worms.*

16 See Munslow (2014) for a wider debate about 'presence' and the past.

17 Making space for each other

Reflections on incommensurate data in interdisciplinary research

David Jeevendrampillai

Introduction

This chapter has its origins in a moment when two social projects met. On the one hand it looks to the *Adaptable Suburbs Project* (*ASP*) at University College London (2010–2014) and on the other to a group called the 'Seething Villagers' who were approached as local enthusiasts to contribute to the *ASP*. Both groups were and are trying to do the same thing: *make better* suburbs. In what follows both groups are referred to as 'social projects', a term derived from the work of Elizabeth Povinelli (2011, 6) who describes them as 'the thick subjective background effects of a life as it has been lived; and these thick subjectivities provide the context of moral and political calculation'. This term is preferred over 'research project' or 'local project' to allude to the ways in which each project has a particular social framing that is a tacit normative position on how *making better* works. The *ASP* encompassed the scale of London, considering suburbs as part of a larger spatial network. It had an advisory board indicative of its policy-facing outputs and deployed objective methods to generate and order data in line with its academic purpose. The Villagers, by contrast, work at a local scale, encouraging face-to-face interaction in fun, creative environments to develop their local community and sense of localness. These two projects are not against each other, indeed they are doing the same thing, making better, though in very different ways. It was, however, through their meeting that their mutual epistemological assumptions became revealed as social projects.

Community mapping in an interdisciplinary research project

Adaptable Suburbs was a four-year project based at University College London's Bartlett Faculty of the Built Environment that aimed at 'developing an understanding of the workings of small scale "below the radar" economic and social sub-urban activities' (Vaughan *et al.* 2010, 1).[1] Citing the lack of policy attention that outer London areas had received, the *ASP*'s 'Case for Support' document, instrumental in securing funding from the Engineering and Physical Sciences Research Council (EPSRC), stated the project aims as being to develop 'an integrated theory of how town centres evolve by developing knowledge on urban sustainability, patterns of social and economic behaviour and how places adapt and change over time'; to improve understanding of the relationship between 'individual actions', 'large scale spatial order' and 'emergent morphological changes'. In this sense the *ASP* is revealed as a distinctly social project; its moral and political calculations steeped in a subjective background that can be traced through the history of city planning and politics of place making (see Rabinow 1989).

The *ASP*'s 'innovative multidisciplinary methodology' aimed at understanding the socio-material landscape of the suburbs as a 'holistic' system and as such it developed an online community map to add a social layer to the data (Vaughan *et al.* 2010). The community map was underpinned by a Geographic Information Systems (GIS) platform, designed to store, represent, analyse and manage its geographic data. In the *ASP* the community map would function as a Public Participation Geographic Information System (PPGIS) in encouraging local residents to contribute of historical and other community-related information as well as providing a method through which to disseminate the project's data to a wider public. The community map was presented as a vital tool through which social data could both contribute to the project's understanding of 'the way in which people use their area', and be layered over other data regarding movement potentials, building type and use. In so doing it helped to reveal the historical relationship between the social needs of an area and its built-environment infrastructure (Vaughan *et al.* 2010).

As an anthropologist on the *ASP* project it was my role to supplement the objective description of the built environment from the network and cartographic analysis with a collection of local testimonies in the oral history tradition. The anthropological research was sketched in the following terms:

> an 'auto-ethnographical' project, in which local inhabitants will be asked to report and tag their local activities and networks on the project website ... [that] will help inform the research project's overall aim of understanding processes of socio-economic adaptation in smaller settlements.[2]

Through this 'auto-ethnographical project', the *ASP* demonstrated its commitment to PPGIS as a democratic approach to research. The implication of the self-ethnography (ethnography meaning the writing of culture) implies that the community map user would be able to select what information was important to them and be able to share it on their own terms. That is, they would be writing about themselves, as *they* liked, collapsing the distance between observers and observed. Thus the 'auto-ethnographic' image of the community map is formed in the radical democratic ideal described by Jürgen Habermas (1990, 44–49) in his description of an 'ideal speech situation' where participants in a social group are able to evaluate each other's decisions and values based on a common sense of reason in an atmosphere free from 'coercive' influences.

PPGIS and community mapping appeal to development and planning discourses due to their promise of facilitating a wider participatory framework. By the 1990s participation had become the 'new orthodoxy' (Henkel and Stirrat 2001, 168). For some, however, the idea of participation as a means and as an end had been conflated (see Parfitt 2004) and by the late 2000s a more critical review of information and communications technologies (ICTs) was emerging. This later literature engages with the ways in which participatory techniques are themselves products of, and permeated by, power relations (Kapoor 2002) that according to Hildyard *et al.* (2001) serve to reduce local opposition to planning and development while creating an appearance of greater democracy. This so-called 'tyranny of participation' (see Cooke and Kothari 2001) highlights a gap between ideal and actual radical democracy. Mouffe (2005, 108) describes how a technocratic move towards a *commensuration* of values 'beyond hegemony' effectively eliminates alternative ideological states and subsumes them into the coherent whole of the collective, an instance suggestive of what Ferguson (1994, 557) in the context of development studies refers to as the 'anti-politics machine'. It is in this sense that a community map, through the visual image that aligns particular aspects of the material form and social value of the built environment, can act to 'perform territory' (see Perkins 2004; Dodge *et al.* 2008). This is because:

> All maps, like all other historically constructed images, do not provide a transparent window on the world. Rather they are signs that present a deceptive appearance of naturalness and transparence [*sic*] concealing an opaque, distorting, arbitrary mechanism of representation, a process of ideological mystification.
>
> (Mitchell 1986, 80; cited in O'Mahony 2014, 50)

In what follows I will examine a particular moment of *incommensurability* that was revealed by the two contrasting social projects that informed the auto-ethnographic component of the *ASP*. The 'moment' occurred when a group of local enthusiasts who call themselves the 'Villagers', in reference to an idealised state of inclusiveness associated with the mythical village of Seething, added 'the wrong type' of information to the community map.

The Seething stories

In order to build on previous research[3] and develop a commensurable data set within the *ASP*, Surbiton was the pre-determined field site for my anthropological enquiry. Whilst identifying people to talk to about the changing landscape of the high street in Surbiton I needed to look further than the traditional historical society and business groups to obtain both the reach and the range of stories the auto-ethnographic method demanded. I was looking for a group that would not have any particular motive or short-term agenda regarding the land use of the built environment nor a group that was narrow and limited to only a few members; at the local studies archives I found a flyer which told me the date, time and location of a parade to remember the 'Legend of Lefi Ganderson'. Over the course of eighteen months' fieldwork I went on to discover that Lefi Ganderson was the central figure in the Seething stories (see Hutchinson 2010a, 2010b, 2011, 2012). These stories form the basis of a number of large events, festivals and parades (around ten to fifteen per year), which hundreds of people attend (Figure 17.1).[4] The Seething stories also create a number of smaller events all of which were, in the words of my informants, 'fun' and, importantly, 'a bit silly'.

At the heart of the Seething stories is Lefi, who being half-boy half-goat was ostracised from a village called Seething. Lefi lived in a cave at the base of a mountain on the top of which a giant, the nasty Thamas Deeton, lived. Nobody would talk to Lefi other than the children of Seething, who would bring him food and play with him. Every year Thamas would come into Seething and terrorise the Seething villagers until one year, Lefi, seeing how frightened this made the villagers, tricked the giant into leaving. By passing milk through a golden ring again and again, Lefi was able to make cheese and fulfil the bet he had made with Thamas that he could live for a

Figure 17.1 A Seething parade fronted by Thamas and a nice man in a hat that represents Lefi

Source: author

whole month on only the food that would pass through that ring. Keeping his promise, Thamas turned from Seething but destroyed Mount Seething as he left. Lefi is remembered in plays and festivals and the spirit of 'doing things for others' is kept alive by a group who call themselves the Seething Villagers.

Seething, the noun, has its origins in the history of the area of Surbiton, relating to a name found on old maps indicating a water spring. The term had, however, fallen out of common usage. Using Seething over Surbiton as a collective name for events and the community allows a sense of belonging beyond the traditional residential or kinship links to locality. If Seething is a 'State of Mind' then anyone can be from Seething.[5] This is where we find the origins of the Surbiton/Seething divide in this chapter. Based on togetherness, fun and happiness, the values spread through Seething events; this inventive play on local history occurs in a suburb making play with the banality of its suburban image. As Wickstead (2013) asserts, it is the cultural image of the suburb as the dormitory of the modern worker; a place, according to *The Big Issue* magazine, which is void of life, 'where the life of the mind curls up in front of the fire in a comfy pair of M&S slippers' which is the heritage of the suburbs.[6] If they don't have history, then they'll make one up; 'no history', then, *is* their history. Through the Seething events a very simple social aim can unfold: the idea of community, togetherness, support and fun. In a very direct way the events bring people together, allow people to sell the products of their craft hobbies, and to learn what skills exist in the community, for example in its musicians and electricians. In so doing new relationships are formed in and around the public spaces of Surbiton that create a sense of local sociability and local economy. People now go to *that* coffee shop as they know the owner. This local sensibility provides particular meanings and values through association and interaction with the local environment. It both works and continues through particular mechanisms such as the performance of local myths, legends and festivals in parks. These ongoing tales and festivals work at a local scale but lose part (though not all) of their efficacy when placed in a context of a citywide audience, such as that which was the imagined audience for the *ASP* community map.

A full account of these rich Seething stories can be found in Jeevendrampillai (2015) However, a few key points can be made here. First, the events are, in the words of my informants, 'stupid, nonsense, daft'. That is, they actively play with the idea of official heritage, history and what can be considered 'proper'. In this sense they can, by definition, be called 'stupid' as they purposefully 'lack common sense'.[7] Rather, they play with what 'common sense' might be with regard to how spaces of the suburbs are used, imagined and lived in. Second, this openness to the stories, to making up history as they go along, permits a form of sovereignty over their claims to be local. Suddenly there are no experts, no official ways and no one person that knows more than the other. In its place comes a mechanism of invention of culture (following Wagner 1981). Third, and key to this mechanism, is a preference for the ambiguities of a stupid history based around fun and a common commitment to developing an inclusive supportive community over 'serious' history that traces facts and traditional tropes of belonging such as kin relationships or time lived in the area.

When told the story of how Thames Ditton Island, a small island in the middle of the nearby Thames river, acquired its name (Thames Ditton is the name that would appear on Google maps and as such on the community map), one is faced with a choice. Is it the place where the body of the giant Thamas Deeton fell whilst turning away from Mount Seething after being hit by a falling boulder or is it not? In this choice one decides either to be in or out. In being fun, a little silly, one goes with it and is drawn into the Seething life of invention, play and making new histories. From here the Seething events, festivals and fun produce an ethical disposition through the affective capacities of the events. That is, the corporeal, emotional and bodily experience draws one into new sets of relations with people and their lives and new sets of relations to the material spaces of the suburbs themselves. With this a sense of localness emerges amid the dancing in parks, parading in streets and putting destroyed mountains on maps.

A moment of interdisciplinary incommensurability

After a few months in the field I presented the *ASP* community map to a few key informants. The long-term idea was that the map would eventually be passed over into the hands of the community to manage its content. The idea of the PPGIS and its use by the community contains within it the ideals of participation, democracy and an 'ideal speech situation' arena which, I have argued, reflect the underlying social project of the *ASP*. In its anthropological component, however, the research project contained an attentiveness to the *process* of radical democracy via the community map. Through ethnography a critical reflection on the mechanisms and workings of the community map as part of this social project was enabled, the results of which are outlined through a description of the incommensurable moment that are the focus of this section.

Each *ASP* suburban case study had its own version of the community map accessible via the project website. It comprised a basic Google API (application program interface) with a specifically designed 'wrap', being a specially tailored interface onto which the user could place a symbol from a limited range that would inform the reader as to the sort of information being shared. These symbols relate to a list of 'categories' which align to the division of interests that a community might hold. To share information a user would register, log-in, upload information via text, photo or video, and locate the information on a point on the map. This information would be listed against a category, seen on the right of Figure 17.2 which shows the community map interface. This interface presents a map of the suburb onto which the user can click and add information in the form of text, photo or video. The user then assigns this information an appropriate category. Initially the categories – historical fact, cycling and transport, green space and so on – were agreed by the project team, and it was intended they would change over the period of consultation with the community to be more relevant to all involved. It was hoped that the community would

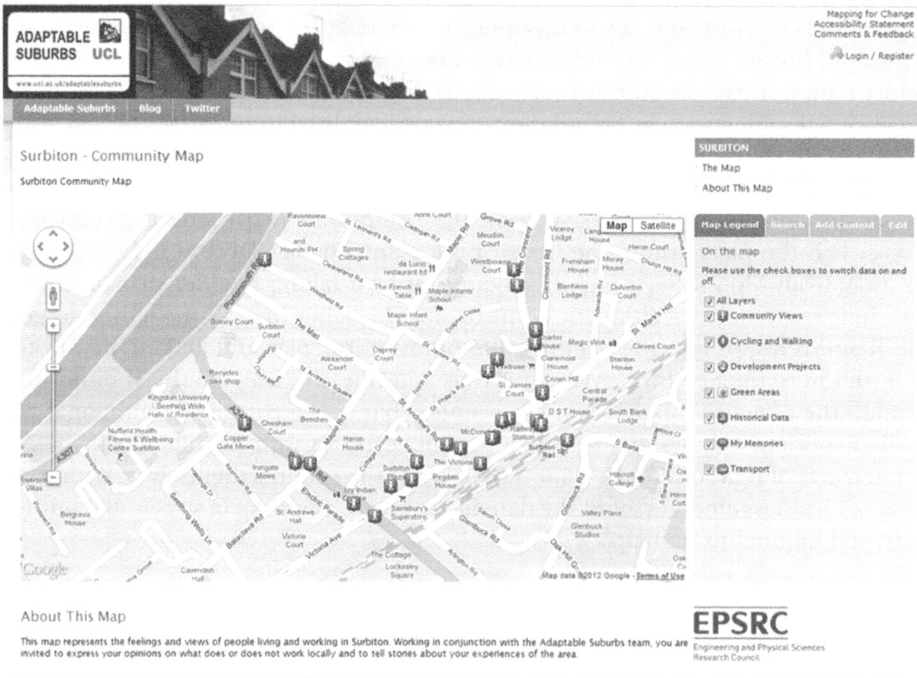

Figure 17.2 The *ASP* Community Map for Surbiton. The visible data points were added by the project in order to test the site, otherwise it is empty

Source: www.ucl.ac.uk/adaptablesuburbs/communitymaps, accessed 3 April 2012; uses map data from Google, © 2012

eventually take over the moderation of the map and that it would be a widely used forum for discussing and displaying oral histories of the area, acting as a living and evolving archive on the community's own terms.

Some time had passed since I had initially asked my informants to try adding information to the map. After introducing the map to many Seethingers at a Free University of Seething[8] (FUS) lecture, I had asked a few Seethingers to test the usability of the site and add information. After some weeks the map remained unedited; initially I assumed enthusiasm for the map was low. However, after asking my informants as to whether they had had time to add information Harry (all names have been anonymised) replied that he had added a point about Mount Seething but it had not yet appeared on the map and he assumed I was yet to moderate the post.

I returned to UCL and spoke with Flo, the Co-Director of *Mapping for Change* (a UCL spin-off company) and key developer of the community mapping platform.[9] Neither Flo, nor *Mapping for Change* were part of the *ASP* directly but Flo was responsible for overseeing the moderation of the map and its long-term community legacy. Flo explained that the post regarding Mount Seething did not pass moderation as it was entered under the label 'historical fact' and yet it was not historical fact as she understood it. For her the presence of a Mountain in Seething was experienced as myth. I shared a working space with Flo and her knowledge of my ethnography came primarily through shared coffee breaks and office chit-chat. She had a vague idea of the weird and wonderful tales of Seething, enough to know their playful nature, but had concerns that the use of the word 'fact' would possibly confuse readers of the map who were not immersed in the Seething tales or folklore. She offered to create a new label, 'myths and legends', on the map so that people could clearly see what was and was not fact, but this offer was refused by the Seethingers as it is precisely the play on the idea of 'fact' which is the productive aspect of Seething tales. This is true to the extent that the Seethingers even created a Seething archaeological dig at the site of Mount Seething, with appropriately dressed 'experts' and scientific process all present at a playful community event.[10] During public readings of the stories, which occurred at most festivals during the 'community play', people collectively shouted 'FACT' after playful assertions. For instance, the etymology of 29 February (Lefi Day) being from the 29 rounds of cheese that Lefi made, 'the 29 that fed you and me' 'FACT'.

This excerpt from a meeting between Seethingers and the *ASP* project Research Assistant (RA) shows the discussion of the notion of fact that occurred after I had been told that only 'historical facts' could be added to the map. We met in a local pub where the landlord (Lee) joins the conversation and provides us with a table and some food. The atmosphere is friendly, relaxed and positive.

RA: just as we have made a Surbiton Mini site we can actually make you a mini site that is called Surbiton historic walk … now we have a fine line with to tread regards what goes on this, if we are working within this platform … and I don't want to open a can of worms here but the head of community maps, the head of *Mapping for Change*, is not too keen on the idea of putting in, as she put it 'false histories' … I really don't want to get into this, I'm not behind any of this, I think technology should be there for the people and however they want to use it this is more a political thing, this is why we have talked about a more disconnected option where people can go out and collect factual information and we can draw a link between the two. We can make a mini site called Seething, as Seething is not a real place …

Table: errrmmmmm …

RA: [quickly] Don't kill the messenger …

Harry: [slowly] No the nice thing is not to be precious about any of it, it's a fascinating idea that from her perspective that if you were in Nottingham and you wrote 'home of Robin Hood' then that would have to be dismissed because there is no actual proof to say that's true or not … Which I bet she wouldn't, she would say that's absolutely right as there is enough folk and legend around it to make that 'a truth' so it's at what point does someone figure out … When you lay it over the business of architectural spaces, etc. then legend and myth are massive influences and if you miss that off then you'll have no concept of what an area is about …

Lee: if she is precious about this legend and myth then let's just use a separate platform.

From here the conversation turned to the use of various alternative technological platforms that would achieve the aim of sharing stories whilst keeping the 'spirit of Seething'. Lee, the landlord, started talking through different platforms and apps he has looked at, including augmented realities, Bluetooth technologies and other emerging ICTs. The Research Assistant attempted to keep the conversation focused on the *ASP* community map, suggesting that Surbiton and Seething have their own sites, but this met with little enthusiasm. Leaving the Free University of Seething, and back at University College London, the RA and I discussed at length how we were unable to find a way to accommodate both the *ASP*'s concerns of 'mis-information' and the Villagers' desire to exchange Seething stories in a Seething style. By the end of the fieldwork period the community map remained little used.

During the fieldwork period my informants rarely asked about the map and showed little interest in it. They did not resist the map, nor did they attempt to engage with it further. Once refused they simply turned away and carried on doing what they were doing in Seething. That is, their social project – the extension of values through particular articulations of social life – endured away from the map. Similarly, the *ASP* seemed little concerned with the continued non-use of the map; it rarely, if ever, came up in meetings or annual reviews. In discussions over the possible development of the map I was told by members of the *ASP* and *Mapping for Change*, that there was neither the money nor the time to develop the platform within the project and that the designer did not 'realise the degree to which anthropologists have spent time in the field with participants', indicating surprise at the degree to which an emic position may be taken towards data. The community map was not considered a failure, however, as a focus on the future potentials of the map allowed the ideals of participation and inclusion to endure through the promise of a technocratic solution in the future, with more time and more money.

A multidisciplinary reading of incommensurability

In the ethnographic data presented above I have described a moment where the democratic ideal of the PPGIS as a research project within the *ASP* met the very local social project of the Seething Villagers. Both projects aimed to make better places through using, showing and articulating the values and potentials of the built environment, but in very different ways. Epseland and Stevens (1998, 314). describe commensality as 'the transformation of different qualities into a common metric'.

An anthropological reading of the data asserts that the moment of incommensurability, the point of refusal whereby a story was not added to the map, was a point of foreclosure. This foreclosure was necessary to maintain the coherence of the values within the community map. The translatability of the stories across a wide, non-local, scale is the essence of the ideal speech situation necessary to the Habermasian notion of democracy based on communication, participation and inclusion. Yet this 'scaling up' results in loss of information whereby complexity gives rise to generalisations and selections (as one scrolls out of a Google Map). Marilyn Strathern (1992, 97) asserts that this loss is necessary and should be considered as part of the data rather than as a loss of data. As such the 'moment' which might look like a failure of an auto-ethnographic tool, can also be seen as the moment when the ethnographer was present to see a 'loss of data' that occurred at a moment of incommensurability. This moment then serves as a useful point through which to identify and assess the normative epistemological frames of two differing social projects.

A social project, in order to maintain coherence and legitimacy, is by necessity a hegemonic order, and it is this necessary production of universality (see Butler *et al.* 2000) which demands commensality in the data set. The Seething story could not be visualised due to its compromising effects on the wider, overarching value of the democratic project. In this sense the act of commensuration is essentially a moral act aimed at eliminating the motives for irrational behaviour, which extend different values. As Epseland and Stevens (1998, 319) note, the 'homogeneity commensuration produces simultaneously diminishes risk and threatens the intensity and integrity of what we value'. It might be then that this is where an anthropological analysis would leave us, with a focus on the mechanisms of power at play in the production of a technocratic solution to the notion of participation in late-liberal democracies. Indeed there is a large body of work concerning planning and mapping as forms of governmentality, territorialisation and regulation inspired in large part by the works of Michel Foucault (see Rainbow 1995). Recently, however, these discussions have also emerged in planning theory, where Yifachel (2000) asserts that discussions within planning theory have been often been overly general about the notion of the 'public' and too concerned with an uncritical engagement with the Habermasian project. In this sense they have been focused on agency within the nexus of participation perhaps at the expense of a structural engagement with what is being participated in. Yifachel critiques the post-political landscape of critical planning literature, where the binaries of inclusion/exclusion, citizen/subject, break down and ideology as difference is replaced with a technocratic problem of participation.

If, as Strathern (1992, 81) asserts, the role of the anthropologist is to identify how one set of values can displace another then anthropologists can, through long-term fieldwork, start to develop attentiveness, not only towards the values being articulated by a group, but also towards those which are not articulated. Such attentiveness allows values and potentials which might be displaced by a particular implementation of PPGIS methods, such as in the *ASP* auto-ethnography of Surbiton, to be articulated by different means. Through these kinds of multidisciplinary engagement anthropologists are starting to find new involvements with GIS. Whereas until recently these had been limited to archaeological mapping and population studies (see Aldenderfer and Maschener 1996), anthropologists have recently started to participate in the process of GIS design. Recent discussions in GIS have seen the question of *who* participates in mapping projects shift to the question of *how* people participate in such projects. New technologies (particularly Web 2.0) are recognised as having the capacity to engage with alternative forms of knowledge production (Tacchi 2012; Haklay *et al.* 2008), whereby design processes are also brought into the process of visualising knowledge. This implicitly recognises that there is no 'one size fits all', with each iteration of PPGIS technologies being tailored to the projects they are meant to engage with.

The extension of the *how* question has meant the emergence of a GIS field which re-frames citizenship as plural and an enlivening of the discussion surrounding the perceived participatory potential of ICTs. The aim of developing a multi-scaled approach is to allow local agency to be built into the project design at all stages. This 'scaling up' speaks directly to questions of radical political change and social justice (see Hickey and Mohan 2004). At its core, however, this move still holds onto the notion of equal knowledge production (see Hacklay 2013), which assumes a process of placing visible values into a common matrix. In essence, the idea of 'equal knowledge production' in PPGIS sits incommensurably with post-structuralist approaches to knowledge in Anthropology. However, this need not be a problem in practice. The research presented in this chapter has shown how disjunctures arising from such a meeting that might be said to 'fail' in an *interdisciplinary* mode (where a new methodology is forged from the various aspects of each discipline) can in fact be productive in a *multidisciplinary* mode, where the disciplines maintain their respective methodologies and compare notes to see the potential of the gap between them.

In conclusion

On the basis of my fieldwork as an anthropologist on the *Adaptable Suburbs Project*, I have argued that it is through *making space for each other* that more productive disciplinary collaborations can occur within research projects. That it is through recognising and respecting disciplinary differences, through reading the particularities of our research practice through each others' disciplines, and through working with different approaches and analyses, that new and productive insights into the normative assumptions of those research approaches can be identified and understood. As such, a critical reflection on the workings of wider social projects can emerge within a rigorous and politically charged academy. This critical practice is one that can span both Anthropology and GIS, each making space for the other, so that both can continue to be innovative in their own ways. These disciplines and others, I have argued, may sometimes be more productive by being attentive and critically engaged in research in an immediate sense, and in relation to wider social projects, than by trying to make commensurate that which simply does not, and sometimes should not, fit.

Acknowledgements

Versions of this chapter were read by Ashley Dhanani and Gillian Conquest whose comments and suggestions were of utmost use. I particularly thank Gillian for her help and advice regarding the literature around PPGIS – see also Conquest (2013). I also thank the *ASP* team and the Seething Villagers for their time effort and support.

The ethnographic research described here formed part of a PhD studentship on the UCL Adaptable Suburbs Project, funded by the UK Engineering and Physical Sciences (EPSRC: project reference EP/I001212/1).

Notes

1 This and other quotatations in this passage from the *Adaptable Suburbs* 'Case for Support' (Vaughan *et al.* 2010). EPSRC reference number EP/I001212/1. 1 November 2010.
2 PhD Studentship details. Taken from the Application documents advertised July 2010. https://uclsstc.wordpress.com/2010/08/13/three-phd-studentships-in-%E2%80%98adaptable-suburbs%E2%80%99/, accessed 2 August 2010.
3 *Towards Successful Suburban Town Centres* 2006–2009, EPSRC EP/D06595x/1. www.sstc.ucl.ac.uk/sstc_index.html, accessed 3 October 2014.
4 See www.seethingwells.org/Welcome_to_Seething_Home.html, accessed 3 October 2014.
5 Seethingers often wear T-shirts that state 'I live in seething – it's a state of mind'.
6 *The Big Issue* (22 February 1999) 'I want the good life'.
7 Oxford Dictionaries. Oxford University Press. www.oxforddictionaries.com/definition/english/stupid, accessed 12 January 2015.
8 A Seething institution established to accommodate the academic collaboration.
9 www.mappingforchange.org.uk/ accessed 3 February 2012.
10 www.youtube.com/watch?v=MGRGv2FbP2Y, accessed 3 October 2014.

18 Reassembling Durkheimian sociology of space

Lasse Suonperä Liebst

Introduction

Over the past ten to fifteen years, sociology has shown growing interest in spatial phenomena, especially in emphasizing space as more than a passive backdrop for social processes: once built, the spatio-material artefact *does* something to social life (Gieryn 2002). Some scholars have suggested that sociology is currently undergoing a 'spatial turn' (Warf and Arias 2008). The French sociologist Bruno Latour (2005) and his actor-network theory (ANT) plays a key role within this emerging research interest in space, in particular ANT's ontological proposition that 'human' and 'non-human' agents hold a symmetrical capacity for agency. Regardless whether space is described in terms of a 'non-representational aspect' (Thrift 2008), 'mobility systems' (Urry 2007), 'connectors' (Yaneva 2009), or 'networked infrastructures' (Graham and Marvin 2001), recent sociology of space is standing on the shoulders of Latour's non-human agents.

The term 'spatial turn' suggests that space should be regarded as a somewhat new dimension of sociology. One might well question, however, the actual novelty of the spatial dimension. Two decades before the current spatial turn, Tonboe's (1993, 524) superb study on the historical lineage of the sociology of space refuted the argument of space being a novel sociological theme. Rather, the spatial dimension tends to be 'lost and found', first then and now again. In the case of the recent spatial turn this manifests itself in an anti-Durkheimian approach adopted from ANT. Specifically, the founding father of French sociology, Émile Durkheim, is criticized for being the originator of sociology's blindness towards non-human things and spatialities (Blok and Jensen 2011). In this critical account of Durkheimian sociology thingness is only something which the sociologist may have epistemological reasons for ascribing to the social, as in the case of Durkheim's methodological dictum to '[c]onsider social facts as things' (Durkheim 1982, 14). Conversely, Latour seeks to breathe ontological life into the non-human things and summarizes accordingly his symmetrical ontology as an inversion of Durkheim's original dictum: 'treat things as social facts' (Latour 1996, 240).

However, as this chapter will argue, this ANT-reading of Durkheim is not theoretically tenable: Latour's rejection of Durkheim is only possible because he neglects the materially-based research programme that Durkheim (1994) established under the label of 'social morphology'. As Lindemann (2011) recently suggested, this programme is in fact closer to a notion of non-human ontologies and agencies than the ANT critics of Durkheim seem aware. With regard to the understanding of space the Latourian and Durkheimian perspectives thus seem to contribute to the same body of cumulative knowledge, regarding space as a key element in the structuring of social life. This chapter suggests that the relevant distinction between the Latourian and Durkheimian sociologies of space does not hinge on the question of whether space shapes social life in certain respects, but rather on the question of which aspects of social life are so spatially shaped. Compared to ANT the Durkheimian tradition allows one to grasp how space shapes the socio-emotional nature and bonds of social life. While Latour's (1996) symmetrical ontology offers a radical theory of inter*objectivity*, Durkheim's social morphological research programme includes a theory of inter*subjectivity* that addresses the emotional-relational processes of social interaction. Insofar as the research interest concerns the interplay between the emotional aspects of social interactions and the physical properties of built space, Durkheim's social morphology offers a rich, but often neglected, sociology of space (Emirbayer 1996).

Durkheim's sociology and social morphology

The social morphological research programme needs reassembling. The elements for a contemporary and coherent sociology of space do exist, but these elements are scattered over numerous texts, two centuries, and different Durkheimian scholars. And such a project of reassembling is also complicated by the fact that generations of sociologists have neglected the spatial contribution of social morphology. Accordingly, scholars cannot currently tap into any existing consensus on what *the* genuine Durkheimian sociology of space looks like and how it may be further developed. This difficulty follows from Tonboe's argument that the space sociological classics tend to be forgotten. In a similar vein Lévi-Strauss expresses regret over the unfulfilled potentials of the morphological tradition:

> It has been Durkheim's and Mauss' great merit to call attention for the first time to the variable properties of space which should be considered in order to understand properly the structure of several primitive societies ... There have been practically no attempts to correlate the spatial configuration with the formal properties of the other aspects of social life.
>
> (Lévi-Strauss 1963, 290–291)

The 'social morphology' constitutes, together with the so-called 'social physiology', the two scientific pillars of Durkheim's 'general sociology'. The distinction between the social morphology and physiology is explained with an analogy to biology: the morphology concerns the external forms or beings of society, while the physiology regards its internal functions or doings (Durkheim 1960). The notion of social morphology suggests that all of social life, notwithstanding its functional variations, is conditioned on a material substratum. Let me quote a key section in which Durkheim presents the social morphology as a separate section in his journal *L'Année Sociologique*:

> Social life rests upon a substratum which is determinate both in its extent and in its form. It is composed of the mass of individuals who comprise the society, the manner in which they are disposed upon the earth, and the nature and configuration of objects of all sorts which affect collective relations. Depending on whether the population is more or less sizeable, more or less dense; depending on whether it is concentrated on cities or dispersed in the countryside; depending on the way in which the cities and the houses are constructed; depending on whether the space occupied by the society is more or less extensive ... the avenues of communication

> which traverse it, and so forth, this social substratum will differ … the constitution of this substratum directly or indirectly affects all social phenomena, just as all psychic phenomena are placed in mediate or immediate relationship to the brain … It is this science which we propose to call *social morphology*.
>
> (Durkheim 1994, 88)

Durkheim regards the social morphology as key to carry out sociology as an explanatory science. The material substratum is ascribed a crucial causal effect on the functions of social life, and accordingly the morphology 'can and must be explanatory' (Durkheim 1994, 88). Moreover, as it appears in the above quotation, the *spatial* properties of society are crucial for the establishment of such morphological causal explanations. The notion of space applied by Durkheim refers equally to macro-scale spatialities and the micro-scale architecture of the built environment, such as '[t]he disposition of streets and squares, the architecture of the houses, and the structure of things made vary from village to town and form the large city to the small one, and so on' (Durkheim 1960, 361).

This outline of the social morphology provides a first piece of evidence that Latour should temper his criticism of Durkheim for neglecting the material properties of social life and how these properties shape sociality. This is not to say that social morphology should simply be considered as a kind of proto-ANT. Durkheim's 'sociocentrism' implies (Durkheim and Mauss 2009, 51), indubitably, a *less* symmetrical understanding of the relationship between human and non-human agents than proposed by Latour. Morphological properties are, as Durkheim remarks, 'the matter to which the vital forces of society are applied, but they do not themselves release any vital forces' (Durkheim 1982, 136). Yet, although suggesting a social ontology less symmetrical than Latour, Durkheim is indeed very explicit about the fact that space and morphology shape societal processes. Once built by society, for example, the spatial architecture starts to act back upon its socio-human creator:

> First, it is not true that society is made up only of individuals; it also includes material things, which play an essential role in the common life. The social fact is sometimes so far materialized as to become an element of the external world. For instance, a definite type of architecture is a social phenomenon; but it is partially embodied in houses and buildings of all sorts which, once constructed, become autonomous realities, independent of individuals … Social life, which is thus crystallized, as it were, and fixed on material supports, is by just so much externalized, and acts upon us from without.
>
> (Durkheim 2005, 278)

Before further considering the role of social morphology, we should briefly consider the social physiology which broadly relates to the functions of society; i.e. moral, religious, etc. institutions which act autonomously as society's collective consciousness, *sui generis*. When elaborating the distinction between social morphology and social physiology, Durkheim is careful to strike the balance between real and ideal socio-ontologies. The social physiology is conditioned on the morphology, but there is a 'variation' and 'complexity' in difference between these societal dimensions; the morphology conditions the emergence of a physiological order whose 'whole very often has very different properties from those which its constituent parts possess' (Durkheim 1978a, 76).

At its most general, Durkheim suggests that study of social morphology is divided into two material components, i.e. a spatial and a population variable: first, '[t]he study of the geographical base of people in relationship to social organization', and second, '[t]he study of population; its volume, density and disposition on earth' (Durkheim 1909, 277). In most cases, Durkheim specifies the population variable as a morphological question of the *dynamic density* of social life. By definition, density (σ) includes a spatial dimension as being defined as the ratio of the numerical mass of individuals (m) to the spatial area (A): $\sigma = m/A$. Furthermore, given that transportation infrastructures shape the possibility of people moving from A to B, space plays a pivotal role in regulating social density levels of spatial regions; for example, in *The Division of Labor in Society*, Durkheim (1984) relates the process of urbanization to the historical emergence of technological infrastructures facilitating urban migration and hence an increasing urban social density.

Reading Durkheim's considerations on dynamic density is hampered by the fact that the dynamic or causal effect that arises from societal densities is attributed two sociologically distinct mechanisms: social conflict or emotional contagion. Most well-known is Durkheim's (1984) argument in his early major work, *Division*, in which he suggests that social densities have social effects as a result of increasing densities being correlated with the conflict level of society. This argument is inspired by Charles Darwin's evolutionary biology: increasing density levels of animals and species within a given ecological niche intensifies the struggle for survival. Such density-conflict mechanism Durkheim (1984: 288) simply identified as the 'law of gravitation in the social world'. Following from this law, the transition from 'mechanic solidarity' (i.e. solidarity arising due to infinite points of resemblance immediately connecting individuals to society) to 'organic solidarity' (i.e. solidarity as a product of the growth of reciprocal independences among differentiated individuals) occurs mechanically as 'a break in the equilibrium of the social mass gives rise to conflict that can only be resolved by a more developed form of the division of labor: this is the driving force for progress' (Durkheim 1984, 212).

Durkheim's (1995) late major work, *The Elementary Forms of the Religious Life*, maintains the analytical emphasis on morphological densities, although the terminology changes and now refers to densities as 'assemblies', 'congregations' and 'concentrations'. What, on the other hand, is abandoned in this work is the conflict-interpretation of densities; instead social densities are seen as related to processes of emotional contagion. That is, the density-conflict mechanism is replaced with a density-contagion mechanism suggesting that emotional contagion arises among co-present or concentrated individuals. Religious rituals are the archetypical example of how emotional contagion may result in an experience of 'collective effervescence', an emotional condition that Durkheim suggests as the foundations of religious ideas and social morals. Effervescent density-contagion processes imply that the social group celebrates themselves and thus become conscious about itself *as* a group. A collective consciousness arises which the individual experiences as being subject to an 'external' and 'constraining' force, be it a transcendental entity or the moral order of society:

> The very fact of the concentration acts as an exceptionally powerful stimulant. When they are once come together, a sort of electricity is formed by their collecting which quickly transports them to an extraordinary degree of exaltation. Every

> sentiment expressed finds a place without resistance in all the minds, which are very open to outside impressions; each re-echoes the others, and is re-echoed by the others … It is in these effervescent social milieu, and indeed from the very effervescence, that a religious idea seems to be born … I have shown how the clan awakens in its members the idea of external forces that dominate and exalt it by the way in which it acts upon its members.
>
> (Durkheim 1995, 217–218, 220–221)

This argument advanced in this chapter that the Durkheimian concept of emotional contagion is morphologically related to the concept of dynamic density is not uncontroversial, however. Some Durkheimian scholars suggest that *Elementary Forms* signifies a rejection of morphological mechanisms and laws expressed in *Division*. This chapter is critical towards the hypothesis that Durkheim simply rejected morphology as his thinking evolved, as suggested by influential scholars such as Steven Lukes (1985), Dénes Némedi (1995) and Jeffrey Alexander (2005). For example, Alexander suggests that 'Durkheim reached his theoretical maturity after a prolonged, if confused, flirtation with materialist forms of structural theory, and eventually a fierce struggle against them' (Alexander 2005, 136). The evidence provided for the existence of such struggle seems flawed. Durkheim (1978b) certainly had a fierce struggle, but if one reads the Durkheimian argument cited by Alexander one realizes that Durkheim specifically refers to and rejects a crude version of Marxian historical materialism, not his own social morphology.

By thus emphasizing social morphology as a continuous Durkheimian line of inquiry, this chapter is on par with the newer Durkheimian studies of Howard Andrews (1993), Ivan Strenski (2006) and Marcel Fournier (2013). These studies concurrently suggest that although social morphology was assigned another, and in certain respects less prominent role in Durkheim's late sociology, morphological themes are found consistently throughout his authorship. Durkheim never rejected social morphology. This is clearly illustrated with the morphological property of dynamic density. Despite using a varying terminology, Durkheim maintained that densities have profound social effects. Moreover, this emphasis of the continuity of Durkheim's thought is also evident if one reads Durkheim in the context of his sociological school; several key members of the school maintained and cultivated the social morphological research programme (Halbwachs 1960). As recently emphasized by Fournier (2013), such contextual reading is consistent with the biographical circumstances of Durkheim and the contribution of his nephew and trusted co-worker, Marcel Mauss:

> There is a collective dimension to Durkheim's approach. *L'Année sociologique* was obviously a collective undertaking. But what about the individual books? Who is the author of *Elementary Forms of Religious Life*? Durkheim or Mauss? The relationship between uncle and nephew was very close – Marcel was Émile's alter ego: 'It is impossible for me to extricate myself from the work of the school', said Marcel Mauss.
>
> (Fournier 2013, 4)

The fact that Durkheim may be more accurately considered as 'Durkheim, Mauss & Co', as Fournier (2013) proposes, allows for a more spatio-morphologically clear-cut interpretation of the late-Durkheimian school, and in particular the density-contagion mechanism. Key in this regard is Mauss' (1979) less known study, *Seasonal Variations of the Eskimo: A Study in Social Morphology*. In this study we find the perhaps strongest formulations of the social morphological programme, despite the fact that the study was published one decade after Durkheim's alleged rejection of social morphology.[1]

Space, density and emotional contagion

Mauss' *Eskimo* is a case study that identifies a methodological opportunity in the way in which the particular seasonal morphology of the Eskimo culture offers a 'favourable condition for a study of how the material form of human groups – the very nature and composition of their substratum – affects different modes of collective activity' (Mauss 1979: 20). The social density of the Eskimos varies profoundly across the seasons. During the summer, they live spatially dispersed in tents and huts across the countryside. This low social density facilitates a social organization with low emotional intensity. This changes dramatically as winter arrives. During the winter, the Eskimos live with an extraordinary high social density in small huts and igloos. The change in the spatio-material substratum of the Eskimos has a profound impact on the social life. As the morphological ratio of the mass of individuals to the spatial area of the dwellings changes, the level of social activities and emotions becomes more intense:

> Winter is a season when Eskimo society is highly concentrated and in a state of continual excitement and hyperactivity. Because the individuals are brought into close contact with each other, their social interactions become more frequent, more continuous and more coherent; ideas are exchanged; feelings are mutually revived and reinforced. By its existence and constant activity, the group becomes more aware of itself and assumes a more prominent place in the consciousness of individuals.
>
> (Mauss 1979, 76)

The fact that we are dealing here with Durkheim, Mauss and Co is supported by Bill Pickering's (1984) suggestion that Durkheim's notion of collective effervescence is likely to have been inspired by Mauss' *Eskimo*. That is to say, Mauss originally outlined the density-contagion mechanism that Durkheim, in turn, included as the key mechanism in *Elementary Forms* explaining how emotionally effervescent group processes develop the collective consciousness of society.

In a space-theoretical perspective it should be emphasized that Mauss specifically suggests the spatial properties of nature and the built environment as the causal force that acts upon and thus intensifies social life. The seasonal morphological rhythm affects the area size of the dwelling which the Eskimos are capable of occupying, and this regulates the dynamic densities of the Eskimo population: 'The seasons are not the direct determining cause of the phenomena they occasion; they act, rather, upon the social density that they regulate' (Mauss 1979, 79). Although Mauss suggests that space has an indirect (i.e. space → density → sociality) rather than a direct (i.e. space → sociality) causal effect on social life, space is attributed a far more prominent role in *Eskimo* compared with *Elementary Forms*. As argued above, in *Elementary Forms* space is mainly found as part of the definition of density, i.e. as the area denominator in the density equation: $\sigma = m/A$. In *Elementary Forms* the group concentrates as a result of purposeful social action, while *Eskimo* redistributes some of this agential capacity to morphological space itself: space may act to densify and thus emotionally intensify social life. Mauss is, so to speak, giving the social morphology a non-human slant. Such a non-human reading of *Eskimo* is not the norm in the Durkheimian literature, however. For example, Mary Douglas advances the opposite argument, suggesting that *Eskimo* illustrates how late-Durkheimian sociology moves beyond any such spatio-morphological causal thinking:

> [*Eskimo*] is an explicit attack on geographical or technological determinism interpreting domestic organization. It demands an ecological approach in which the structure of ideas and of society, the mode of gaining livelihood and the domestic architecture are interpreted as a single interaction whole in which no element can be said to determine the others.
>
> (Douglas 1972, 513–514)

Indeed, Mauss (2007, 23) and Durkheim (1899) are opposed to any such crude spatial determinism; for example, Mauss emphasizes that spatial factors should always be considered in the context of society's complex totality. But Douglas misses the fact that Mauss is indeed applying a spatio-morphological causal mechanism in *Eskimo,* suggesting that space acts as the factor triggering the changing nature and lived intensity of the Eskimos.

Putting this discussion of historical sociological thought briefly to one side, I will highlight one of the analytical strengths of applying Durkheim, Mauss and Co's density-contagion mechanism in empirical studies within urban spatial sociology. While the Durkheimian tradition suggests that social density is correlated with experiences at the positive end of the emotional spectrum, a persistent urban sociological research tradition has, as documented by Boyko and Cooper (2011), overemphasized the 'dark sides' of social density and its correlate with negative emotions. Extremely influential in this regard is Georg Simmel's (1976) argument that urbanism has a rhythm and density that intensifies life to the point of over-stimulation. Following this argument, Louis Wirth (1938) proposes that high urban social density leads to mental stress, social conflict and moral anomie. And in a similar vein, Stanley Milgram (1974) argues that the dense living conditions of the city result in a sensory overload that, in turn, dehumanizes urban social interactions. As argued by Jonathan Freedman (1975), the problem with this line of inquiry is that it overlooks the positive feelings which are also, and perhaps more commonly, stimulated by the density of urban life. A key analytical argument for reassessing the Durkheimian sociology of space is, accordingly, how this morphological tradition, as in the case of *Eskimo,* has theorized the nexus between space, density and the positive emotions of contagion. Spatial mechanisms acting to densify individuals into a mass may shape the emergence of positive emotions given the Durkheimian fact that being in a densely crowded mass gives some sense of being 'where the action is' (Collins 2004, 82; Goffman 1967).

Cumulation contra anti-Durkheimianism

The outline of the Durkheimian tradition's morphologically-based sociology of space serves to highlight the one-sidedness of Latour's critique of Durkheim for missing the spatial thingness of society. The uncompromising nature of Latour's argument may best be understood in the light of his anti-cumulative epistemology. Contemporary sociology suffers, according to Randall Collins (1999), from a conspicuous tendency to misrecognize sociology's own actual knowledge accumulation, and Latour should be considered one of the sociologists that contribute to this unfortunate process (Osborne 1996). This gives rise to the question: perhaps Latour's rejection of Durkheim simply reflects the fact that he misrecognizes how the Durkheimian and Latourian traditions actually have contributed to the same cumulative knowledge base regarding the importance of spatial objects and things. Let us consider the following quote in which Latour defines ANT and its non-human perspective in contrast to any sort of Durkheimian sociology of 'the social'. Such sociology allegedly misses the fact that 'the social' only acquires *durability* and *extension* as connected with non-human things:

> ANT is not the empty claim that objects do things 'instead' of human actors: it simply says that no science of the social can even begin if the question of who and what participates in the action is not first of all thoroughly explored, even though it might mean letting elements in which, for lack of a better term, we would call *non-humans*. This expression, like all the others chosen by ANT is meaningless in itself. It does not designate a domain of reality. It does not designate little goblins with red hats acting at atomic levels, only that the analyst should be prepared to look in order to account for the durability and extension of any interaction. The project of ANT is simply to extend the list and modify the shapes and figures of those assembled as participants and to design a way to make them act as a durable whole.
>
> (Latour 2005, 72)

First we should note that it isn't crucial whether or not one applies the term 'non-human' (a term which the Durkheimians do not apply). Rather, ANT suggests an analytical openness towards who and what is taking part in social actions and relations. Without such ontological sensitivity sociology has difficulties grasping the durability and extension of interactions. The quotation given above does not explicitly mention spatial objects, but space seems to be touched upon indirectly with the terms 'shapes' and 'figures'. ANT has elsewhere been related to architectural space (Latour and Yaneva 2008). Among others, Albena Yaneva argues that non-human design and architecture should be understood as 'connectors' allowing durability and extension to social interactions and thus suggests that space 'shapes, conditions, facilitates and makes possible everyday sociality' (Yaneva 2010, 273).

Notwithstanding the different theoretical terminology, the spatio-material conclusions found in Durkheim and Mauss are in many respects converging with those of ANT. In suggesting this, the argument proposed here parallels Lindemann's (2011, 93) suggestion that Latour, despite his contrary intentions, offers a 'brilliant proof of Durkheim's theory of the morphology of social facts'. As already mentioned, the emphasis of morphological space as a force structuring social life is evident in

Durkheim's *Division* and Mauss' *Eskimo*. In *Division*, for example, Durkheim suggests that '[t]he communication network forcibly prescribes the direction of internal migrations or commercial exchanges, etc., and even their intensity' (Durkheim 1984, 58). The transition from pre-modernity to modernity occurred as the undifferentiated population segments of pre-modernity gradually expand their social relations through space; that is to say, the historical process of modernity was morphologically shaped by the emergence of infrastructural and technological relations facilitating more extensive forms of interaction, communication and mobilities (Durkheim 1984, 58). Essentially, such morphological considerations coincide with Latour's definition of non-human agents as allowing an extension of societal interaction patterns.

Next, in terms of the durability property of the non-human agents, Latour himself involuntarily lends my argument a helping hand. Under the title 'Durkheim having a Tardian moment', Latour (2005, 38) quotes a key section from *Elementary Forms*, and adds the following (ironically intended) comment: 'Does the totem express the group, facilitate its cohesion, or is it what allows the group to exist as a group?' Let us quote from the same section from *Elementary Forms*:

> Without symbols, moreover, social feelings could have only an unstable existence. Those feelings are very strong so long as men are assembled, mutually influencing one another. But when the gathering is over, they survive only in the form of memories that gradually dim and fade away if left to themselves … But if movements by which these feelings have been expressed eventually become inscribed on things that are durable, then they too become durable.
>
> (Durkheim 1995, 232–233)

What Latour scoffs at is the fact that Durkheim here seems compelled to do what Gabriel Tarde (proclaimed by Latour (2002) as the founding father of ANT) criticizes Durkheim for not doing; i.e. to realize that 'the group' or 'the social' does not explain anything, but on the contrary is what needs sociological explanation. According to Latour a proper sociological explanation of 'the social' should include the non-human objects. The fact that Durkheim is forced into having a 'Tardian moment' in which he includes a notion of totem objects into his sociology allegedly proves the limits of Durkheim's sociology. Otherwise, Durkheim would not have been able to explain how the social groups become durable.

The problem with Latour's argument, however, is that in the quoted passage the notion of totem objects does *not* represent a momentary lapse in Durkheim's sociological reasoning. Rather, Durkheim (1995, 208) consistently discusses and defines totems as materially anchored symbols: 'The totem is above all a symbol, a tangible expression of something else', i.e. the group celebrating itself which charges the totem emblem with emotions and thus turns this object into being the 'flag' of the group. As Bjørn Schiermer (2011, 86) puts it: 'In a word, Durkheim's sociology of religion is not simply about *social* things but also about social *things*; about sacred objects.' Symbols and emotions which are not anchored in objects are experienced as being more abstract, less durable, and thus act as weaker representations of the group's collective consciousness. Additionally, the tangible nature of the totem emblem provides social interactions with an important durability, in particular as these are extended across time and space. A good example might be the sacred amulet which is emotionally charged in the emplaced ritual, but evokes similar effervescent emotions in the believer even if removed from the ritual site. Thus, essentially, the Durkheimian totem object serves the same social function as Latour's 'immutable mobile' (i.e. an object that may be transported, but has some permanence to it; cf. Latour 1990) allowing a durable extension of scientific ideas (as suggested by Latour) and religious ideas and emotions (as proposed by Durkheim).

Durkheim and Mauss also provide several space-morphological examples concerning how spatial objects stabilize social interaction. For example, in *Primitive Classification*, Durkheim and Mauss (2009) argue that the mental classification of the world of tribe cultures ('cosmic space') reflects and corresponds with the spatio-morphological organization of the tribes ('tribal space'): 'Cosmic space and tribal space are thus only very imperfectly distinguished, and the mind passes from one to the other without difficulty, almost without being aware of doing so' (Durkheim and Mauss 2009, 7). The classification system of the tribes is modelled on the 'morphological organization' of the tribe, and according to Durkheim and Mauss such structured correspondence between classification and morphology supports the reproduction of the social world. In a nutshell, the spatial morphology helps in maintaining the durability of the social interactions.

Closing remarks

In this chapter I have identified and criticized two conventional readings of the Durkheimian tradition, both of which have contributed to the marginalization of social morphology from the space-sociological literature. First, this is attributable to the line of Durkheimian scholars who argue that late-Durkheimian thought signifies a clear-cut rejection of social morphology; and second, Latour's anti-Durkheimianism has been accepted as part of the ANT foundations of the recent spatial turn in sociology. Note the paradox that these two, respectively, sympathetic and critical readings of Durkheimian sociology have had the same detrimental effect on social morphology.

However, one should not jump to the conclusion that these two readings are solely responsible for the reduced status of social morphology; this also has a technical explanation in the underdeveloped status of the quantitative methodologies available to Durkheim and Mauss. This deficiency made it difficult for them to conduct space-morphological studies that met Durkheim's positivist dictum to consider social facts *as* things. The qualitative case study method applied in *Eskimo* provided one possible methodological strategy to measure morphological space as a thing. But Durkheim (1982) and Mauss (2003) emphasized that morphology should also be made subject to quantification and statistical analysis. Statistics allow one to 'cancel out' the impact of individual effects, and statistics thus offer a privileged strategy to examine collective social facts: 'Basically, every social problem is a statistical problem', as Mauss (2003, 69) summarized the methodological rule of thumb.

The fact that the lack of quantitative tools constrained social morphology from being carried out as an empirical research program is expressed in connection to Mauss' reflections upon the Durkheimian tradition in the aftermath of Durkheim's death. Here, Mauss emphasizes that the measurement of the morphological properties of space, numbers and time 'demands the use of quantitative methods. To refer to material and social structures and the movements of these structures is to refer to things which can be measured'. And then he adds with regret that such quantification lags behind: 'Alas! Even in *L'Année sociologique*, we are far from what is desirable.' Yet, Mauss is a methodological optimist on behalf of the future generations of scholars as these most likely will be better technically 'armed than we were' (Mauss 2003, 69–70).

Mauss' reflections are prescient, both with regards to the argument that morphological properties pose a technical challenge to sociological analysis, and his intuition that the future generations will be better armed to face this challenge. This appears evident if we position Mauss' reflections in relation to the space-morphological methodology of space syntax developed by Bill Hillier, Julienne Hanson and colleagues, and which notably draws inspiration from Durkheim's social morphology (Hillier and Hanson 1984). Here, a key methodological insight is that the relationship between space and society presents, above all, a measurement challenge that calls for the development of specialized quantitative tools:

> [T]here is a gap between social theory and spatial design, and that the inability of contemporary theory to join them is not a function of their lack of relationship, but instead a function of our lack of ability to measure them properly and thus construct evidence-based theories and testable hypothesis. The gap between space and society, in other words, is a measurement challenge first and foremost.
> (Hillier and Raford 2010, 266)

The reason for highlighting space syntax in this context is due to the fact that this methodology has made remarkable progress in the development of techniques of measuring and modelling morphological space. The key methodological achievement of space syntax is the articulation of space as a network of relations that can be objectively measured by the mathematical tools of graph theory (Hillier and Raford 2010). Thus, consistent with Durkheim's methodological dictum to consider social facts as things, space syntax allows sociologists to measure the non-discursive regularity of space as a social fact, i.e. as an exterior and restraining 'thing', acting upon us from without. Space syntax should accordingly be considered part of the technically advanced future that Mauss placed his trust in. So far, unfortunately, space syntax techniques have not been adopted by the Durkheimian literature, and this may be ascribed to the fact that space syntax and sociology are practised within the almost entirely unrelated paradigms of architectural theory and sociology. I have recently examined the potentials of bridging the interdisciplinary gap between the fields of Durkheimian sociology and space syntax (Liebst 2014). In my view, such interdisciplinary inquiry holds both the prospect of further sharpening the sociological sensitivity of space syntax and – as this chapter suggests – the prospect of providing Durkheimian sociology with a methodological tool that matches up to the space-sociological potential of the social morphology.

Note

1 Alexander suggests that Durkheim's rupture occurred in 1896, i.e. ten years *before* the 1906 publication of *Eskimo*.

19 Movement in Adaptive Architecture

Holger Schnädelbach

Introduction: Adaptive Architecture

Hale (2012) argues that everything people use to engage with the world can be described as technology, whether it is language, tools or architecture (Hale 2012). This chapter adopts this strategy in arguing that all human engagement with the world – and therefore also with architecture – is through movement. This association becomes particularly interesting now that architecture itself has started to place greater emphasis on mobility, a phenomenon that is well exemplified in the emerging field of Adaptive Architecture. This chapter is intended to stimulate further discussion into the under-investigated subject of how to conceptualize movement in Adaptive Architecture. It draws on a number of examples from current practice and thought experiments to make its point.

I have been drawn to consider human movement in relation to the built environment by recent work in Adaptive Architecture. In one sense all architecture is *adaptable*. We can change many aspects of our built environment from the décor to the layout, services and external appearances. To what extent this is possible has been summarised by Duffy (1990) and Brand (1994), the latter presenting a continuum from site to objects. This chapter is, however, primarily concerned with Adaptive Architecture, architecture which is specifically designed to be adaptive to its environment and to its inhabitants (Schnädelbach 2010). Adaptive Architecture covers a wide range of conceptual, prototypical and everyday built examples. These might include buildings that 'talk to us', change colour, are more environmentally friendly and those that adapt to different spatial uses – to name just a few (Fox and Kemp 2009). Some buildings have the capacity to change their shape, internal spatial topology, orientation or even location, although spatially dynamic buildings remain relatively rare, despite some early developments detailed by Cook (1970).[1]

In the past, the label of 'smart buildings' has been applied to a subset of the work mentioned above and this is how Adaptive Architecture has entered the public imagination. Smart buildings promised to be our servants, knowing what room temperature we prefer, what music we like to listen to and when we might want a hot bath (Harper 2003). Such innovations were mainly driven by technological developments, rather than by an analysis of what people actually desired, as pointed out by Chan *et al.* (2008). More recently, however, with greater maturity of the technology, specific uses for home automation have been identified, for example in practically supporting religious practice (Woodruff *et al.* 2007).

In my own work, I have sought to better understand how people interact and live with Adaptive Architecture, specifically when it is underpinned by *computational means* rather than by the *interpretation* of human behaviours. Affective Computing (Picard 1997), concerned with measuring people's emotional states, remains a fascinating research area in this regard. It is also clear, however, that there are much simpler and more direct forms of interaction between inhabitant and adaptive environments that are relatively under-explored. Adaptive Architecture can be driven by either environmental data or it can be driven by inhabitant behaviour – the latter being the principal focus here. When Adaptive Architecture responds spatially to inhabitants it does so through manual interaction (for example, moving a sliding wall) and through automatic interactions (for example, reacting to a Passive Infrared (PIR) sensor to open an automatic door). In an earlier publication (Schnädelbach 2012), I experimented with the construction and evaluation of 'hybrid-spatial topologies', combining physical and virtual spaces that are manually driven by human movement and interaction. This research has demonstrated the ways in which hybrid spatial topologies underpin social interaction at a distance (cf. Schnädelbach *et al.* 2006). My recent work has concentrated on directly mapping human movement to architectural movements through sensor technologies (cf. Schnädelbach *et al.* 2010). Here the concern has been to explore the feedback loops that emerge between inhabitants and the built environment, for example observing how it is possible to trigger changes to people's respiratory behaviour through exposure to a biofeedback environment, as we outline in Schnädelbach *et al.* (2012). This work deliberately avoids attempts to automatically interpret inhabitant intent, i.e. to create what might be called a 'smart' or 'intelligent' building. In similar vein, this chapter focuses on one particular area of human – buildings interaction: namely how human movement and movement in architecture become related to each other.

Human movement

Solodkin *et al.* (2007, 507) state that:

> The behavioral repertoire of humans is broad, extending from simple behaviors such as sensory perception to more complex cognitive behaviors like language or creativity. Interestingly, no matter how simple or complex are these behaviors, they share without exception the common feature that their expression is a motor act.

From this one can deduce the value of considering human movement to be much more than simply concerned with movement through space, which is the principal consideration of architects. All human expressions of behaviour are motor acts and therefore result in muscle movements somewhere in the body. Of course, not all behaviours become expressed and there are 'behaviours' that do not involve muscle movements such as thinking. It is worth reflecting briefly on the word 'expression'. Expression, in the sense it is employed here, is used to indicate a communicative act from human being to human being; it assumes a human audience. From an Adaptive Architecture perspective, this audience or 'target' might be a technical system, which is capable of detecting movement. To develop the argument linking human movement to movement in architecture, it is useful to briefly consider the movement range produced by humans. For brevity this is necessarily a highly generalised description. The human body has more than 600 muscles, which can be divided into skeletal muscles, smooth muscles and cardiac muscles. The resulting movements actuated from these muscles can be subdivided into *internal* and *external*, reflecting their potential for movement-expression, rather than any medical definitions.

First, there are muscle movements that are typically considered to be internal. They are usually not directly visible on the outside, i.e. they are not necessarily communicative expressions of behaviour. They can, however, (literally) surface and become expressive. Examples would be the heart muscle driving internal blood flow, which then becomes visible under certain circumstances on the skin (for example, following physical activity), or the internal muscles involved in respiration changing the size of the thoracic cavity, thus clearly producing an externally visible movement. Second, there are also movements which are external, very clearly visible and in some sense communicative: facial muscles are used for talking; there are muscles involved in eye movement; the muscles in the arms and hands are used to execute fine-grained grasping actions. These kinds of movement are larger scale than the previous (internal) set, but the human actor may yet remain stationary when executing them. Finally, there are the largest scale movements that a single human can produce, which are typically intended to move our bodies through the world, for example by walking, running, jumping and climbing.

Frequencies and rhythms

Muscle movements occur at different frequencies, partly defined by the respective size and inertia of parts of the human body. This can be understood through frequency analysis, which is commonly used in physiology to understand, for example, the various rhythms present in the heartbeat (Cacioppo *et al.* 2007, 194). Through this technique, certain lower frequencies in heartbeat patterns can be associated with respiration patterns through the effect called Respiratory Sinus Arrhythmia (RSA). Without recourse to such analysis people have the apparently innate capacity to actively synchronise respiration with another body movement, for example in jogging, to line up breaths with steps. People may also synchronise movements with other people, for example in dance, or more subtly through deliberate mimicry of the body posture of another person to establish rapport. This theme recalls Lefebvre's project of *Rhythmanalysis* (which I shall return to later) in setting the internal rhythms of people into relation with those found in the environment, putting human rhythms at the centre of this worldview (Lefebvre 2013, 29).

Control of movements

Some movement behaviours are controlled sub-cortically (that is, unconsciously through the agency of the body), for example rhythmic movements or locomotion patterns. Even when they are triggered unconsciously, control and feedback can still be local to the movement location. The triggering of pre-defined motor patterns can account for breathing but also swimming, for example. Other movements, typically more complex ones in reaction to a changing environment, may be a grasping gesture or pointing at a moving target; these require much more centralised control. The planning, control and execution of voluntary movements is driven by the motor cortex of the brain, for example when we consciously control fine-grained interactional movements as Solodkin *et al.* (2007) discuss. It is, of course, important to acknowledge that there are many reasons for an individual's control over movement and therefore their movement range to be limited. Such reasons include a person's age as well their physical and mental well-being. For the purposes of this chapter, however, it is sufficient to assume an average movement range and control over movements.

Detection of movements

To establish the average range of human movement expressions with actuation potential in Adaptive Architecture it is essential that movements in that range can be detected. Detection ultimately depends on the way movement phenomena present themselves to ourselves, to others and to technical systems.

Ourselves

As humans we are not consciously aware of all of the movements (whether expressive or not) that we produce ourselves, while we do principally have the perceptual instruments to be aware of most of them. We certainly have more than the often-assumed five senses of sight, hearing, smell, touch and taste. Other sense organs allow us, for example, to sense temperature differences or the relative spatial positions of our limbs. This sensory range should allow us in principle to detect and experience our own movements, even if they are not voluntary. For example, in certain body positions, one can feel the heart beat in one's chest. One can feel one's own pulse as it lifts up the skin. One can feel one's eyelids close and one can know the positions of one's limbs in relation to each other. We might generally proclaim that we remain unaware of those phenomena, but this is because we just do not, on the whole, direct our attention to all of them very frequently, as there is no need for us to do so.

Others

The human organism can also sense the movements produced by others, whether those are on the larger or on the smaller scales mentioned. At the larger scale, we see the movements of others' limbs and we see movements of others through space. People are extremely good at *doing something* with this information: for example, judging people's spatial behaviour, maintaining a spatial position, navigating in space, accessing contested physical resources and interpreting culturally informed spatial behaviours. At the smaller scale, people can see and hear somebody else breathe. People can see somebody's pulse on the surface of their body, typically in circumstances where they have just exercised. In addition, people's reactions to others result in mirroring behaviours. This happens mostly unconsciously in an embodied fashion. Fuchs (2013) describes the bodily and interbodily resonance patterns of this built-in detection and feedback mechanism. The ability to detect the movement of others underpins how people determine and control their own movement behaviours, whether this is concerned with stepping close to someone in conversation or whether it is the synchronisation of respiration patterns to establish rapport.

Systems – feeding into Adaptive Architecture

Numerous sensor and associated infrastructure technologies are available to detect human movement, converting movement into data readable by a machine. On a smaller scale, systems can detect physiological data. For this purpose, sensor technologies are commonly (but not always) mounted on or inside the human body. For example, electrodes are mounted on the skin to detect the electrical activity signifying the phases of our heartbeat. Through electromyography, the electrical signature of muscle activity closer to the surface can be detected, for example to evaluate whether people are smiling or frowning; much simpler measurement belts can be use to detect people's respiration amplitude and frequency. At a larger scale, movement data is concerned with the direction of our gestures, our orientation towards others and the environment, and our location. Mechanisms to detect movements at this larger scale are frequently installed in the environment (they can also be found on our bodies). Vision techniques drawing on data from various types of cameras allow the construction of live point-clouds of people in space, providing data on location, orientation and body position. Through Global Positioning Systems (GPS), which are distributed in the environment and on people, the near-precise position of people and objects can be determined in relation to a global frame of reference.

Considering the human capacity to detect movement in ourselves and in others, it is worth highlighting the potential for exploiting this capacity *within* technical systems. People are already very good at detecting movement and far better than machines at interpreting human movement. Until systems improve, there therefore remains an argument for people to contribute sensor data if it benefits them and others, in order to complement data about human movement gathered through technology. Regardless of how it was acquired, whether through human or technological sensing, human movement can be made available as data. In fact, Thrift (2006) argues that as both smaller scale and larger scale human movement can be translated into data, everything becomes addressable. In this new 'movement space' human activities are permanently accompanied by distributed calculations occurring in embedded computing. This gives rise to a relative space of people and objects based on the hard absolute space of locations anchored by positioning systems. It is not difficult to see how this data stream of relative locations will in the future extend to information about more or even all of our movements and therefore be increasingly sensitive to information about our movement-expressions at all scales. Such data would serve technically to underpin a general link between those expressions and the built environment.

Human movement relating to Adaptive Architecture

It now remains to put human movements in relation to those movements possible in the architectural domain. In the broader context of this endeavour, Lefebvre (2013) details how nothing in our environment is immobile. Over time and at different frequencies all physical things move, including buildings, even if their movements are imperceptible to us. Such movements occur on different scales, from micro- to macro- movements, which in turn can be directly related to the increasing scale of Brand's (1994) layers from objects to sites. Lefebvre's *Rhythmanalysis* is akin to, but much broader than, the previously mentioned technical frequency analysis used in physiology. He associates human rhythms, whether physiological or cultural, with those of the wider environment, scaling up this approach even to the analysis of the city. Experiencing these interwoven rhythms of people and environment offers up a new form of seeing, in which the reference point is constituted by the rhythms present within the observer (Lefebvre 2013, 29).

From a phenomenological perspective, our bodies and our environment must be seen as a continuum. Every movement that appears on the human 'surface', no matter how small, 'protrudes into' or perturbs the environment in some way, with the skin being a tenuous and porous membrane between the world and us. We cannot be separated from the environment: we are *embodied and embedded* to use Wheeler's (2005) terminology. Human movement pushing against 'empty' space has no visible effect. When our movement hits upon resistance, however, it is stopped or parts of the environment are marked. This process, similar to what Varela (1979) refers to as 'structural coupling', means we cannot help leaving traces in our environment. Through continued use, our lives become imprinted on the spaces that we occupy, a process that Richard Long (e.g. in his performance of 'A Line made by Walking' in 1967) has articulated in his environmental art.

On the other hand, there is a more immediate link, characterised by direct feedback. This link might cause simple movements of artefacts, furniture and adaptive features of designed space, which go beyond mark making. Experimental Adaptive Architecture, as already mentioned, is exploring how architectural elements and buildings begin to move themselves. Once architecture has the capacity to spatially change, beyond the generally imperceptible movements discussed by Lefebvre, linking human movement to architectural movement becomes evidently possible. Such a link might affect the topology, orientation, location and configuration of architectural elements. Development in the discipline of Ubiquitous Computing as originally framed by Weiser (1991), which – among other things – has seen information technology become embedded into the built environment, has enabled another set of opportunities. Particularly relevant for this chapter is the possibility of transforming human movement (for example by amplifying it from a small scale to the larger scale of architecture) so that it can become actuated in the environment. To discuss this transformation between human and architectural movements, I will first reflect on the nature of human movement as it has been outlined above and then secondly, refer to a number of exemplar architectural projects, both implemented and imaginary, to illustrate the argument.

First, it is worth summarising the nature of the human movements, as they might become relevant to architecture. Along an axis of scale, one can see human movement as a continuum from very small movements to the largest movements we can produce, see Figure 19.1.

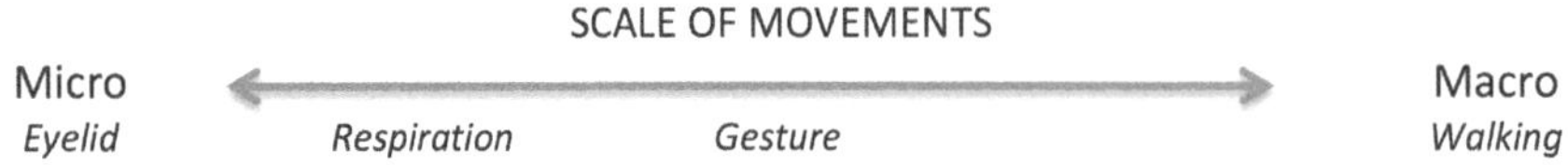

Figure 19.1 Scale of movements
Source: author

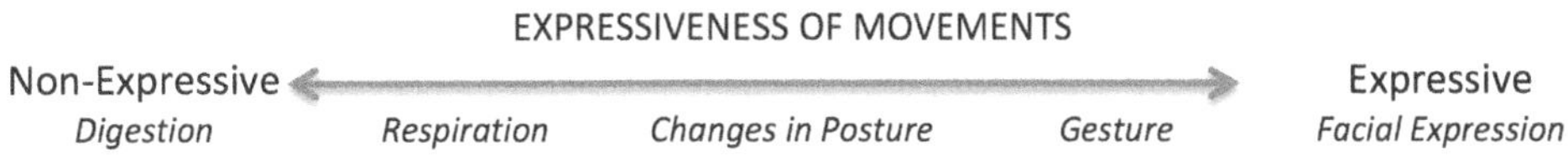

Figure 19.2 Expressiveness of movements
Source: author

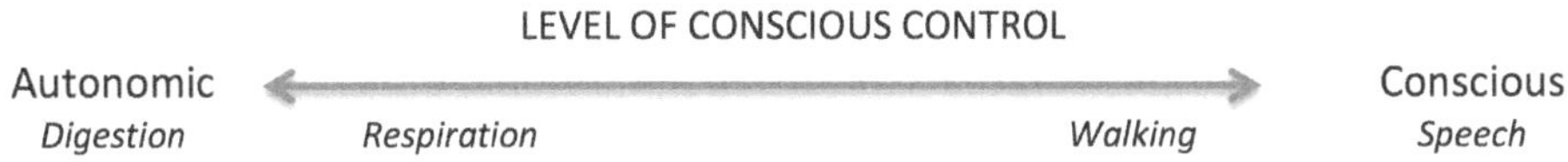

Figure 19.3 Level of conscious control
Source: author

Movements can be more or less expressive, in other words they can serve more or less effectively as communication medium to other humans. Again, this might usefully be thought of as a continuum, as illustrated in Figure 19.2. The idea is to evaluate the expressiveness of the actual movement, which is why speech does not appear at the most expressive end of the scale. Although it is a motor act, speech is mostly expressive because of the sound produced (discounting for now the possibility of lip-reading).

Our movements are more or less under conscious control, which presents itself as a continuum from autonomic to conscious, as illustrated in Figure 19.3. There is a very interesting transition space, where our movements are controlled by a mixture of conscious control and unconscious bodily stimuli. Respiration is typically put in this space; we can never stop breathing but we can finely regulate how we breathe through respiration training for example (Montgomery 1994) with the aim to relax.

The following actual and proposed examples illustrate how these dimensions of human movement can, in some sense, be actuated in the built environment. The survey covers five ways of making links with these dimensions and I will briefly reflect on each.

Rietveld's *Schröder House* (Kronenburg 2007, 26), Holl's *Fukuoka Housing* (Kronenburg 2007, 52) and Ban's *The Naked House* (Kronenburg 2007, 170), all offer internal architectural features that can be physically adapted. Sliding and hinged walls allow people to re-configure the spatial topology in the first two, while the ability to move rooms around in a larger enclosure is a feature of the third example. This is actuated directly through human movement with a one-to-one mapping between that movement and the architectural movement, where the direction of that movement remains intact. There is no manipulation of the human effort required (attenuation or amplification). It requires physical force and effort to trigger. Reflecting on the three continua listed above, it can also be said that movement is medium to large scale, that it is expressive and that it is voluntarily controlled.

Moving on from entirely manual engagement, both Studio Gang's *Starlight Theatre* (Studio Gang Architects 2009) and DRMM's *Sliding House* (DRMM 2009) are examples in which the larger scale movement of architectural elements have been implemented. A button pressed by the human agent (this action is assumed and it could be a slider for example) is used to trigger the operation of a motor (or any other suitable actuation driver) to operate parts of the building, to open or move out of the way. Bodily movement is thus translated from a very small-scale movement to the large-scale movement that is visible architecturally. It is amplified in scale and power through technical systems, consisting of a simple sensor, some logic and a simple actuator. The movement that we employ is relatively low in expressiveness and it is entirely conscious.

Movement relationships are becoming more subtle and coupled in experimental architecture such as TU Delft's *Muscle Tower* (Hubers 2004) and Ruariri Glynn's *Reciprocal Space* (Glynn 2005). Here architectural movement is programmed to react to the proximity of its inhabitants to actuate (at least parts of) the built structure. Movement through physical space, which happens to be interactionally enclosed by those prototypes, triggers responses in their architectural elements. Translations are introduced to map between human and architectural movement, where it seems certain that human movement becomes amplified in the *Muscle Tower* while it might be attenuated in *Reciprocal Space*. In these examples, human movements are on the largest scale we can produce; they are of medium expressiveness and they are conscious.

In my own prototype *ExoBuilding* (Schnädelbach *et al.* 2010) but also in Jacob and Findley's *Breathe* (Jacobs and Findley 2001), architectural elements respond to the respiratory behaviour of inhabitants. Respiration and, in a later mapping also heartbeat, is amplified to such an extent that it can be mapped to a room-sized enclosure that moves in synchronisation with people's physiological behaviour. In other words the movement produced to keep the blood oxygenated and to supply all parts of the body with oxygenated blood. As the feedback from many study inhabitants showed, this direct and direction-consistent mapping provides for highly immersive and embodied experiences. The human movements detected and employed are of a small to medium scale, they have low to medium expressiveness and they can be both voluntarily controlled while autonomic control has the upper hand.

As the final example in this loose taxonomy of the relationship of human and architectural movement, it is worth speculating about the things that have not been tried. For various practical reasons, the mappings described remain on a small scale and sometimes involve very few concurrent inhabitants, which is clearly not how we inhabit architecture. An example with a larger scale would be an adaptive city, which changes its physical topology to adapt to our aggregate movement. Here human movement would be aggregated, increasing the scale that we can individually produce to an urban-scale space. Such movement would, for example, automatically be mapped to routes opening or closing, to bridges turning to connect different places and to levels changing on public squares. Movement would be of a larger scale than individually possible, it would be of medium expressiveness and it would be voluntarily controlled or be 'autonomic', to the extent that, as space syntax (Hillier 1984) research maintains, urban-scale aggregate movement patterns are determined to a large degree by spatial topology.

A spatial culture of movement

The above examples have demonstrated how the latest developments in experimental Adaptive Architecture begin to offer a different, perhaps more malleable but also more complex type of relationship with human movement. Architectural movement can be physically actuated or digitally controlled. Human movements become mapped directly, or in amplified or attenuated fashions. The coupling between human movement and that in Adaptive Architecture could persist or couplings could be dynamic, giving control over what is mapped to what in different circumstances. To enable human movement to have an instrumental role in Adaptive Architecture, this expression of behaviour must be translated to (digital) information streams, abstracting it from physical space into malleable data.

The overview provided in this chapter has attempted to sketch the broad design-research space in this area. When designing Adaptive Architecture, when making links between human movement and movement in architecture, it is essential that architects consider the entire range of people's movement capacity, as all expressive behaviours are movements. Reflecting on these design possibilities returns us to *Rhythmanalysis* as proposed by Lefebvre (2013). In mapping human movements to those of architecture, architects can construct the type of *polyrhythmic* relationship that Lefebvre sought to analyse and that he observed as already present in our environment (Lefebvre 2013, 77). Looking at Adaptive Architecture from Lefebvre's standpoint and considering the multiple rhythms that permeate it, it is clear that the rhythms of a more general architecture continue to persist: the longer term physical and cultural rhythms of the site and ground as well as the physical and cultural rhythms of our embodied selves. In addition, however, the architect picks a set of movements in ourselves and in the environment to 'match' to each other. In the examples of architectural projects included above, the matching has typically resulted in *isorhythmia*, where human and architectural movements are occurring at the same frequency, such as in the previously presented manually triggered movements but also in the physiological mappings of ExoBuilding. *Eurhythmia*, the harmonious co-existence of multiple rhythms can also be found in the designed relationships of human movement and architectural movement in Adaptive Architecture. One example is constituted by the experiments with ExoBuilding where participants' respiration was linked to environmental movement and their skin conductance (influenced by levels of activation and relaxation) was linked to environmental lighting. The design introduced multiple related physiological rhythms into the environment in this way, as presented by Schnädelbach *et al.* (2010). When *arrhythmia* has arisen in our work, multiple rhythms were present in the environment but they were dissonant; inhabitants felt uneasy and uncomfortable.

This research points to the fundamental relationship between movement and internal, including mental, states. While the correlations and causalities of that relationship remain under investigation, a number of techniques have been proposed for people to use movement to positively influence their internal state. Country walking, as described by Edensor (2000), is designed to imprint itself on us through a higher state of consciousness and sensory awareness, though the author also cautions that it is heavily framed by cultural references, which might make this effect difficult to achieve. Mindfulness training offers a practical strategy to support mental well-being which begins by gaining more control over physiological movements (e.g. via respiration training; cf. Kabat-Zinn 2005). Respiration training, outside the mindfulness context, is now an accepted coping mechanism for example to reduce anxiety (cf. Clark and Hirschman 1990). Adaptive Architecture can be supportive of such strategies, helping people to move in certain ways that have been shown to be beneficial or that they feel are beneficial. Technically, it can also be set in contrast to human behaviour, where arrhythmia present in the relationship between people and their environment is likely to cause ill-health. All this presents a specific design challenge for architects who wish to make links between human movement and movement in architecture: it is essential they consider the resulting isorhythmic, eurythmic and arrhythmic relationships in order to interpret possible inhabitant reactions in such actuated environments.

Information about human movement must also be seen as part of other types of personal information that are becoming technically available. Such information, for example, includes social networking data, the data present in personal file stores and the messages people exchange electronically. The current interest in 'Big Data' captures an aspect of our present reality in which electronic data sources are not only measured for immediate re-use, as for example in the way that the output of an external temperature sensor influences the heating of a building at a given movement. Once data sources have been stored by technology giants such as Google, Microsoft and Apple, the data becomes available for data mining, and the creation of semantically enriched meta-data. Once there is record of personal data, cross-references can be made between behaviours over time and the behaviour of others. While none of the Adaptive Architecture examples introduced above relies on Big Data, this is an implicit context of this research. Via the Big Data cloud, individual adaptive buildings can become integrated into the 'smart city' and interaction with them can draw on accumulated records of personal information, rather than only being focused on the present.

As Morozov (2014) argues, once personal information has been accessed by organisations (for example, insurance companies or the state), people might more easily be rewarded or indeed penalised for their behaviour. With increasing computing power embedded in the environment (and increasingly in people for that matter), responses to behaviours become automated, in the sense that support is offered to decision makers automatically (for example, to the credit card fraud detection team). 'Algorithmic Regulation', as proposed by O'Reilly (Goldstein and Dyson 2013, chapter 22), takes this yet another step further with the aim of driving up the efficiency of decision making. The Big Data context of Adaptive Architecture presents the second specific design challenge that needs to be addressed: it will be critical for architects working in this area to consider the most appropriate interfaces to give people control over their personal data, i.e. to let them have a say in what is captured, stored, how it is mined and by whom.

Conclusions

In conclusion, I would like to reflect on some limitations of the present discussion. My own experimentation with ExoBuilding has highlighted the possibilities that emerge when human and architectural movement are tightly coupled, in the way it allows for highly immersive, embodied experiences under certain circumstances to emerge (cf. Schnädelbach *et al.* 2012). That work has also emphasised the potential for architecture to trigger movement in humans, an aspect that is not covered in this chapter. ExoBuilding influences human movement through biofeedback, whereas Khan's *Open Columns Homeostat* (Khan 2011), for example, is designed to trigger human movement by occupying more or less physical space. There seems no reason to think it should not be possible to classify architectural movement based on scale, expressiveness and level of autonomy in a similar way to the scale of human movements proposed here. Finally, as already mentioned, humans produce behaviours that do not necessarily manifest themselves as human movement, for example thoughts. With constantly developing sensor technologies such behaviours might become detectable by machines, however, and they will then be available to architects to map architectural behaviours against.

Note

1 For a detailed exploration of which architectural elements (e.g. internal partitions or roof elements) can possibly be actuated in an adaptive building 'to make it move', the reader is referred to Schnädelbach *et al.* (2010).

20 Mobilities Design

Towards an experimental field of research and practice

Ole B. Jensen and Ditte Bendix Lanng

Introduction

Contemporary society is defined by the ways in which mobile goods, bodies, vehicles, objects and data are organized, moved and staged. Against the background of the 'mobilities turn' in social theory[1] this chapter proposes a further development of the perspective in the direction of a material and design oriented turn. In order to fulfil this purpose we articulate an emerging research field of 'mobilities design'. In our understanding the time has come to articulate 'Mobilities Design' as a dedicated research field in and of its own. There is a pressing need for research targeting the material, physical and design-oriented dimensions of multiple mobilities from the local to the global. Despite its cross-disciplinary identity the 'mobilities turn' has not sufficiently capitalized its potential for exploring issues of material design and physical form. The exchange value with design is twofold: first it means getting closer to the 'material' which is needed if mobilities research can claim to have understood contemporary mobilities; second, it means that the creative, explorative and experimental approaches of the design world are put within reach of mobilities researchers, increasing the potential for innovation in this area. Design research, on the other hand, might also enter into a fruitful relationship with mobilities research, offering a distinctively 'mobile' lens on design objects and issues. The emerging field of 'mobilities design' explores the borders between architecture, urban design, urban planning, and infrastructure design. The field is intended to address a research gap in areas of increasing societal importance whether inequality, environmental sustainability, or the meaning of vernacular mobilities design in everyday life. 'Mobilities design' opens up an architectural research agenda examining infrastructure spaces as 'spaces of everyday life' for billions of people. This chapter is structured in three parts: after the general introduction we present the mobilities theory perspective of 'Staging Mobilities' and connect this to the empirical phenomenon of parking lots and their design. The chapter ends with some concluding remarks.

Mobilities design and parking lots

In this section we will briefly introduce some of the key theoretical and conceptual framings underpinning our understanding of 'mobilities design'. As previously mentioned this takes as its point of departure the now well established so-called 'mobilities turn' (Urry 2000, 2007). However, we believe that the time has now come to focus more closely on the materialities of mobilities. Much of the existing mobilities

Figure 20.1 Staging Mobilities
Source: Jensen (2013, 6)

research has proven to be very valuable but it has connected insufficiently to ideas of architecture, design and physical intervention.[2] At the same time the exploration of material design and physical intervention has not sufficiently connected to the mobilities literature.[3] Therefore we aim to establish a link between theories coming out of the mobilities turn and the material and design-oriented insights and perspectives related to architecture and urban design. Theoretically, we start out from a normative perspective that recognizes the importance of mundane and ordinary everyday life situations. Within the conceptual framework of 'Staging Mobilities' as defined by Jensen (2013) it is shown that everyday life situations are 'staged' from above through planning, design and regulation as well as staged and 'acted out' from below by consociates in interactions. Such 'mobilities in situ' may be explored from the intersection of three analytical dimensions: material spaces and design, social interactions and the embodied performances (see Figure 20.1).

Such an analytical perspective is related to what we may term 'Critical Mobilities Thinking'. By this we mean that theorizing and reflection should both focus on 'problems' (for example, social exclusion and environmental stress) as well as 'potentials' (for example, seeing how generic transit spaces may afford new experiences and social interaction). In this chapter we are particularly interested in developing the notion of these potentials by examining a very mundane and ordinary place: the parking lot. On the basis of the 'Staging Mobilities' framework with its emphasis on situational mobilities and its explicit concern with material space and design, we intend to link this to the discussion of underdeveloped potentials in transit spaces such as parking lots. This is done in order to achieve two things: first, we believe that so-called 'non-places' as described by Augé (1995) should be understood as more than merely instrumental and generic sites and we think the emerging field of mobilities design may facilitate this development. Second, by exploring its relationship to material space and design we argue that our conceptual and theoretical understanding of mobilities becomes more detailed as well as more concrete and less abstract.

The situational perspective on everyday life mobilities thus opens up an informed analysis and interest into the 'materialities of mobilities'. We find these in multiple incremental decisions and interventions across a vast number of professions and disciplines. We embark on this research with a very open – some may say too open – understanding of what would qualify as 'mobilities design'. However, in order not to reproduce the understanding of 'design' as neither a mystical creative force nor a simple and rational endeavour, as well as not to confine it to any particular profession, we ask the pragmatic question: *what enables, affords or prevents this particular mobile situation?* This question is a direct outcome of using a situational analytical frame and is well suited to a research agenda more interested in how actions are becoming possible than in determining to which fields of thought and academic disciplines they might 'belong'. In other words, we want to foreground the situational affordances creating specific mobile practices. In order to do so we will open up 'design' to include the decisions taken by the traffic engineer on how long the green period of traffic lights should be, as well as the service design interventions creating way-finding systems or ticket machine systems. Since the authors' background is within urban studies and urban design, the theory and practices of architecture and urban design will also be included. In this way we aim to define a rather wider conception of 'design' practice than is conventionally recognized, a step which may seem offensive to custodians guarding the traditional disciplinary boundaries of architectural design. In this respect we agree with American urban designer and planner Kevin Lynch when he states that:

> Design is the playful creation and strict evaluation of the possible forms of something, including how it is to be made. That something need not be a physical object, nor is design expressed only in drawings. Although attempts have been made to reduce design to completely explicit systems of search and synthesis, it remains an art, a peculiar mix of rationality and irrationality. Design deals with qualities, with complex connections, and also with ambiguities.
>
> (Lynch 1981, 290)

From theories we want to move towards specific and empirical examples of design that affords or prevents situational mobilities.

The parking lot as illustrative case

Parking lots are ubiquitous and monofunctional; their design tends to be generic and they are often derided as people-hostile asphalt environments of automobility. Nonetheless, our point of departure in this study is that parking lots hold underused potential for social exchange and sensorial pleasure. We will use the situational 'Mobilities Design' thinking outlined above to consider the potential for a selected parking lot to become a cherished public space which welcomes heterogenous mobile situations. Through this case study we intend to develop a single tangible example of the ways of thinking and doing inflected in 'Mobilities Design'. We set out by outlining the conventional instrumentality of parking lot design standards, including critiques of the resulting material environments. This is what we term a *mode 1 design* of parking lots. After this, we will exemplify a *mode 2 design* – Mobilities Design – in which we synthesize the functional-programmatic design considerations of mode 1 with a situational Mobilities Design thinking, encompassing the identification of heterogenous mobile situations that occur and *could* occur in the parking lot. For this purpose we will use an experimental study of selected parking lots in the 1970s suburban district of Aalborg East, Denmark. The aim of the study is not to provide a solution for the optimal design of parking lots; rather it is to linger in a state of exploration which allows questions to be asked, facilitates the imagining of an alternative future for parking lot design, and inclines us to open the agenda for another mode of consideration to be included in the design of some of our most mundane, yet also most utilized public spaces.

Designing a standard parking lot

The parking lot in our focus here (Figure 20.2) is a vivid material manifestation of functionalist principles of traffic planning, as they appear in, for example, the Swedish SCAFT guidelines from 1968, which were widely used in traffic planning in Scandinavia (cf. Harder 2002). The SCAFT guidelines provide a generic set of principles for the organization and design for traffic. Forms, spaces and materials are kept simple and with a uniform expression. The motive behind these design guides is safety: the traveller's overview of the traffic situation and a minimum of disturbances to keep concentration on the traffic (Hagson 2000, 16).

MIT professor Ben-Joseph asserts how optimized space usage and vehicular flow have often been achieved through design standards, which come packaged even as mathematical equations (Ben-Joseph 2012, 89). In the United States, which is the focus of Ben-Joseph's research, city ordinances for parking lot design from the 1960s specified accurate minimum requirements for dimensions, construction, ingress and egress, while landscaping, accessibility criteria and environmental concerns only vaguely and gradually have been integrated in formal regulations. Not least, as Ben-Joseph remarks, 'Planners, designers, developers, and the public rarely pay attention to the design of parking lots' (Ben-Joseph 2012, xx). The ubiquitous landscapes of parking have become naturalized, absorbed into everyday life, without being questioned as a space for other activities than its primary function.

In Denmark, contemporary principles for parking lot planning and design reveal that *mode 1* functional-programmatic considerations continue to dominate the agenda for the materialization of the widespread phenomenon of surface parking. Danish design of parking lots is mainly regulated by municipal authorities through local planning ordinances. A set of national instructive guidelines have been formulated by the Road Directorate (cf. Danish Road Directorate 2011, 2012). These do contain considerations for elements such as vegetation, lighting and double use, however only in vague and minimal terms. In the local plan for the area discussed

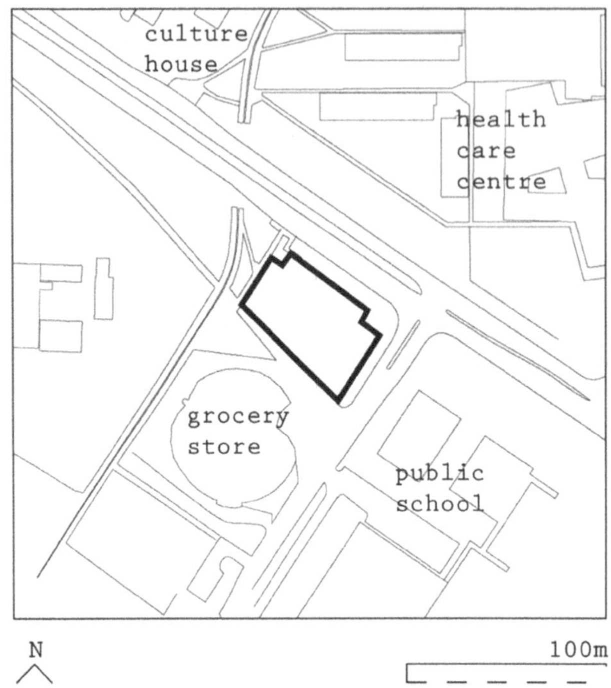

Figure 20.2 Parking lot site plan
Source: authors

not much is specified except that the unbuilt areas should be of 'proper appearance' (Aalborg Municipality 1980). However, newer municipal documents address the issue in more precise terms. The most distinct example with regards to our case study is the municipal Architectural Policy from 2004. It contains a section on Traffic Spaces, and points out both a dilemma and an underused potential for the design of these spaces:

> The challenge lies in creating cohesion between infrastructure, urban and rural spaces, and creating original and functional solutions that benefit both the traffic and the surroundings … The city's different types of traffic spaces shall in general be infused with more aspects in relation to usage and experience so that their interaction with the buildings and structures in the surrounding district is improved. This can be achieved through a very deliberate choice of vegetation, physical cross section, construction materials, lighting and other urban furniture.
> (Aalborg Municipality 2004, 12; our translation)

In spite of their good intentions these guidelines and regulations leave the parking lot designers with very sparse formal knowledge, tools and power to include sensorial and social considerations in parking lot design. Only functional standards for parking norms, principles of circulation, dimensioning of booths, area-optimized 90-degree parking schemes, etc., have been formalized and quantified. These can be found in municipal ordinances and in, for example, the internationally known *Architect's Data* (Neufert and Neufert 2000).

The minor degree of interest in the design of parking lots, and the standardized or deficient design guidelines only poorly address the plurality of mobile situations which unfold – or could be unfolded – on the parking lot: the people who perform mobilities in variations of modalities, rhythms, routes, social and sensorial engagements. The generic forms have been heavily criticized for being exclusively (mono-) functional and lacking sensorial and social qualities (see Jacobs 1961 or Sennett 1994). Such transit spaces tend to facilitate a distanced form of car-driving only, thus 'locking-in' our social life to automobility as Urry remarks (2007). Asphalt, the prevalent paving material of the extensive parking lot surfaces, has in itself gained considerable attention and critique (such as Zardini 2005), in that most often it does not comply with environmental issues and fails to invite social exchange and the sensorial pleasure of travelling.

Mobilities Design – affording social and sensorial mobile situations

Following the points above, our exploration of the parking lot is spurred by the hypothesis that our most mundane transit spaces might hold underused potential. Ben-Joseph reasserts this point, when he sets out his interrogation of the culture and design of parking by asking: 'Can't the mundane also be valued, appreciated, or become spectacular in its own right?' (Ben-Joseph 2012, 4). Our purpose in this section is to search for redesign potential in the mobile practices of parking lot situations. One of the sparse, trivial and – apparently – desensitizing transport situations, a journey across the parking lot, will be unpacked to articulate potentials for the *mode 2* design thinking.[4]

Several researchers of mobilities advocate that making a journey is a matter of social, cultural and experiential complexity. Urry emphasizes that '[m]ovement often involves an embodied experience of the material and sociable modes of dwelling-in-motion' (Urry 2007, 11). According to Vannini people are not just passively *getting transported*. Instead they are wayfarers who *perform journeys*: 'a mundane but meaningfully ritualistic and artful practice which creates occasions for unique interaction settings and relationships' (Vannini 2012, 162). A manifold of embodied practices and experiences are inherent in making ordinary journeys, and wayfarers encounter the world in various ways: 'physical travel involves lumpy, fragile, gendered, racialized bodies encountering other bodies, objects and the physical world multi-sensuously' (Urry 2007, 272). Likewise diverse modalities provide us with different 'kinaesthetic investments' (Sheller 2004): walking, bicycling and car-driving orient us towards the material affordances of the world in different ways. Also cultural and historical dimensions are key in shaping the journey experience, not least with regard to the actual material environment that the wayfarer inhabits on the journey. Here we should clarify that in accordance with Ingold (2011) we use the term 'wayfarer' as well as the term 'journey' to capture the phenomenological and evocative dimension of ordinary everyday life mobility. So instead of reserving these notions to exotic long-distance movement we want to use these terms to draw attention to the proposition that mundane mobility practices also have effects on identity, belonging and understanding of place (as is often understood to be the case of exotic journeys to far-away destinations). In other words there is an analytical point in using the notions of 'wayfarer' and 'journeys'.

We are not, of course, seeking to reimagine mundane mobility practices as 'exotic' so much as to utilize the ethnographic discourse of mobilities in order to highlight that the everyday mobility has its own 'magic' and importance. The terminology used is a deliberate attempt to counter the neglect of the everyday life and its importance.

The material environment of mundane journeys is our focus here: a parking lot where wayfarers 'stitch together' their embodied experiences and interactions and enact the mundane transit space as a sensuous and ambiguous geography, which in turn must organize and orchestrate those journeys. When approaching the design of the parking lot, then, we suggest that in spite of its abundance of mundane mobilities, we should not think of it as a 'non-place' of detachment (see Augé 1995). Rather we should stimulate the search for potential by unfolding it as a public space animated and co-produced by those embodied practiced mobilities. In the words of Cresswell and Merriman (2011, 7), such spaces are 'continually practiced and performed through the movement and enfolding of a myriad of people and things'.

In travelling by foot across the parking lot – from its far western corner to the far eastern – we experience one of its regularly traversed flow routes; the observations were conducted at recurrent field visits between 2011 and 2013. The journeys along this route are performed by wayfarers who arrive by bus from the city centre and head for the grocery shop, school or day care institutions south of the parking lot. Taking part in and observing this ordinary journey allows us to get a sense of the manifold nature of wayfaring practices in the parking lot. Though it is a place with a relatively low density of people, each journey is performed within a pulsating network of co-existing mobile situations. The parking lot is an arena of changing formations of mobile and (temporarily) immobile people and things which fosters subtle 'negotiations in motion' (Jensen 2010a, 389; 2010b, 344). Such mobile negotiations happen when pedestrians arrive by bus, trot across the asphalt surface and exchange delicate signs with car-drivers about turns and halts (see Figure 20.3). At localities where routes intersect or overlap they shape temporary social condensers for wayfarers who flow in and out of different 'mobile withs' (Jensen 2013, 4). These include, for example, small groups of fellow travellers, such as two kids who playfully stumble across the kerb side of the parking lot, and an unplanned meeting between acquaintances on their way.

When performing a journey by foot across the parking lot our attention might also be directed towards the sensorial enlacement of wayfarers and environment. Our everyday mobile engagements with this space is multisensory and highly variable:

> Urban spaces provide numerous ambiances to be felt with all the senses. Whether we think of a lively outdoor marketplace or an ordinary parking lot, an attractive historical center or an accessible subway station, the very way we relate to these places is based on the sensory experience they provide. It is a matter of light and colour, sound, smell, touch and heat, as well as the manner in which we walk and talk, move and look, relate and behave. In other words, urban ambiances always create a subtle interweaving of synaesthesia and kinaesthesia, a complex mixture of percepts and affects, a close relationship between sensations and expressions.
> (Thibaud 2011, 1)

During our walk or drive across the parking lot, we unfold a multisensorial journey. We move, and so do many of the people and things around us. We move, for example, from the enclosement of the car out into the sun, where sounds are diffuse,

Figure 20.3 Co-existence of modalities at the parking lot
Source: authors

and we might suddenly hear clearly the wheels of shopping carts rolling over the asphalt. Maybe we find ourselves inclined to get away as fast as possible, or we might enjoy a breath of air, a hint of a smell of new mown grass, or a chance encounter with a friend. Delving into these commonplace mobile practices and experiences points to the social and sensorial heterogeneity of wayfaring and to the ambiguities of transit spaces which come to life through such heterogeneous journeys. In this view the parking lot should not be confined to a strict functional categorization but acknowledged for the possible social and sensorial richness of its mobilities. It possesses a design potential to cultivate this ambiguity in order to greater facilitate and value wayfarers' mundane encounters and diverse experiences in their own right.

Parking lot materialities perform mobile affordances in relation to the natural environment and transport technologies, thereby conditioning practices and experiences. What we are aiming at here is to work with the relationship between the design of the lot and the wayfarers, focusing on the properties of materialities (see Ingold 2011) as these afford mundane, everyday practices. We should not understand the parking lot as a stable artefact in which design can inscribe specified uses and experiences only. Instead, the parking lot might be occupied by a plurality of other ways of mobile inhabitation than the prescribed ones. The actual accomplishment of this plurality lies beyond the control of mobility designers. The design of the parking lot must, therefore, be open to an unknown future, which cannot be determined by the singularity of design, however prescriptive its form-making seems to be. In our parking lot study this leaves the mobility designer with the task of working with the plurality and uncertainty of embodied mobilities, and allowing design to be a mediator of the yet-to-be.[5]

Our suggestion for a way into exploring these potentials is to acknowledge the many mobile situations which take place in the parking lot, as well as those which *could* take place, thereby contributing in various ways to the character of the parking lot. Ben-Joseph provides us with a sparkling directory of uses observed on American parking lots, including the more exotic ones such as *movie-on-the-lot* and *tailgating*. These discovered uses transform the mundane asphalt surfaces into an official and festive part of American football, where people gather around the car prior to the game with barbecues, chairs and canopies for eating, drinking and socializing (Ben-Joseph 2012, 28). One could imagine identifying a catalogue of mundane, yet heterogenous mobile situations, which could occur in the parking lot. Such mobile situations from a car-driver's perspective could range from a straightforward one where someone enters the parking lot by car, parks it, walks to the shop, returns by foot, and exits the parking lot by car; to a situation which includes a stop to chat, perhaps with another car driver, window open; and finally to a linger on the lot, stretching legs and enjoying a sandwich. A pedestrian would maybe carry his heavy grocery bags to the bus stop across the parking lot, being forced to make several stops to rest his arms and hands from the load. Other wayfarers would possibly be involved in ludic activities of wheeling or rolling across the smooth asphalt surface, perhaps during a less busy time of day, outside school and opening hours. Together a vast array of mobile situations shapes the Mobilities Design programme for the redesign of the parking lot. The array of possible situations reveals that the parking lot is a space of diverse modalities. As described above plenty of practices are not carried out from within the 'iron cage' of the car, but on foot, bike, moped, even in a wheelchair or on rollerblades. These situations also point to a wealth of different routes and rhythms in which small places, which might not have seemed important, could stand out as significant. Plurality, however, lies not only in this range of diverse situations; as we saw above even each quotidian mobile situation unfolds a heterogeneity of practices and experiences, when one pays close attention. How, then, could the parking lot afford such mobile situations, its redesign welcoming 'little practices' and pleasurable, embodied experiences of mobilities?

This question lies at the heart of the 'Mobilities Design' approach to the design of a mundane transit space. As such, it is an approach that overlays the technical and instrumental mode 1 design of facilities for our daily life with greater sensitivity for the everyday practices and social and sensorial dimensions of mobilities. Public space is not necessarily constrained to traditional urban squares, and so we have suggested to approach the parking lot as a 'mobile agora' (Jensen 2009, 151) – a place of movements, connections and in-betweenness, where design considerations for human interaction and sensorial pleasure might support the relinking of the dissected zones of modern planning (see Hajer and Reijndorp 2001). A mundane parking lot might be a location loaded with possibilities for wayfarers to cultivate the 'precious gift of travel time' and develop meaningful social practices and joyful experiences on their way, like Vannini illustrates with the case of the ferry line-up (Vannini 2012, 192).

Conclusions

At the onset of this chapter we made the distinction between 'Mode 1' and 'Mode 2' design as a way of addressing the shift of mind and practice that we are articulating with the emergence of 'Mobilities Design' as both an academic subfield to the 'mobilities turn', but also as a new way of engaging with the cross-disciplinary set of professions that are taking design decisions and executing design interventions that together create the practices and pragmatism behind all situational mobilities. We intend that this 'concluding' discussion should be regarded as provisional since this stage of the articulation of 'Mobilities Design' remains a work-in-progress. Clearly in this chapter we have focussed more on the research part than the design practice. However, this economy is necessary since it would require more space and visual material to establish the full potential of 'Mobilities Design' to the practice of parking lot design than we have available here.

Coming back to the contributions to the 'mobilities turn' that we see emerging out of the field of 'Mobilities Design' we may start by stating that sites of mobilities are richer in their potentials than the 'non-place' discourse has given them credit for. Moreover, that engaging in what has been termed 'Critical Mobilities Thinking' means challenging the self-evident, innovation, seeking solutions and recognizing that all problems have a 'potential side' and all potentials have a negative side. It is precisely here that we see the added value of articulating a notion such as 'Mobilities Design' in order to open up this agenda: to become sensitive to the materialities of mobilities. In short, we need a new 'material turn' in mobilities research! In general, mobilities often take place in what some term 'non-places' and its effects beyond simple physical movement are often invisible. The research into design and mobilities opens up to a material pragmatism that seeks to understand and provide languages for these mundane, but increasingly important, practices. 'Mobilities design' research explores the question: '*What's the name of my mobility experience, what does it mean, and how does it materialize?*'

As this is a field in the making we do not mean to close off any additional discussion about its potential directions or more precise content. However, as we (to our knowledge at least) are the only ones who have made the mobilities/design nexus explicit we do want to make a statement on the potential layout of a research programme or agenda for 'Mobilities Design'.

In terms of 'focus of studies', research in Mobilities Design would focus on situated practices, doings, acts and interactions. It would contemplate the rich and complex relationships between multiple objects, artefacts, systems, technologies, spaces, as well as the relationship between the human and non-human dimensions of situational mobilities. Mobilities Design research would be particularly interested in exploring the sensed and embodied mobile practices (i.e. kinesis). When it comes to themes such as 'methods and approaches' a Mobilities Design research agenda will address methods and approaches that are experimental, creative and performative. Moreover they would be situated and empirical (and at times involve one-on-one interventions in the city). In terms of the epistemologies and frames of thinking underpinning this research agenda we see intellectual legacies (and this list is definitely not all-embracing) pointing towards ideas of actor-networks, post-phenomenology and pragmatism. Much inspired by the 'more than representational' or at times 'non-representational' (as is the case with Thrift (2008)) perspectives we could see a particular role for re-reading and rethinking the tradition of pragmatism in particular which seems to have been neglected in the general re-inventions of classic streams of thought within the social sciences. The link between pragmatism and Mobilities Design requires more explanation than we have been able to offer here, but it constitutes a necessary part of the future work in this field.

Next to the theoretical and methodological issues connecting the emerging discipline of Mobilities Design to existing research within mobilities, there lies a more grounded discussion about the future of parking lot in urban design and planning. Our research opens up one possible angle for discussing and analysing the policy question 'what is the future of the parking lot?' This should be seen as an important issue in urban design, partly due to the massive investment in areas already occupied by this particular artefact, but also in the light of possible 'post-car' scenarios as developed by Dennis and Urry (2009). Perhaps the parking lot of the near future will have to accommodate for more experiences and more sensations than simply parked cars. Mobilities design as a research field establishes such connections as issues for future research.

This chapter has reported on an initial effort to connect the academic analysis of the 'mobilities turn' and the interventionist practices of the various built-environment professions dealing with the material design of mobilities and place. We have, at least superficially, been able to open up both the theoretical and the design-oriented dimensions of Mobilities Design. Further effort is now needed to construct a more robust and coherent theoretical framing as well as greater exploration into how different design elements afford or inhibit mobile practices. As this is only the early days of articulating 'Mobilities Design' as an experimental field of research and practice we are, however, rather confident that it will become more widely acknowledged as a novel and necessary field of study.

Notes

1 For short review papers on this see Sheller (2011) and Vannini (2010).
2 Exceptions are, for example, Merriman's (2007) study of motorways and Adey's work in airports (2010).
3 Exceptions are for example Houben and Calabrese' (2003) book on motorways as everyday life spaces and Hoete's (2003) reader on the 'aesthetics of mobility'.
4 See also Lanng *et al.* (2012) for a slightly more detailed account of such a journey.
5 See the works of for example Yaneva (2009) and Dyrssen (2011) for an exploration of design as 'mediator'.

Bibliography

Aalborg Municipality. 1980. *Lokalplan 08–009 for institutionsområde Tornhøj*. Aalborg: Aalborg Kommune.

Aalborg Municipality. 2004. *Architectural Policy: For an Attractive and Vigorous Urban Environment.* Aalborg: Aalborg Kommune.

Ackerman, D. 1990. *A Natural History of the Senses.* New York: Vintage.

Addams, Jane. 1972 [1909]. *The Spirit of Youth and the City Streets.* Chicago: University of Illinois Press.

Adey, Peter. 2010. *Mobility.* London: Routledge.

Agamben, Giorgio. 2009. *What is an Apparatus? And Other Essays.* Stanford: Stanford University Press.

Agbe-Davies, A.S. 2010. 'Concepts of community in the pursuit of an inclusive archaeology'. *International Journal of Heritage Studies* 16 (6): 373–389.

Aldenderfer, M. and H.D.G. Maschener. 1996. *Anthropology, Space, and Geographic Information Systems.* Oxford: Oxford University Press.

Alexander, Jeffery C. 2005. 'The inner development of Durkheim's sociological theory: from early writings to maturity'. In *The Cambridge Companion to Durkheim,* edited by J.C. Alexander and P. Smith. Cambridge: Cambridge University Press, 136–159.

Allemandy, Victor Henry. 1914. *Enfield: Past and Present.* Enfield: Meyers, Brooks and Co.

Allison, P., ed. 1997. 'Why do excavation reports have finds' catalogues?' In *Not so Much a Pot, More a Way of Life,* edited by C.G. Cumberpatch and P.W. Blinkhorn. Oxford: Oxbow, 77–84.

Allison, P. 1999. *The Archaeology of Household Activities.* London: Routledge.

Allison, P. 2005. *Pompeian Households: An Analysis of the Material Culture.* Los Angeles: Cotsen Institute of Archaeology, University of California.

Amin, Ash. 2008. 'Collective culture and urban public space'. *City* 12 (1): 5–24.

Anderson, Benedict. 1991. *Imagined Communities: Reflections on the Origin and Spread of Nationalism.* London: Verso.

Anderson, Benedict. 2010. 'Preemption, precaution, preparedness: anticipatory action and future geographies'. *Progress in Human Geography* 34 (6): 777–798.

Anderson, Stanford, ed. 1978. *On Streets.* Cambridge, MA and London: MIT Press.

Andrews, Howard. 1993. 'Durkheim and social morphology'. In *Emile Durkheim: Sociologist and Moralist,* edited by S. Turner. London and New York: Routledge, 111–138.

Appadurai, Arjun. 1996. *Modernity At Large: Cultural Dimensions of Globalization.* Minneapolis: University of Minnesota Press.

Aron, Jean Paul, Paul Dumont and Emmanuel Le Roy Ladurie. 1972. *Anthropologie du conscrit français d'après les comp.* Paris: Mouton.

Aucott, Paula, Alexander von Lünen and Humphrey Southall. 2009. 'Exposing the history of Europe: the creation of a structure to enable time-spatial searching of historical resources within a European framework'. *OCLC Systems & Services* 25 (4): 270–286.

Auge, Marc. 1995. *Non-Places: An Introduction to Supermodernity.* London: Verso.

Augustine. 1853. *The Confessions*. Edited and translated by E. Pusey. Oxford: John Henry Parker.

Ault, B.A. and L. Nevett. 1999. 'Digging houses: archaeologies of Classical and Hellenistic Greek domestic assemblages'. In *The Archaeology of Household Activities*, edited by P. Allison. London: Routledge, 43–56.

Bachelard, Gaston. 2000. *The Dialectics of Duration*. London: Clinamen Press.

Bafna, Sonit. 2003. 'Space syntax: a brief introduction to its logic and analytical techniques'. *Environment and Behaviour* 35 (1): 17–29.

Baggs, A.P., Diane K. Bolton, M.A. Hicks and R.B. Pugh. 1980. *A History of the County of Middlesex*, vol. 6, *Friern Barnet, Finchley, Hornsey with Highgate*, edited by T.F.T. Baker and C.R. Elrington. Accessed 30 November 2014. www.british-history.ac.uk/source.aspx?pubid=88.

Baggs, A.P., Diane K. Bolton, Eileen P. Scarff and G.C. Tyack. 1976. *A History of the County of Middlesex*, vol. 5, *Hendon, Kingsbury, Great Stanmore, Little Stanmore, Edmonton Enfield, Monken Hadley, South Mimms, Tottenham*, edited by T.F.T. Baker and R.B. Pugh. Accessed 30 November 2014. www.british-history.ac.uk/source.aspx?pubid=87.

Baird, J. and A. Pudsey. 2015. *Between Words and Walls*. Cambridge: Cambridge University Press.

Baird, Jennifer and C. Taylor. 2010. *Ancient Graffiti in Context*. Abingdon: Routledge.

Baker, Alan R.H. and Mark Billinge. 1982. *Period and Place: Research Methods in Historical Geography*. Cambridge: Cambridge University Press.

Baker, N. 2010. *Shrewsbury: An Archaeological Assessment of an English Border Town*. Oxford: Oxbow.

Banister, David. 2008. 'The sustainable mobility paradigm'. *Transport Policy* 15: 73–80.

Barley, M.W., ed. 1975. *The Plans and Topography of Medieval Towns in England and Wales*. London: Council for British Archaeology.

Barrell, John. 2006. *The Spirit of Despotism: Invasions of Privacy in the 1790s*. Oxford: Oxford University Press.

Barron, C.M. 2005. *London in the Later Middle Ages: Government and People 1200–1500*. Oxford: Oxford University Press.

Barthes, Roland. 1977 [1967]. 'The death of the author'. In *Image, Music, Text*. London: HarperCollins, 142–148.

Barthes, Roland. 1982. *Empire of Signs*. Translated by R. Howard. New York: Anchor.

Baudelaire, Charles. 1999. *Les Fleurs du Mal*. Paris: Edition Livre de Poche.

Bawa-Cavia, Anil. 2011. 'Sensing the urban: using location-based social network data in urban analysis'. Paper presented at the Pervasive PURBA workshop. Accessed 1 November 2014. www.cavvia.net.

Beed Davey, Gwenda, Kate Darian-Smith and Carla Pascoe. 2013. 'Playlore as cultural heritage: traditions and change in Australian children's play'. In *Children, Childhood and Cultural Heritage*, edited by Kate Darian-Smith and Carla Pascoe. London and New York: Routledge, 40–54.

Behrens, Moritz, Ava Fatah gen. Schieck, Efstathia Kostopoulou, Steve North, Wallis Motta, Lei Ye and Holger Schnädelbach. 2013. 'Exploring the effect of spatial layout on mediated urban interactions'. In *Proceedings of the 2nd ACM International Symposium on Pervasive Displays*. Mountain View: ACM.

Belk, Russell W. 1988. 'Possessions and the extended self'. *Journal of Consumer Research* 15 (2): 139–168.

Bell, A. 1997. 'Cicero and the spectacle of power'. *Journal of Roman Studies* 87: 1–22.

Bell, A. 2004. *Spectacular Power in the Greek and Roman City*. Oxford: Oxford University Press.

Benjamin, Walter. 2000. *Berliner Kindheit um Neunzehnhundert*. Giessener Fassung. Frankfurt am Main: Suhrkamp.

Ben-Joseph, Eran. 2012. *ReThinking a Lot: The Design and Culture of Parking*. Cambridge, MA: MIT Press.

Bergson, Henri. 1999. *Duration and Simultaneity*. Manchester: Clinamen Press.

Berlo, David. 1960. *The Process of Communication*. New York: Holt, Rinehart and Winston.

Berry, J. 1997. 'Household artefacts: towards a re-interpretation of Roman domestic space'. In *Domestic Space in the Roman World: Pompeii and Beyond*, edited by R. Laurence and A. Wallace-Hadrill. Portsmouth, RI: Journal of Roman Archaeology, L.L.C, 183–195.

Berry, J. 2007. *The Complete Pompeii*. London: Thames & Hudson.

Berry, J. and R. Laurence. 2001. *Cultural Identity in the Roman Empire*. London: Routledge.

Betts, E. 2011. 'Towards a multisensory experience of movement in the city of Rome'. In *Rome, Ostia, Pompeii: Movement and Space*, edited by R. Laurence and D. Newsome. Oxford: Oxford University Press, 118–132.

Birley, Andrew. 2010. *The Nature and Significance of Extramural Settlement at Vindolanda and Other Selected Sites on the Northern Frontier of Roman Britain*. Leicester: University of Leicester, School of Archaeology and Ancient History.

Birley, Eric et al. 1933. 'Excavations at Housesteads in 1932'. *Archaeologia Aeliana* 10: 82–96.

Bissell, David. 2010. 'Passenger mobilities: affective atmospheres and the sociality of public transport'. *Environment and Planning D* 28: 270–289.

Blake, E. 2004. 'Space, spatiality, and archaeology'. In *A Companion to Social Archaeology*, edited by R.W. Preucel and L. Meskell. Oxford: Blackwell, 230–254.

Blok, Anders and Torben E. Jensen. 2011. *Bruno Latour: Hybrid Thoughts in a Hybrid World*. London: Routledge.

Blunt, Alison. 2003. 'Geography and the humanities tradition'. In *Key Concepts in Geography*, edited by Sarah L. Holloway, Stephen P. Rice and Gill Valentine. London: SAGE, 73–91.

Bodenhamer, David J. 2013. 'The spatial humanities: space, time and place in the new digital age'. In *History in the Digital Age*, edited by Toni Weller. London and New York: Routledge, 23–38.

Borges, Jorge Luis. 1956. 'El advenimiento de Buenos Aires'. *Crítica, Suplemento Literario Letras Hispano-Americanas*, 30 November.

Borsay, Peter, ed. 1990. *The Eighteenth-Century Town: A Reader in English Urban History, 1688–1820*. London: Longman.

Borsay, Peter. 1991. *The English Urban Renaissance: Culture and Society in the Provincial Town, 1660–1770*. Oxford: Clarendon Press.

Bourdieu, Pierre. 1977. *Outline of the Theory of Practice*. Cambridge: Cambridge University Press.

Bourdieu, Pierre. 1984. *Distinction: A Social Critique of the Judgment of Taste*. Boston: Harvard University Press.

Bourdieu, Pierre. 1990. *The Logic of Practice*. Translated by R. Nice. Stanford: Stanford University Press.

Bowsher, D., T. Dyson, N. Holder and I. Howell. 2007. *The London Guildhall: An Archaeological History of a Neighbourhood from Early Medieval to Modern Times*. London: Museum of London Archaeology.

Boyer, Carl B. and Uta C. Merzbach. 1991. *A History of Mathematics*, 2nd edn. New York: Wiley.

Boyko, Christopher T. and Rachel Cooper. 2011. 'Clarifying and re-conceptualising density'. *Progress in Planning* 76 (1): 1–61.

Braidotti, Rosie. 1996. 'Cyberfeminism with a difference'. *New Formations* 29: 9–25.

Brand, Stewart. 1994. *How Buildings Learn: What Happens After They're Built*. London and New York: Viking.

Braudel, Fernand. 1983. *Civilization and Capitalism, 15th–18th Centuries*, vol. 1, *The Structures of Everyday Life*. London: Harper & Row.

Brown, Chloë, Vincenzo Nicosia, Salvatore Scellato, Anastasios Noulas and Cecilia Mascolo. 2012. 'The importance of being placefriends: discovering location-focussed online communities'. In *Proceedings of the 2012 ACM Workshop on Online Social Networks*, edited by Stefan Saroiu and Ben Zhao, 31–36. Accessed 16 March 2015. http://conferences.sigcomm.org/sigcomm/2012/wosn12.php.

Brown-May, Andrew. 1995. *The Highway of Civilisation and Common Sense: Street Regulation and the Transformation of Social Space in 19th and Early 20th Century Melbourne*. Canberra: Australian National University.

Bryan, Dominic. 2000. *Orange Parades: The Politics of Ritual, Tradition and Control.* London: Pluto Press.

Burch, M. and P. Treveil. 2011. *The Development of Early Medieval and Later Poultry and Cheapside.* London: Museum of London Archaeology.

Burgess, Ernest W. 1967 [1925]. 'The growth of a city: an introduction to a research project'. In *The City*, edited by Robert E. Park, Ernest W. Burgess and Roderick D. McKenzie. Chicago and London: University of Chicago Press, 47–62.

Burton, Paul. 2009. 'Growing pains: adolescent urbanism on the Gold Coast'. In *Proceedings of the 4th National Conference on the State of Australian Cities*, edited by Paul Maginn and F. Haslam-Mackenzie. Accessed 14 November 2014. www.researchgate.net/publication/44001094_Growing_Pains_Adolescent_Urbanism_on_the_Gold_Coast/links/0912f50f75b3b8747d000000.

Busch, Alexandra. 2011. *Militär in Rom.* Wiesbaden: Deutsches Archäologisches Institut.

Butler, Judith, Ernesto Laclau and Slavoj Zizek. 2000. *Contingency, Hegemony, Universality: Conte mporary: Dialogues on the Left.* London and New York: Verso.

Byrne, Denis and Maria Nugent. 2004. *Mapping Attachment: A Spatial Approach to Aboriginal Post-Contact Heritage.* Hurstville: Department of Education and Conservation (NSW).

Cacioppo, John T., Louis G. Tassinary and Gary G. Berntson. 2007. *Handbook of Psychophysiology*, 3rd edn. Cambridge: Cambridge University Press.

Caillois, Roger. 1961. *Man, Play and Games.* New York: Free Press of Glencoe [English translation, first published in French 1958].

Cairncross, Frances. 2001. *The Death of Distance: How the Communications Revolution is Changing Our Lives.* Boston: Harvard Business School Press.

Camille, M. 2000. 'Signs of the city: place, power, and public fantasy in medieval Paris'. In *Medieval Practices of Space*, edited by B.A Hanawalt and M. Kobialka. Minneapolis: University of Minnesota Press, 1–36.

Capp, B. 2003. *When Gossips Meet: Women, Family, and Neighbourhood in Early Modern England.* Oxford: Oxford University Press.

Carcopino, Jérôme. 1962 [1941]. *Daily Life in Ancient Rome: The People and the City at the Height of the Empire.* Middlesex: Penguin Books.

Carey, Edward. 2003. *Alva & Irva: The Twins Who Saved a City.* Basingstoke and Oxford: Picador.

Carmona, M., M. Davis, F. Scott, J. Gort and H. Meltis. 2010. *High Street London.* London: Design for London.

Castells, Manuel. 1996. *The Information Age: Economy, Society and Culture*, vol. 1, *The Rise of the Network Society.* London: Blackwell.

Caygill, Howard. 1995. *A Kant Dictionary.* Oxford: Blackwell.

Celik, Z., D. Favro and R. Ingersoll, eds. 1994. *Streets: Critical Perspectives on Public Space.* Berkeley: University of California.

Certeau, Michel de. 1984. *The Practice of Everyday Life.* Berkeley: University of California Press.

Cessford, C. 2009. 'Post-1550 urban archaeology in a developer funded context: an example from Grand Arcade, Cambridge'. In *Crossing Paths or Sharing Tracks? Future Directions in the Archaeological Study of Post-1550 Britain and Ireland*, edited by A. Horning and M. Palmer. Woodbridge: Boydell & Brewer, 301–321.

Chadwick, E. 1842. *The Sanitary Condition of the Labouring Population of Great Britain.* London: Poor Law Commission Office.

Chan, Marie, Daniel Estève, Christophe Escriba and Eric Campo. 2008. 'A review of smart homes–present state and future challenges'. *Computer Methods and Programs in Biomedicine* 91 (1): 55–81.

Chatterton, Paul. 2010. 'The urban impossible: a eulogy for the unfinished city'. *City* 14 (3): 234–244.

Chauncey, George. 1994. *Gay New York: Gender, Urban Culture and the Making of the Gay Male World, 1890–1940.* New York: Basic.

Childe, V.G. 1951. 'The urban revolution'. *The Town Planning Review* 21 (1): 3–17.

Chomley, H.B. 1944. 'Reminiscences of an early colonist'. *British Australian and New Zealander*, 22 January.

Christesen, Paul and Donald G. Kyle. 2014. *A Companion to Sport and Spectacle in Greek and Roman Antiquity*. Malden and Oxford: Wiley-Blackwell.

Clark, Michael E. and Richard Hirschman. 1990. 'Effects of paced respiration on anxiety reduction in a clinical population'. *Applied Psychophysiology and Biofeedback* 15 (3): 273–284.

Clarke, D.L. 1977. *Spatial Archaeology*. London: Academic Press.

Classen, C. 1993. *Worlds of Sense: Exploring the Senses in History and Across Cultures*. London: Routledge.

Classen, C. 1997. 'Foundations for an anthropology of the senses'. *International Social Science Journal* 153: 401–412.

Coleman, Kathleen. 2000. 'Entertaining Rome'. In *Ancient Rome: The Archaeology of the Eternal City*, edited by Jon Coulston and Hazel Dodge. Oxford: Oxford University School of Archaeology Monograph 54, 210–258.

Collingwood History Committee. 1979. *In Those Days: Collingwood Remembered*. Melbourne: Carringbush Regional Library.

Collins, Randall. 1999. 'Socially unrecognized cumulation'. *The American Sociologist* 30 (2): 41–61.

Collins, Randall. 2004. *Interaction Ritual Chains*. Princeton: Princeton University Press.

Collins, Robert. 2012. *Hadrian's Wall and the End of Empire: The Roman Frontier in the 4th and 5th Centuries*. London: Routledge.

Connerton, Paul. 1989. *How Societies Remember*. Cambridge: Cambridge University Press.

Conquest, G. 2013. 'Dodging silver bullets: opportunities and challenges for an "Extreme Citizen Science" approach to forest management in the Republic of the Congo'. Unpublished MA dissertation. London: University College London.

Cook, Peter. 1970. *Experimental Architecture*. New York: Universe Books.

Cooke, B. and U. Kothari. 2001. 'The case for participation as tyranny'. In *Participation the New Tyranny*, edited by B. Cooke and U. Kothari. London: Zed Books, 1–15.

Cooley, A. and M. Cooley. 2004. *Pompeii: A Sourcebook*. London: Routledge.

Cope, Meghan and Sarah Elwood, eds. 2009. *Qualitative GIS: A Mixed Method Approach*. London: SAGE.

Corbett, Jon, Giacomo Rambaldi, Peter Kyem, Dan Weiner, Rachel Olson, Julius Muchemi, Mike McCall and Robert Chambers. 2006. 'Overview: mapping for change – the emergence of a new practice'. *Participatory Learning and Action* 54: 13–20.

Corfield, Penelope J. 2007. *Time and the Shape of History*. New Haven and London: Yale University Press.

Cornish, Henry. 1975 [1880]. *Under the Southern Cross*. Melbourne: Penguin.

Cosgrove, Denis. 1984. *Social Formation and the Symbolic Landscape*. Madison: Croom Helm.

Cosgrove, Denis. 2005. 'Mapping/cartography'. In *Cultural Geography: A Critical Dictionary of Key Concepts*, edited by David Atkinson, Peter Jackson, David Sibley and Neil Washbourne. London and New York: Tauris, 27–33.

Couchman, Harriet. 1910. *Reminiscences of Tottenham*. Tottenham: Crusha and Son.

Couchman, Sophie. c. 2009. 'Miss Suey Land and Ron Wong Loy: the children of Little Bourke Street and the entertainment industry'. Accessed 20 November 2014. https://arrow.latrobe.edu.au/store/3/4/5/5/1/public/stories/suey_land.htm.

Cowman, Krista. 2001. 'The battle of the boulevards: class, gender and the purpose of public space in later Victorian Liverpool'. In *Identities in Space: Contested Terrains in the Western City since 1850*, edited by Simon Gunn and Robert J. Morris. Aldershot: Ashgate, 152–164.

Crawford, Michael. 1996. *Roman Statutes*. London: Institute of Classical Studies.

Cresswell, Tim and Peter Merriman, eds. 2011. *Geographies of Mobilities: Practices, Spaces, Subjects*. Farnham: Ashgate.

Crosby, George. 1849. *Crosby's Parliamentary Record of Elections in Great Britain and Ireland.* Leeds: George Crosby.

Crouch, David. 1998. 'The street in the making of popular geographical knowledge'. In *Images of the Street: Planning, Identity and Control in Public Space,* edited by Nicholas R. Fyfe. Abingdon and New York: Routledge, 160–175.

Cumberpatch C.G. and P.W. Blinkhorn. 1997. *Not so Much a Pot, More a Way of Life.* Oxford: Oxbow.

Cunningham, Gail. 2000. 'The riddle of suburbia: suburban fictions at the Victorian fin de siècle'. In *Expanding Suburbia: Reviewing Suburban Narratives,* edited by Roger Webster. Oxford: Berghahn Books, 51–70.

Danish Road Directorate. 2011. *Byernes trafikarealer: Hæfte 9. Anlæg for parkering og standsning i byer.* Copenhagen: Vejdirektoratet.

Danish Road Directorate. 2012. *Håndbog i trafikplanlægning i byer.* Copenhagen: Vejdirektoratet.

Davies, Andrew. 2009. *The Gangs of Manchester: The Story of the Scuttlers, Britain's First Youth Cult.* Preston: Milo.

Davis, Martin. 2000. *The Universal Computer: The Road from Leibniz to Turing.* New York and London: W.W. Norton.

Day, J. 2013. *Making Senses of the Past: Towards a Sensory Archaeology.* Illinois: Southern Illinois University Press.

Deegan, Mary J. 2001. 'The Chicago school of ethnography'. In *The Handbook of Ethnography,* edited by Paul Atkinson, Amanda Coffey, Sara Delamont, John Lofland and Lyn Lofland. London: SAGE, 11–25.

Deleuze, Gillles. 1992. 'Post-script on the societies of control'. *October* 59: 3–7.

Dennis, Kingsley and John Urry. 2009. *After the Car.* Cambridge, MA: Polity.

Dennis, Richard. 1984. *English Industrial Cities of the Nineteenth Century: A Social Geography.* Cambridge: Cambridge University Press.

Dennis, Richard. 2008. *Cities in Modernity: Representations and Productions of Metropolitan Space, 1840–1930.* Cambridge: Cambridge University Press.

Department of Transport, Planning and Local Infrastructure, State Government of Victoria, Australia. 2014. 'Victorian State Planning Policy Framework', amended 30 May 2014. Accessed 25 November 2014. http://planningschemes.dpcd.vic.gov.au/schemes/vpps/11_SPPF.pdf.

Descartes, Rene. 1985 [1644]. 'Principles of philosophy'. In *The Philosophical Writings of Descartes,* vol. 1, edited by John Cottingham, Robert Stoothoff and Dugald Murdoch. Cambridge: Cambridge University Press, 177–291.

De Souza e Silva, Adriana. 2006. 'From cyber to hybrid: mobile technologies as interfaces of hybrid spaces'. *Space & Culture* 9 (3): 261–278.

de Vries, S. 2010. 'Nearby nature and human health: looking at mechanisms and their implications'. In *Innovative Approaches to Researching Landscape and Health; Open Space: People Space 2,* edited by C. Ward Thompson, P. Aspinall and S. Bell. London: Routledge, 77–96.

Dobbins, J. and P. Foss, eds. 2007. *The World of Pompeii.* London: Routledge.

Dodge, M., R. Kitchen and C. Perkins, eds. 2011. *The Map Reader: Theories of Mapping Practice and Cartographic Representation.* New York: Wiley-Blackwell.

Douglas, Mary. 1972. 'Symbolic orders in the use of domestic space'. In *Man, Settlement and Urbanism,* edited by P.J. Ucko, R. Tringham and G.W. Dimbleby. London: Duckworth, 513–521.

Driessen, Mark. 2005. 'Unifying aspects of Roman forts'. In *TRAC 2004: Proceedings of the Fourteenth Annual Theoretical Roman Archaeology Conference, the University of Durham 26–27 March 2004,* edited by J. Croxford Bruhn, B. and D. Grigoropoulos. Oxford: Oxbow Books, 157–162.

DRMM. 2009. 'Sliding House'. Accessed 3 March 2015. http://drmm.co.uk/projects/view.php?p=sliding-house.

Drobnik, J., ed. 2006. *The Smell Culture Reader*. Oxford: Berg.

Duffy, Francis. 1990. 'Measuring building performance'. *Facilities* 8 (5): 17–20.

Durkheim, Émile. 1899. 'Friedrich Ratzel, Anthropogéographie'. *l'Année sociologique*, 3: 550–558.

Durkheim, Émile. 1909. 'Sociologie et sciences sociales'. In *De la méthode dans les sciences*. Paris: Félix Alcan, 259–285.

Durkheim, Émile. 1960. 'Sociology and its scientific field'. In *Emile Durkheim, 1858–1917: A Collection of Essays*, edited by K.H. Wolff. Columbus: The Ohio State University Press, 354–375.

Durkheim, Émile. 1978a. 'Sociology and the social sciences'. In *Emile Durkheim on Institutional Analysis*, edited by M. Traugott. Chicago: University of Chicago Press, 71–87.

Durkheim, Émile. 1978b. 'Antonio Labriola, essays on the materialist conception of history (review)'. In *Emile Durkheim on Institutional Analysis*, edited by R. Traugott. Chicago: University of Chicago Press, 131–138.

Durkheim, Émile. 1982. *Rules of Sociological Method*. New York: The Free Press.

Durkheim, Émile. 1984. *The Division of Labor in Society*. New York: The Free Press.

Durkheim, Émile. 1994. 'Note on social morphology'. In *On Institutional Analysis*, edited by M. Traugott. Chicago: University of Chicago Press, 88–90.

Durkheim, Émile. 1995. *The Elementary Forms of Religious Life*. New York: Free Press.

Durkheim, Émile. 2005. *Suicide: A Study in Sociology*. London and New York: Routledge.

Durkheim, Émile and Marcel Mauss. 2009. *Primitive Classification*. London: Taylor & Francis e-Library.

Durrschmidt, J. 2000. *Everyday Lives in the Global City: The Delinking of Locale and Milieu*. London: Routledge.

Dyos, Harold J. 1966. *Victorian Suburb: A Study of the Growth of Camberwell*. Leicester: Leicester University Press.

Dyrssen, Catharina. 2011. 'Navigating in heterogeneity: architectural thinking and art-based research'. In *The Routledge Companion to Research in the Arts*, edited by Michael Biggs and Henrik Karlsson. London: Routledge, 223–239.

Edensor, Tim. 2000. 'Walking in the British countryside: reflexivity, embodied practices and ways to escape'. *Body & Society* 6 (3/4): 81–106.

Edwards, E. and K. Bhaumik, eds. 2008. *Visual Sense: A Cultural Reader*. Oxford: Berg.

Elden, Stuart. 2001. *Mapping the Present: Heidegger, Foucault and the Project of a Spatial History*. London: Continuum.

Elliott, Anthony and John Urry. 2010. *Mobile Lives*. London: Routledge.

Ellis, C., T.E. Adams and A.P. Bochner. 2010. 'Autoethnography: an overview'. *Forum Qualitative Sozialforschung / Forum: Qualitative Social Research* 12 (1). Accessed 24 February 2015. www.qualitative-research.net/index.php/fqs/article/view/1589/3095.

Emirbayer, Mustafa. 1996. 'Useful Durkheim'. *Sociological Theory* 14: 109–130.

Entwistle, Dorothy. 2012. 'The Whit walks of Hyde: glorious spectacle, religious witness, and celebration of a custom'. *Journal of Religious History* 36: 204–233.

Epstein, James. 2003. *In Practice: Studies in the Language and Culture of Popular Politics in Modern Britain*. Redwood City: Stanford University Press.

Eriksen, Jørgen W. 2010. 'Should soldiers think before they shoot?' *Journal of Military Ethics* 9: 195–218.

Espeland, W.N. and M.L. Stevens. 1998. 'Commensuration as a social process'. *Annual Review of Sociology* 24: 313–343.

Ethington, Phillip J. 2000. 'Los Angeles and the problem of historical knowledge: a multimedia essay to accompany the December issue of *The American Historical Review*'. Accessed 19 December 2014. www.usc.edu/dept/LAS/history/historylab/LAPUHK/.

Ethington, Phillip J. 2007. 'Placing the past: "groundwork" for a spatial theory of history'. *Rethinking History* 11 (4): 465–493.

Everdell, William R. 1997. *The First Moderns: Profiles in the Origins of Twentieth-Century Thought.* Chicago: University of Chicago Press.

Farinelli, Franco. 2013. 'Im Anfang war die Karte'. In *Die Zukunft der Kartographie. Neue und nicht so neue epistemologische Krisen,* edited by Marion Picker, Véronique Maleval and Florent Gabaude. Bielefeld: transcript, 257–277.

Fatah gen Schieck, A., Holger Schnädelbach, Wallis Motta, Moritz Behrens, Steve North, Lei Ye and Efstathia Kostopoulou. 2014. 'Screens in the wild: exploring the potential of networked urban screens for communities and culture'. In *Proceedings of the International Symposium on Pervasive Displays.* Copenhagen: ACM, 166–167.

Favro, Diane. 1996. *The Urban Image of Augustan Rome.* Cambridge: Cambridge University Press.

Favro, Diane. 2006. 'In the eyes of the beholder: virtual reality re-creations and academia'. In *Imaging Ancient Rome: Documentation – Visualisation – Imagination,* edited by L. Haselberger and J. Humphrey. Portsmouth, RI: Journal of Roman Archaeology, L.L.C, 321–334.

Feld, S. and K. Baso. 1996. *Sense of Place.* Santa Fe: School of American Research Press.

Fellman, Rudolf. 1958. *Die Principia des Legionslagers Vindonissa und das Zentralgebäute der römischen Lager und Kastelle.* Brugg: Vindonissa Museum.

Ferguson, J. 1994. *The Anti-Politics Machine: 'Development,' Depoliticization, and Bureaucratic Power in Lesotho.* London: University of Minnesota Press.

Fink, Robert O. 1971. *Roman Military Records on Papyrus.* Cleveland: Press of Case Western Reserve University.

Fisher, N. and H. Van Wees, eds. 2011. *Competition in the Ancient World.* Wales: Classical Press of Wales.

Fisk, Frederic. 1913. *The History of Tottenham, in the County of Middlesex.* Tottenham: Fred Fisk.

Fletcher, Harry and A. F. Maxwell. 1989. *Air Force Bases Volume II, Active Air Force Bases outside the United States of America on 17 September 1982.* Alabama: Office of Air Force History.

Florida, Richard. 2011. 'The paradox of unlimited choice: place matters more than you think in e-commerce'. *Martin Prosperity Insights,* 22 February 2011. Accessed 28 April 2014, http://martinprosperity.org/2011/02/22/the-paradox-of-unlimited-choice-place-matters-more-than-you-think-in-e-commerce/.

Flower, H. 1999. *Ancestor Masks and Aristocratic Power in Roman Culture.* Oxford: Oxford University Press.

Forman Cody, Lisa. 2003. 'Every lane teems with instruction, and every alley is big with erudition'. In *The Streets of London: From the Great Fire to the Great Exhibition,* edited by Tim Hitchcock and Heather Shore. London: Rivers Oram Press, 82–100.

Foster, John. 1977. *Class Struggle and the Industrial Revolution: Early Industrial Capitalism in Three Towns.* Abingdon: Routledge.

Foucault, Michel. 1972. *The Archaeology of Knowledge.* London: Tavistock.

Foucault, Michel. 1986. 'Of other spaces'. *Diacritics* 16: 22–27.

Foucault, Michel. 1995. *Discipline and Punish: The Birth of the Prison,* 2nd edn. New York: Vintage Books.

Fournier, Marcel. 2013. *Emile Durkheim: A Biography.* London: Polity Press.

Fox, Michael Aug and Miles Kemp. 2009. *Interactive Architecture.* New York: Princeton Architectural Press.

Frederick, Dominic. 2003. 'Architecture and surveillance in Flavian Rome'. In *Flavian Rome: Culture, Image, Text,* edited by A.J. Boyle and W.J. Dominik. Leiden and Boston: Brill, 199–227.

Freedman, Jonathan L. 1975. *Crowding and Behavior.* Oxford: W.H. Freedman.

French, K.L. 2001. *The People of the Parish: Community Life in a Late Medieval English Diocese.* Philadelphia: University of Pennsylvania Press.

Frere, Sheppard and Roger Tomlin. 1992. *The Roman Inscriptions of Britain / Vol.2, Instrumentum domesticum (Personal belongings and the like). Fascicule 4, Wooden barrels, stilus-tablets, objects of wood, leather, oculists' stamps, wallplaster, mosaics, handmills, stone tablets, stone balls, stone pebbles, small stone votives, miscellaneous objects of stone, jet, clay figurines, clay objects, antefixes, tile-stamps of Legion II Augusta, of Legion VI Victrix, of Legion IX Hispana, of Legion XX Valeria Victrix, tile-stamps of the auxiliaries (RIB 2442–2480).* Gloucester: Published for the Administrators of the Haverfield, Bequest by Alan Sutton.

Frere, Sheppard and Roger Tomlin. 1993. *The Roman Inscriptions of Britain / Vol.2, Instrumentum domesticum (Personal belongings and the like). Fascicule 6, Dipinti and graffiti on amphorae, dipinti and graffiti on mortuaria, inscriptions in white barbotine, dipinti on coarse pottery, Samian barbotine or moulded inscriptions (RIB 2492–2500).* Gloucester: Published for the Administrators of the Haverfield, Bequest by Alan Sutton.

Frere, Sheppard and Roger Tomlin. 1995. *The Roman Inscriptions of Britain / Vol.2, Instrumentum domesticum (personal belongings and the like). Fascicule 7, Graffiti on samian ware (terra sigillata) (RIB 2501).* Stroud: Published for the Administrators of the Haverfield, Bequest by Alan Sutton Publishing.

Frézouls, Edmond. 1987. 'Rome Ville Ouverte'. In *L'Urbs Espace Urbain et Histoire.* Rome: L'École Française de Rome, 373–392.

Frontius, Sextus Julius. 1950 [1925]. *Stratagems. Aqueducts of Rome.* Reprint. Translated by C.E. Bennett. London and New York: Loeb Classical Library.

Fuchs, T. 2013. 'The phenomenology of affectivity'. In *The Oxford Handbook of Philosophy and Psychiatry,* edited by K.W.M, Fulham, M. Davies, G. Graham, J.Z. Sadler, G. Stanghellini, R.G.T. Gipps and T. Thornton. Oxford: Oxford University Press, 612–631.

Furlong, Kathryn. 2011. 'Small technologies, big change: rethinking infrastructure through STS and geography'. *Progress in Human Geography* 35: 460–482.

Gane, Douglas. 1886. *New South Wales and Victoria in 1885.* London: Sampson Low, Marston, Searle and Rivington.

Gardner, Andrew. 2007. *An Archaeology of Identity: Soldiers and Society in Late Roman Britain.* Walnut Creek: Left Coast Press.

Garfinkel, Harold. 1963. 'A conception of, and experiments with, "trust" as a social condition of stable concerted actions'. In *Motivation and Social Interaction,* edited by O.J. Harvey. New York: Ronald Press, 187–238.

Garrioch, D. 1986. *Neighbourhood and Community in Paris, 1740–1790.* Cambridge: Cambridge University Press.

Gaukroger, Stephen. 1992. 'The nature of abstract reasoning: philosophical aspects of Descartes' work in algebra'. In *The Cambridge Companion to Descartes,* edited by John Cottingham. Cambridge: Cambridge University Press, 91–114.

Gazda, E. 1991. *Roman Art in the Private Sphere.* Michigan: Michigan University Press.

Gehl, Jan. 1987. *Life Between Buildings: Using Public Space.* Translated by Jo Koch. New York: Van Nostrand Reinhold.

Gell, Alfred. 1998. *Art and Agency: An Anthropological Theory.* Oxford: Oxford University Press.

General Register Office. 1872. *Census of England and Wales, 1871. Population Tables. Area, Houses, and Inhabitants,* vol. II, *Registration or Unions Counties.* London: Her Majesty's Stationery Office.

General Register Office. 1883. *Census of England and Wales. 1881,* vol. I, *Area, Houses, and Population. Counties.* London: Her Majesty's Stationery Office.

General Register Office. 1912. *Census of England and Wales. 1911. Area, Families or Separate Occupiers, and Population,* vol. I, *Administrative Areas. Counties, Urban and Rural Districts, & c.* London: His Majesty's Stationery Office.

General Register Office. 1922. *Census of England & Wales 1921. County of London. Tables (Part I).* London: His Majesty's Stationery Office.

General Register Office. 1925. *Census of England & Wales 1921. Workplaces.* London: His Majesty's Stationery Office.

George, M., ed. 1999. *Constructions of Space IV: Further Developments in Examining Social Space in Ancient Israel.* New York: LHBOTS 569.

George, M. 2002. 'Review of *Reading Space: Social Interaction and Identity in the Houses of Roman Pompeii*, by Grahame, M.'. *Journal of Roman Studies* 92: 238–239.

Georgiou, Dion. 2014. '"The drab suburban streets were metamorphosed into a veritable fairy-land": spectacle and festivity in the Ilford hospital carnival, 1905–1914'. *London Journal* 39 (3): 227–248.

Georgiou, Myria. 2013. *Media and the City*. London: Polity.

Gering, Axel. 2004. 'Plätze und Straßensperren an Promenaden. Zum Funktionswandel Ostias in der Spätantike'. *Mitteilungen des Deutschen Archäologischen Instituts, Römische Abteilung* 111: 299–382.

Gibson, William. 2010. 'Quote of the day'. *The Brooks Review*, 29 November 2010. Accessed 24 July 2012. https://brooksreview.net/2010/11/gibson-qotd/.

Giddens, Anthony. 1984. *The Constitution of Society: Outline of the Theory of Structuration*. Cambridge: Polity.

Gieryn, Thomas. 2000. 'A space for place in sociology'. *Annual Review of Sociology* 26: 463–496.

Gieryn, Thomas F. 2002. 'What buildings do'. *Theory and Society* 31 (1): 35–74.

Gilbert, Pamela K. 2004. *Mapping the Victorian Social Body*. Albany: State University of New York Press.

Giles, K. 2000. *An Archaeology of Social Identity: Guildhalls in York, c. 1350–1630*. British Archaeological Reports 315. Oxford: John and Erica Hedges Ltd and Archaeopress.

Gill, Tim. 2007. *No Fear: Growing Up in a Risk Averse Society*. London: Calouste Gulbenkian Foundation.

Gilliam, J. Frank. 1954. 'The Roman military feriale'. *The Harvard Theological Review* 47: 183–196.

Gillis, John R., 2008. 'Epilogue: the islanding of children – reshaping the mythical landscapes of childhood'. In *Designing Modern Childhoods: History, Space, and the Material Culture of Children*, edited by Marta Gutman and Ning de Coninck-Smith. New Brunswick and London: Rutgers University Press, 316–330.

Ginzburg, Carlo. 1976. *The Cheese and the Worms: The Cosmos of a Sixteenth-Century Miller*. London: Routledge.

Glynn, Ruariri. 2005. 'Reciprocal space'. Accessed 3 March 2015. www.futurefeeder.com/2005/06/reciprocal-space/.

Goffman, Erving. 1967. *Interaction Ritual: Essays on Face-to-Face Behavior*. New York: Pantheon Books.

Goffman, Erving. 2010 [1959]. *The Presentation of Self in Everyday Life*. New York: Doubleday.

Goheen, Peter G. 1990. 'Symbols in the streets: parades in Victorian urban Canada'. *Urban History Review* 18 (3): 237–243.

Goheen, Peter G. 1993. 'The ritual of the streets in mid-19th century Toronto'. *Environment and Planning D: Society and Space* 11: 127–145.

Goldbeck, F. 2010. *Salutationes: die Morgenbegrüßungen in Rom in der Republik und der frühen Kaiserzeit*. Berlin: Klio.

Goldberg, P.J.P. 1996. *Women, Work, and Life Cycle in a Medieval Economy*. Oxford: Clarendon Press.

Goldberg, P.J.P. 2008. 'The fashioning of bourgeois domesticity in later medieval England: a material culture perspective'. In *Domesticity: Home, Housing and Household in Medieval England*, edited by P.J.P Goldberg and M. Kowaleski. Cambridge: Cambridge University Press, 66–91.

Goldstein, Brett and Lauren Dyson. 2013. *Beyond Transparency: Open Data and the Future of Civic Innovation*. San Francisco: Code for America Press.

Goldsworthy, Adrian Keith, Ian Haynes and C.E.P. Adams, eds. 1999. *The Roman Army as a Community: Including Papers of a Conference Held at Birkbeck College, University of London on 11–12 January, 1997*. Portsmouth, RI: Journal of Roman Archaeology, L.L.C.

Goodson, C., A.E. Lester and C. Symes, eds. 2010. *Cities, Texts and Social Networks, 400–1500*. Farnham: Ashgate.

Gosden, C. 2004. 'The past and foreign countries: colonial and post-colonial archaeology and anthropology'. In *A Companion to Social Archaeology*, edited by R.W. Preucel and L. Meskell. Oxford: Blackwell, 161–178.

Grady, Kevin. 1989. *Georgian Public Buildings of Leeds and the West Riding.* Leeds: Thoresby Society.

Graham, Stephen and Simon Marvin. 2001. *Splintering Urbanism: Networked Infrastructures, Technological Mobilities and the Urban Condition.* London: Routledge.

Grahame, M. 1993. 'Reading the Roman house: the social interpretation of spatial order'. In *TRAC 1993,* edited by A. Leslie. Oxford: Oxbow, 48–74.

Grahame, M. 1997. 'Public and private in the Roman house: the spatial order of the Casa del Fauno'. In *Domestic Space in the Roman World: Pompeii and Beyond,* edited by R. Laurence and A. Wallace-Hadrill. Portsmouth, RI: Journal of Roman Archaeology, L.L.C., 137–164.

Grahame, M. 1998. 'Material culture and Roman identity: the spatial layout of Pompeian houses and the problem of ethnicity'. In *Cultural Identity in the Roman Empire,* edited by J. Berry and R. Laurence. London: Routledge, 156–178.

Grahame, M. 2000. *Reading Space: Social Interaction and Identity in the Houses of Roman Pompeii (A Syntactical Approach to the Analysis and Interpretation of Built Space).* Oxford: Oxbow.

Graves, P. 1989. 'Social space in the English medieval parish church'. *Economy and Society* 18: 297–322.

Green, Adrian. 2003. 'Houses in north-eastern England: regionality and the British beyond, c.1600–1750'. In *Archaeologies of the British: Explorations of Identity in Great Britain and its Colonies, 1600–1945,* edited by Susan Lawrence. London: Routledge, 55–75.

Greenfield, P.M. 1984. 'A theory of the teacher in the learning activities of everyday life'. In *Everyday Cognition: Its Development in Social Context,* edited by B. Rogoff and J. Lave. Cambridge, MA: Harvard University Press.

Greenstein, D. 1997. 'Bringing bacon home: the divergent progress of computer-aided historical research in Europe and the United States'. *Computers and the Humanities* 30: 351–364.

Grenfeld, Bernard Pyne and Arthur Hunt. 1916. *The Oxyrhynchus Papyri Part XII.* London: Egypt Exploration Fund.

Grenville, J. 1996. *Medieval Housing.* London: Leicester University Press.

Grenville, J. 2004. 'The archaeology of the late and post-medieval workshop – a review and proposal for a research agenda'. In *The Vernacular Workshop: From Craft to Industry, 1400–1900,* edited by P. Barnwell, M. Palmer and M. Airs York. Council for British Archaeology, 28–37.

Griffin, Carl J. 2010. 'The violent Captain Swing'. *Past and Present* 209: 149–180.

Griffin, Carl J. and Adrian B. Evans. 2008. 'On historical geographies of embodied practice and performance'. *Historical Geography* 36: 5–16.

Griffiths, Sam. 2008. 'Historical space and the interpretation of urban transformation: the spatiality of social and cultural change in Sheffield c1770–1910'. PhD diss., University of London.

Griffiths, Sam. 2009. 'Persistence and change in the spatio-temporal description of Sheffield parish 1770–1910'. In *Proceedings of the 7th International Space Syntax Symposium,* edited by Daniel Koch, Lars Marcus and Jesper Steen. Stockholm: KTH Royal Institute of Technology, 1–15.

Griffiths, Sam. 2011. 'Temporality in Hillier and Hanson's theory of spatial description: some implications of historical research for space syntax'. *Journal of Space Syntax* 1 (2): 73–96.

Griffiths, Sam. 2012. 'The use of space syntax in historical research: current practice and future possibilities'. In *Proceedings of the Eighth International Space Syntax Symposium,* edited by Margarita Greene, José Reyes and Andrea Castro. Santiago, Chile: PUC, 1–26.

Griffiths, Sam. 2013. 'GIS and research into historical "spaces of practice": overcoming the epistemological barriers'. In *History and GIS: Epistemologies, Considerations and Reflections,* edited by Alexander von Lünen and Charles Travis. Dordrecht: Springer, 153–172.

Griffiths, Sam. 2015a forthcoming. 'Manufacturing innovation as urban spatial practice: Sheffield's cutlery and metals industries c.1750–1900'. In *Unscrewing the Creative City: The Historical Fabrication of Cities as Agents of Economic Innovation and Creativity,* edited by Ilya Van Damme, Bruno Blondé and Andrew Miles. London: Routledge.

Griffiths, Sam. 2015b forthcoming. 'The high street as a morphological event'. In *Surburban Urbanities: Suburbs and the Life of the High Street.* London: UCL Press.

Griffiths, Sam and Tom Quick. 2005. 'How the individual, society and space become structurally coupled over time'. In *Proceedings of the 5th Space Syntax Symposium, June 2005*, vol. 2, edited by Akkelie van Nes. Amsterdam: Techne Press, 447–458.

Grindr. 2015. 'Learn more'. Accessed 1 January 2015. http://grindr.com/learn-more.

Groenman van Vateeringe, Willy. 1986. 'The Horrea of Valkenburg ZH'. In *Studien zu dem Militärgrenzen Roms III. 13. Internationaler Limeskongress Aalen 1983 Vorträge.* Stuttgart: Konrad Theiss Verlag, 159–168.

Gros, Pierre. 2005. 'Le role du people de Rome dans la definition, l'organisation et le déplacement des lieux de la convergence sous l'empire'. In *Popolo e Potere nel Mondo Antico,* edited by Gianpaolo Urso. Pisa: Edizioni ETS, 191–214.

Gunn, Simon. 2000a. *The Public Culture of the Victorian Middle Class: Ritual and Authority and the English Industrial City, 1840–1914.* Manchester and New York: Manchester University Press.

Gunn, Simon. 2000b. 'Ritual and civic culture in the English industrial city, c.1835–1914'. In *Urban Governance in Britain and Beyond Since 1750,* edited by Robert Morris and Richard Trainor. Aldershot and Burlington: Ashgate, 227–241.

Gunn, Simon. 2001. 'The spatial turn: changing histories of space and place'. In *Identities in Space: Contested Terrain in the Western City Since 1850,* edited by Simon Gunn and R.J. Morris. Farnham: Ashgate, 1–14.

Gunn, Simon and R.J. Morris. 2001. *Identities in Space: Contested Terrain in the Western City Since 1850.* Farnham: Ashgate.

Gutman, Marta and Ning de Coninck-Smith, eds. 2008. *Designing Modern Childhoods: History, Space, and the Material Culture of Children.* New Brunswick and London: Rutgers University Press.

Habermas, J. 1990. *Discourse Ethics: Notes on a Program of Philosophical Justification. Moral Consciousness and Communicative Action.* Translated by C. Lenhart and S.W. Nicholson. Cambridge, MA: MIT Press.

Hägerstrand, Torsten. 1970. 'What about people in regional science?' *Papers of the Regional Science Association* 24 (1): 6–21.

Haggerty, K.D. and Ericson, R.V. 2007. 'The surveillant assemblage'. *British Journal of Sociology* 51 (4): 605–622.

Hagson, Anders. 2000. *Stads- och trafikplaneringens paradigm.* Göteborg: Chalmers Tekniska Högskola.

Hajer, Maarten and Arnold Reijndorp. 2001. *In Search of New Public Domain.* Rotterdam: NAI Publishers.

Haken, Hermann and Juval Portugali. 2003. 'The face of the city is its information'. *Journal of Environmental Psychology* 23 (4): 385–408.

Haklay, M. 2013. 'Citizen science and volunteered geographic information – overview and typology of participation'. In *Crowdsourcing Geographic Knowledge: Volunteered Geographic Information (VGI) in Theory and Practice,* edited by D. Sui, S. Elwood and M. Goodchild. Berlin: Springer, 105–122.

Haklay, M., A. Singleton and C. Parker. 2008. 'Web mapping 2.0: the neogeography of the GeoWeb. Geography'. *Compass* 2 (6): 2011–2039.

Halbwachs, Maurice. 1960. *Population and Society: Introduction to Social Morphology.* Glencoe: Free Press.

Hale, Jonathan. 2012. 'Architecture, technology and the body: from the prehuman to the posthuman'. In *The SAGE Handbook of Architectural Theory,* edited by S. Cairns, C.G. Crysle and H. Heynen. London: SAGE.

Hall, R.A. and K. Hunter-Mann. 2002. *Medieval Urbanism in Coppergate: Refining a Townscape.* York: Council for British Archaeology.

Hall, Stuart. 2001. 'Foucault: power, knowledge and discourse'. In *Discourse Theory and Practice: A Reader*, edited by Margaret Wetherell, Stephanie Taylor and Simeon J. Yates. London: SAGE, 72–81.
Halvorson, M.J. and K.E. Spierling. 2008. *Defining Community in Early Modern Europe*. Aldershot: Ashgate.
Hamilakis, Y. 2013a. *Archaeology and the Senses: Human Experience, Memory and Affect*. Cambridge: Cambridge University Press.
Hamilakis, Y. 2013b. 'Afterword: eleven theses on the archaeology of the senses'. In *Making Senses of the Past: Towards a Sensory Archaeology*, edited by J. Day. Illinois: Southern Illinois University Press, 409–419.
Hamilakis, Y., M. Pluciennik and S. Tarlow, eds. 2002. *Thinking Through the Body: Archaeologies of Corporeality*. New York: Kluwer.
Hamilton, S. and R. Whitehouse. 2006a. 'Three senses of dwelling: beginning to socialise the neolithic ditched villages of the Tavoliere, southeast Italy'. *Journal of Iberian Archaeology* 8: 159–184.
Hamilton, S. and R. Whitehouse (with K. Brown, P. Combes, E. Herring and M. Seager Thomas). 2006b. 'Phenomenology in practice: towards a methodology of a "subjective" approach'. *European Journal of Archaeology* 9 (I): 31–71.
Hanawalt, B. and M. Kobialka. 2000. *Medieval Practices of Space*. Minnesota: University of Minnesota Press.
Hanson, J. 2003 [1998]. *Decoding Homes and Houses*. Cambridge: Cambridge University Press.
Hanson, William S. 2007. *Elginhaugh: A Flavian Fort and Its Annexe*, vol. 1. Oxford: Oxbow.
Hapgood, Lynne. 2005. *Margins of Desire: The Suburbs in Fiction and Culture, 1885–1925*. Manchester and New York: Manchester University Press.
Haraway, Donna. 1991. 'A cyborg manifesto: science, technology, and socialist-feminism in the late twentieth century'. In *Simians, Cyborgs and Women: The Reinvention of Nature*. New York: Routledge, 149–181.
Harder, Henrik. 2002. *Vejplanlægning, politik og praksis*. PhD thesis. Aalborg: Aalborg University.
Harder, Henrik. 2003. 'A "network" and a "dictum"'. Paper presented at the AESOP/ACP Third Joint Conference, Leuven.
Harding, V. 2002. 'Space, property, and propriety in urban England'. *Journal of Interdisciplinary History* 32 (4): 549–569.
Harley, J. B. 1989. 'Deconstructing the map'. *Cartographica* 26 (2): 1–20.
Harman, Graham. 2010. *Towards Speculative Realism: Essays and Lectures*. Winchester: Zero Books.
Harper, Richard. 2003. *Inside the Smart Home*. London and New York: Springer.
Harrison, Mark. 1988. *Crowds and History: Mass Phenomena in English Towns 1790–1835*. Cambridge: Cambridge University Press.
Hart, Roger. 1979. *Children's Experience of Place*. New York: Irvington.
Hartnett, J. 2008. '*Si quis hic sederit*: streetside benches and urban society in Pompeii'. *American Journal of Archaeology* 112: 91–119.
Hartnett, J. 2011. 'The power of nuisances on the Roman street'. In *Rome, Ostia, Pompeii: Movement and Space*, edited by R. Laurence and D. Newsome. Oxford: Oxford University Press, 135–159.
Hartshorne, P.B. 2004. *The Street and the Perception of Public Space in York, 1476–1586*. Unpublished PhD thesis. York: University of York.
Harvey, David. 2006. 'Space as a key word'. In *David Harvey: A Critical Reader*, edited by Noel Castree and Derek Gregory. Oxford: Wiley-Blackwell, 270–294.
Harvey, Francis, Mei-Po Kwan and Marianna Pavlovskaya. 2005. 'Introduction: critical GIS'. *Cartographica* 40 (4): 1–5.
Haselberger, Lothar. 2007. *Urbem Adornare*. Portsmouth, RI: Journal of Roman Archaeology, L.L.C.
Haselberger, L. and J. Humphrey, eds. 2006. *Imaging Ancient Rome: Documentation – Visualisation – Imagination*. Portsmouth, RI: Journal of Roman Archaeology, L.L.C.

Hayles, N. Katherine. 1999. *How We Became Posthuman.* Cambridge, MA: MIT Press.
Hayles, N. Katherine. 2005. *My Mother Was a Computer: Digital and Literary Texts.* Chicago and London: University of Chicago Press.
Haynes, Ian. 2013. *Blood of the Provinces: The Roman Auxilia and the Making of Provincial Society from Augustus to the Severans.* Oxford: University of Oxford Press.
Hebbert, Michael. 2005. 'The street as locus of collective memory'. *Environment and Planning D: Society and Space* 23 (4): 581–596.
Heidegger, Martin. 1962. *Being and Time.* New York: Harper & Row.
Heidegger, Martin. 1977. *The Question Concerning Technology and Other Essays.* Edited by William Lovitt. New York: Harper Torchbooks.
Heidegger, Martin. 1992 [1924]. *The Concept of Time.* Translated by William McNeill. Oxford: Blackwell.
Heidegger, Martin. 1996. *Being and Time: A Translation of Sein und Zeit.* Translated by Joan Stambaugh. Albany: State University of New York Press.
Heidegger, Martin. 2000. *Introduction to Metaphysics.* London: Yale University Press.
Heidegger, Martin. 2008. 'The end of philosophy and the task of thinking'. In *Basic Writings, from 'Being and Time' (1927) to 'The Task of Thinking' (1964),* edited by David Farrell. London and New York: Routledge, 311–325.
Heller, Michael. 2014. 'Suburbia, marketing and stakeholders: developing Ilford, Essex, 1880–1914'. *Urban History* 41 (1): 62–80.
Hellström Reimer, M. 2012. 'Christiania Copenhagen: a common out of the ordinary'. In *Urban Wildscapes,* edited by A. Jorgensen and R. Keenan. London and New York: Routledge, 120–130.
Henkel, H. and R. Stirrat. 2001. 'Participation as spiritual duty; empowerment as secular subjection'. In *Participation the New Tyranny,* edited by B. Cooke and U. Kothari. London: Zed Books, 168–184.
Hewitt, Lucy E. 2012. 'The civic survey of Greater London: social mapping, planners and urban space in the early twentieth century'. *Journal of Historical Geography* 38 (3): 247–262.
Hickey, S. and G. Mohan. 2004. 'Towards participation as transformation: critical themes and challenges'. In *Participation: From Tyranny to Transformation? Exploring New Approaches to Participation in Development,* edited by S. Hickey and G. Mohan. London: Zed Books, 159–175.
Hicks, D. and M.C. Beaudry, eds. 2006. *The Cambridge Companion to Historical Archaeology.* Cambridge: Cambridge University Press.
Hicks, D. and A. Horning. 2006. 'Historical archaeology and buildings'. In *The Cambridge Companion to Historical Archaeology,* edited by D. Hicks and M.C. Beaudry. Cambridge: Cambridge University Press, 273–293.
Hildyard, N. *et al.* 2001. 'Pluralism, participation and power: joint forest management in India'. In *Participation the New Tyranny,* edited by B. Cooke and U. Kothari. London: Zed Books, 55–71.
Hill, Albert. 1899. *Seventy-Two Years in Tottenham: Reminiscences.* London: Thos. De la Rue and Co.
Hill, Kate. 2001. '"Roughs of both sexes": the working class in Victorian museums and art galleries'. In *Identities in Space: Contested Terrain in the Western City Since 1850,* edited by Simon Gunn and R.J. Morris. Farnham: Ashgate, 190–203.
Hillier, Bill. 1989. 'The architecture of the urban object'. *Ekistics* 56 (334/33): 5–21.
Hillier, Bill. 1996. *Space is the Machine: A Configurational Theory of Architecture.* Cambridge: Cambridge University Press.
Hillier, Bill. 2003. 'The architectures of seeing and going: or, are cities shaped by bodies or minds? And is there a syntax of spatial cognition?' In *Proceedings to the 4th International Space Syntax Symposium,* edited by Julienne Hanson. London: Space Syntax Laboratory, 6.1–34.
Hillier, Bill. 2005. 'Between social physics and phenomenology: explorations towards an urban synthesis?' In *Proceedings of the 5th Space Syntax Symposium, June 2005,* vol. 1, edited by Akkelie van Nes. Amsterdam: Techne Press, 3–23.

Hillier, Bill. 2007. *Space is the Machine*. Accessed 1 November 2014. www.ninsight.at/ak_stdb/SpaceIsTheMachine.pdf.

Hillier, Bill. 2009. 'Spatial sustainability in cities: organic patterns and sustainable forms'. In *Proceedings of the 7th International Space Syntax Symposium*, edited by Daniel Koch, Lars Marcus and Jesper Steen. Stockholm: KTH Royal Institute of Technology, K01–20.

Hillier, Bill and Julienne Hanson. 1984. *The Social Logic of Space*. Cambridge: Cambridge University Press.

Hillier, Bill and Shinichi Iida. 2005. 'Network effects and psychological effects: a theory of urban movement'. In *Spatial Information Theory*, edited by A. Cohn and D. Mark, *Lecture Notes in Computer Science* 3603. Berlin, Heidelberg and New York: Springer, 473–490.

Hillier, Bill and Noah Raford. 2010. 'Description and discovery in socio-spatial analysis: the case of space syntax'. In *The SAGE Handbook of Measurement*, edited by G. Walford, E. Tucker and M. Viswanathan. London: SAGE, 265–281.

Hillier, Bill, and Laura Vaughan. 2007. 'The city as one thing'. *Progress in Planning* 67 (3): 205–230.

Hillier, Bill, Adrian Leaman, Paul Stansall and Michael Bedford. 1976. 'Space syntax'. *Environment and Planning B* 3: 147–185.

Hillier, Bill, Alan Penn, Julienne Hanson, T. Grajevski and J. Xu. 1993. 'Natural movement – or, configuration and attraction in urban pedestrian movement'. *Environment and Planning B* 20 (1): 29–66.

Hitchcock, Tim. 2004. *Down and Out in Eighteenth-Century London*. London: Hambledon.

Hobsbawm, Eric J. 1992. 'Introduction: inventing traditions'. In *The Invention of Tradition*, edited by Eric J. Hobsbawm and Terence Ranger. Cambridge: Canto, 1–15.

Hodder, I. 1977. *The Spatial Organisation of Culture*. London: Duckworth.

Hodder, I. 2004. 'The "social" in archaeological theory: an historical and contemporary perspective'. In *A Companion to Social Archaeology*, edited by R.W. Preucel and L. Meskell. Oxford: Blackwell, 23–42.

Hodgson, Nick. 2003. *The Roman Fort at Wallsend (Segedunum): Excavations in 1997–8*. Newcastle upon Tyne: Tyne & Wear Museums and the Arbeia Society.

Hodgson, Nick and Paul Bidwell. 2004. 'Auxiliary barracks in a new light: recent discoveries on Hadrian's Wall'. *Britannia* 35: 121–157.

Hoete, Anthony, ed. 2003. *ROAM: Reader on Aesthetics of Mobility*. London: Black Dog Publishing.

Hopkin, D. and P. Denley. 1991. 'History, historians and the new technology'. In *Scholarship and Technology in the Humanities*, edited by May Katzen. London: Bowker-Saur, 63–76.

Hopkins, Nick. 2012. 'Camp Bastion'. *Guardian*, 8 September: 7.

Hornsey, Richard. 2012. 'Listening to the Tube map: rhythm and the historiography of urban map use'. *Environment and Planning D: Society and Space* 30 (4): 675–693.

Houben, Francine and Luisa Maria Calabrese, eds. 2003. *Mobility: A Room with a View*. Rotterdam: NAi Publishers.

Houston, S. and K. Taube. 2000. 'An archaeology of the senses: perception and cultural expression in Ancient Mesoamerica'. *Cambridge Archaeological Journal* 10: 261–294.

Howe, Nicholas, ed. 2007. *Ceremonial Culture in Pre-Modern Europe*. Chicago: University of Notre Dame Press.

Howell, M. 2000. 'The spaces of late medieval urbanity'. In *Shaping Urban Identity in Late Medieval Europe*, edited by M. Boone and P. Stabel. Leuven: Garrant, 3–19.

Howes, D. 1991. *The Varieties of Sensory Experience: A Sourcebook in the Anthropology of the Senses*. Toronto: University of Toronto Press.

Howes, D. 2003. *Sensual Relations: Engaging the Senses in Culture and Social Theory*. Michigan: University of Michigan Press.

Howes, D, ed. 2005. *Empire of the Senses: The Sensual Culture Reader*. Oxford: Berg.

Howes, D, ed. 2009. *The Sixth Sense Reader*. Oxford: Berg.

Hubers, J.C. 2004. 'Muscle Tower II'. Accessed 1 November 2013. www.bk.tudelft.nl/index.php?id=16060&L=1.

Huizinga, Johan. 1955. *Homo Ludens: A Study of the Play Element in Culture*. Boston: The Beacon Press.

Hunt, Henry. 1820. *Memoirs of Henry Hunt Esq*. London: T. Dolby.

Husserl, Edmund. 1970. *The Crisis of European Sciences and Transcendental Phenomenology: An Introduction to Phenomenological Philosophy*. Translated (with an introduction) by David Carr. Evanston: Northwestern University Press.

Husserl, E. 2001 [1900]. *Logical Investigations*. London: Routledge.

Hutchinson, R. 2010a. *The Legend of Lefi Ganderson*. Seething: Homage Publishing.

Hutchinson, R. 2010b. *The King's Soup*. Seething: Homage Publishing.

Hutchinson, R. 2011. *Jack and the Golden Egg, or, How the Seething Community Sports Day Started*. Seething: Homage Publishing.

Hutchinson, R. 2012. *The Last Sardines*. Seething: Homage Publishing.

Hüttenmoser, Marco. 1995. 'Children and their living surroundings: empirical investigations into the significance of living surroundings for the everyday life and development of children'. *Children's Environments* 12 (4): 403–413.

Huxley, M. and O. Yiftachel. 2000. 'New paradigm or old myopia? Unsettling the communicative turn in planning theory'. *Journal of Planning Education and Research* 19 (4): 333–342.

Ihde, Don. 2010. *Heidegger's Technologies: Postphenomenological Perspectives*. New York: Fordham University Press.

Ingleby, Matthew. 2010. 'Building plots: metropolitan fiction, 1848–1897, and the conception of urban sprawl'. *Literatur in Wissenschaft and Unterricht* 43 (2–3): 127–141.

Ingold, Tim. 2000. *The Perception of the Environment: Essays on Dwelling, Livelihood and Skill*. London: Routledge.

Ingold, Tim. 2011. *Being Alive: Essays on Movement, Knowledge and Description*. New York: Routledge.

Inwood, Michael. 2000. *Heidegger: A Very Short Introduction*. Oxford: Oxford University Press.

Ivanov, A.B. 2011. 'Euclidean geometry'. *Encyclopedia of Mathematics*. Accessed 1 February 2015. www.encyclopediaofmath.org/index.php?title=Euclidean_geometry&oldid=34034.

Jackson, Peter. 1989. *Maps of Meaning*. London and New York: Routledge.

Jacobs, Jane. 1961. *The Death and Life of Great American Cities*. New York: Random House.

Jacobs, Jane. 1962. *The Death and Life of Great American Cities*. London: Jonathan Cape.

Jacobs, Jane. 1993. *The Death and Life of Great American Cities*. New York and Toronto: Random House.

Jacobs, M. and J. Findley. 2001. 'Breathe'. Accessed 3 March 2015. www.sonicribbon.com/sonicribbon/breathe/.

James, Simon. 1999. 'The community of the soldiers: a major identity and centre of power in the Roman Empire'. In *TRAC 98: Proceedings of the Eighth Annual Theoretical Roman Archaeology Conference, which took place at the University of Leicester, April 1998*, edited by P. Baker, S. Jundi and R. Witcher. Oxford: Oxbow, 14–25.

James, Simon. 2011. *Rome and the Sword: How Warriors and Weapons Shaped Roman History*. London: Thames & Hudson.

Jameson, Frederic. 1984. 'Postmodernism, or the cultural logic of late capitalism'. *New Left Review* 146: 53–92.

Jardine, Lisa. 2010. 'The aftermath of *Homo Ludens*: from Huizinga to Zemon Davis and beyond'. Huizinga Lecture given at Leiden University, 10 December 2010.

Jensen, Ole B. 2009. 'Flows of meaning, cultures of movement – urban mobility as meaningful everyday life practice'. *Mobilities* 4: 139–158.

Jensen, Ole B. 2010a. 'Negotiation in motion: unpacking a geography of mobility'. *Space and Culture* 13: 389–402.

Jensen, Ole B. 2010b. 'Erving Goffman and everyday life mobility'. In *The Contemporary Goffman*, edited by Michael Hviid Jacobsen. New York: Routledge, 333–351.

Jensen, Ole B. 2013. *Staging Mobilities*. London: Routledge.

Jensen, Ole B. and Malene Freudendal-Pedersen. 2012. 'Utopias of mobilities'. In *Utopia: Social Theory and the Future*, edited by Michael Hviid Jacobsen. Farnham: Ashgate, 197–217.

Jerram, Leif. 2013. 'Space: a useless category for historical analysis?' *History and Theory* 52: 400–419.

Jessop, Bob, Neil Brenner and Martin Jones. 2008. 'Theorizing sociospatial relations'. *Environment and Planning D* 26: 389–401.

Jeevendrampillai, D. 2013. 'Unpacking the digital silence: failure in public participation GIS'. In *Proceedings of the 112th American Anthropological Association*. San Francisco.

Jeevendrampillai, D. 2015. *Making the Local*. Forthcoming PhD thesis. London: University College London.

Johnson, M. 1999. *Archaeological Theory: An Introduction*. Oxford: Blackwell.

Johnson, M. 2010. *English Houses 1300–1800: Vernacular Architecture, Social Life*. Harlow: Longman.

Jones, A. and G. Macgregor, eds. 2002. *Colouring the Past*. Oxford: Berg.

Jones, K. and J. Wills. 2005. *The Invention of the Park: Recreational Landscapes from the Garden of Eden to Disney's Magic Kingdom*. Cambridge, MA: Polity.

Jorgensen, A. and R. Keenan, eds. 2012. *Urban Wildscapes*. London and New York: Routledge.

Josephus, Flavius. 1970. *The Jewish War*. Translated by G.A. Williamson. Harmondsworth: Penguin.

Joyce, Patrick. 2003. *The Rule of Freedom: Liberalism and the Modern City*. London: Verso.

Joynt, W.D. 1979. *Breaking the Road for the Rest*. Melbourne: Hyland House.

Justinian I. 1998. *The Digest of Justinian*. Translated by A. Watson. Philadelphia: University of Pennsylvania Press.

Jütte, R. 2005. *A History of the Senses from Antiquity to Cyberspace*. Cambridge: Cambridge University Press.

Kabat-Zinn, Jon. 2005. *Coming to Our Senses: Healing Ourselves and the World Through Mindfulness*. New York: Hyperion.

Kaiser, Alan. 2011. 'Cart traffic in Pompeii and Rome'. In *Rome, Ostia, Pompeii: Movement and Space*, edited by R. Laurence and D. Newsome. Oxford: Oxford University Press, 174–193.

Kapchan, Deborah A. 1995. 'Performance'. *The Journal of American Folklore* 108 (430): 479–508.

Kapoor, I. 2002. 'The devil's in the theory: a critical assessment of Robert Chambers' work on participatory development'. *Third World Quarterly* 23: 101–117.

Karskens, G. 1999. *Inside the Rocks*. Erskineville: Hale and Iremonger.

Keene, D. 1996. 'Landlords, the property market and urban development in medieval England'. In *Power, Profit and Urban Land: Landownership in Medieval and Early Modern Northern European Towns*, edited by F.E. Eliassen and G.A. Ersland. Aldershot: Scholar Press, 93–119.

Kellett, John R. 1966. *The Impact of Railways on Victorian Cities*. London: Routledge, Kegan and Paul.

Kelman, Herbert C. and Lee Hamilton. 1989. *Crimes of Obedience: Toward a Social Psychology of Authority and Responsibility*. Yale: Yale University Press.

Kent, S., ed. 1990. *Domestic Architecture and the Use of Space: An Interdisciplinary, Cross-Cultural Study*. Cambridge: Cambridge University Press.

Khan, Omar. 2011. 'Open Columns | A CO2 Actuated Responsive Architecture'. Accessed 3 March 2015. http://cast.b-ap.net/opencolumns/.

King, C. 2006. *House and Society in and English Provincial City: The Archaeology of Urban Household in Norwich, 1370–1700*. Unpublished PhD thesis. Reading: University of Reading.

Kingston, Ralph. 2010. 'Mind over matter? History and the spatial turn'. *Cultural and Social History* 7 (1): 111–121.

Klynne, A. 2004. 'The laurel grove of the Caesars: looking in and looking out'. In *Roman Villas around the Urbs. Interaction with Landscape and Environment. Proceedings of a Conference held at the Swedish Institute in Rome, September 17–18, 2004*, edited by A. Klynne and B. Santillo Frizel. Rome. Accessed 1 March 2015.www.isvroma.it/public/villa/print/klynne.pdf.

Knowles, Anne. 2014. 'The contested nature of historical GIS'. *International Journal of GIS* 28 (1): 206–211.

Koch, Daniel. 2012. 'Isovists revisited: egocentric space, allocentric space, and the logic of the mannequin'. In *Proceedings of the Eighth International Space Syntax Symposium*, edited by Margarita Greene, José Reyes and Andrea Castro. Santiago: PUC, 1–23.

Kockel, V. and B.F. Weber. 1983. 'Die Villa delle colonne a mosaico in Pompeji'. *Mitteilungen des Deutschen Archäologischen Instituts, Römische Abteilung* 90 (I): 51–89.

Korsmeyer, C., ed. 2005. *The Taste Culture Reader: Experiencing Food and Drink*. Oxford: Berg.

Kroll, W. 1933. *Die Kultur der ciceronischen Zeit*. Leipzig: Dieterich.

Kronenburg, Robert. 2007. *Flexible: Architecture That Responds to Change*. London: Laurence King.

Kuhlenbeck, Britta. 2010. *Re-Writing Spatiality: The Production of Space in the Pilbara Region of Western Australia*. Münster: Lit Verlag.

Kumin, B. 1999. *The Shaping of a Community: The Rise and Reformation of the English Parish c. 1400–1560*. Aldershot: Ashgate.

Langford, Paul. 1998. *A Polite and Commercial People: England, 1727–1783*. Oxford: Oxford University Press.

Lanng, Ditte Bendix, Henrik Harder and Ole B. Jensen. 2012. 'Towards urban mobility designs: en route in the functional city'. In *Selected Proceedings from the Annual Transport Conference at Aalborg University*. Aalborg: Aalborg University.

Larmor, D.H.J. and D.J. Spencer. 2007. *The Sites of Rome: Time, Space, Memory*. Oxford: Oxford University Press.

Lasdun, S. 1991. *The English Park: Royal, Private and Public*. London: Deutsch.

Latour, Bruno. 1990. 'Drawing things together'. In *Representation in Scientific Practice*, edited by M. Lynch and S. Woolgar. Cambridge, MA: MIT Press, 19–68.

Latour, Bruno. 1996. 'On interobjectivity'. *Mind, Culture, and Activity* 3 (4): 228–245.

Latour, Bruno. 1999. 'On recalling ANT'. In *Actor-Network Theory and After*, edited by John Law and John Hassard. Oxford: Blackwell, 15–25.

Latour, Bruno. 2002. 'Gabriel Tarde and the end of the social'. In *The Social in Question: New Bearings in History and the Social Sciences*, edited by Patrick Joyce. London: Routledge, 117–132.

Latour, Bruno. 2004. *Reassembling the Social: An Introduction to Actor-Network Theory*. Oxford: University Press.

Latour, Bruno. 2005. *Reassembling the Social: An Introduction to Actor-Network Theory*. Oxford: Oxford University Press.

Latour, Bruno and Albena Yaneva. 2008. 'Give me a gun and I will make all buildings move'. In *Explorations in Architecture: Teaching, Design, Research*, edited by Teto Geiser. Basel: Birkhäuser, 80–89.

Laurence, Ray. 1994. 'Rumour and communication in Roman politics'. *Greece and Rome* 41: 62–74.

Laurence, Ray. 1997. 'Space and text'. In *Domestic Space in the Roman World*, edited by Ray Laurence and Andrew Wallace-Hadrill. Portsmouth, RI: Journal of Roman Archaeology, L.L.C., 7–15.

Laurence, Ray. 2001. 'Roman Italy's urban revolution'. In *Modalità Insediative e Strutture Agrarie nell'Italia Meridionale in Età Romana*, edited by Elio Lo Cascio and Afredina Storchi Marino. Bari: Edipuglia, 593–611.

Laurence, Ray. 2008. 'City traffic and the archaeology of Roman streets'. In *Stadtverkehr in der Antiken Welt*, edited by D. Mertens. Wiesbaden: Deutsches Archäologisches Institut, 87–106.

Laurence, Ray. 2010. *Roman Pompeii: Space and Society*, 2nd edn. London: Routledge

Laurence, Ray. 2011. 'Literature and the spatial turn'. In *Rome, Ostia, Pompeii: Movement and Space*, edited by David Newsome and Ray Laurence. Oxford: Oxford University Press, 81–99.

Laurence, Ray. 2013a. 'Streets and facades'. In *The Wiley-Blackwell Companion to Roman Architecture*, edited by Caroline Quenemoen and Roger Ulrich. Oxford: Wiley-Blackwell, 399–411.

Laurence, Ray. 2013b. 'Traffic and land transportation in and near Rome'. In *The Cambridge Companion to Ancient Rome*, edited by Paul Erdkamp. Cambridge: Cambridge University Press, 246–261.

Laurence, Ray. 2013c. 'Roads and bridges'. In *The Wiley-Blackwell Companion to Roman Republican Archaeology*, edited by Jane DeRose Evans. Oxford: Wiley-Blackwell, 296–308.

Laurence, Ray. 2015. 'Towards a history of mobility in Ancient Rome (300 BC to AD 100)'. In *The Moving City: Processions, Passages and Promenades in Ancient Rome*, edited by Jonas Bjørnebye, Simon Malmberg and Ida Östenberg. London: Bloomsbury.

Laurier, Eric and Hayden Lorimer. 2012. 'Other ways: landscapes of commuting'. *Landscape Research* 37: 207–224.

Law, John. 2007. 'Actor network theory and material semiotics', version from 25 April 2007. Accessed 18 May 2007. www. heterogeneities. net/publications/Law2007ANTandMaterial Semiotics.pdf.

Lawrence, R.J. 1990. 'Public collective and private space: a study of urban housing in Switzerland'. In *Domestic Architecture and the Use of Space: An Interdisciplinary, Cross-Cultural Study*, edited by S. Kent. Cambridge: Cambridge University Press, 73–91.

Lawson, G. and C. Scarre, eds. 2006. *Archaeoacoustics*. Cambridge: Cambridge University Press.

Le Bohec, Yann. 1994. *The Imperial Roman Army*. London: Batsford.

Lefebvre, Henri. 1991 [1974]. *The Production of Space*. Translated by Nicholson-Smith. Oxford: Oxford University Press.

Lefebvre, Henri. 2004. *Rhythmanalysis: Space, Time and Everyday Life*. London: Continuum.

Lefebvre, Henri. 2013. *Rhythmanalysis: Space, Time and Everyday Life*. Translated by Stuart Elden and Gerald Moore. London: Bloomsbury.

Lemonnier, Pierre. 1993. *Technological Choices: Transformation in Material Cultures Since the Neolithic*. Oxon: Routledge.

Lemos, André. 2010. 'Post-mass media functions, locative media, and informational territories: new ways of thinking about territory, place, and mobility in contemporary society'. *Space and Culture* 13 (4): 403–420.

Leroi-Gourhan, Andre. 1993. *Gesture and Speech*. Boston: MIT Press.

Lester, A.E. 2010. 'Crafting a charitable landscape: urban topographies in charters and testaments from medieval Champagne'. In *Cities, Texts and Social Networks 400–1500*, edited by C. Goodson, A.E. Lester and C. Symes. Farnham: Ashgate, 125–148.

Levi, Giovanni. 1991. 'On microhistory'. In *New Perspectives on Historical Writing*, edited by Peter Burke. Cambridge: Polity, 93–113.

Lévi-Strauss, Claude. 1963. *Structural Anthropology*. New York: Basic Books.

Levick, Barbara. 1990. *Claudius*. London: Routledge.

Levinson, David. 1998. 'Accessibility and the journey to work'. *Journal of Transport Geography* 6 (1): 11–21.

Liebst, Lasse Suonperä. 2014. *Staged, yet Unstaged. Sociological Inquiries into Space and Micro-Interaction*. PhD thesis. Copenhagen: University of Copenhagen.

Lilley, K. 2002. *Urban Life in the Middle Ages: 1000–1450*. London: Palgrave.

Lilley, K. 2009. *City and Cosmos*. London: Reaktion Books.

Lindemann, Gesa. 2011. 'On Latour's social theory and theory of society, and his contribution to saving the world'. *Human Studies* 34 (1): 93–110.

Livingstone, Sonia. 2009. 'On the mediation of everything: ICA presidential address 2008'. *Journal of Communication* 59 (1): 1–18.

Lock, G. and B. Molyneaux, eds. 2006. *In Confronting Scale in Archaeology: Issues of Theory and Practice*. New York: Springer.

Lott, Bert. 2004. *The Neighbourhoods of Augustan Rome*. Cambridge: Cambridge University Press.

Loudon, J.C. 1829. 'Hints on breathing places for the metropolis, and for country towns and villages, on fixed principles'. *Gardener's Magazine* 4: 686–690.

Loughlin, Elizabeth. 1988. *Among the Terraces: A Carlton Street*. Melbourne: Carlton Forest Project.

Low, Setha M. 2009. 'Towards an anthropological theory of space and place'. *Semiotica* 175: 21–37.

Low, Setha and Denise Lawrence-Zúñiga. 2003. *The Anthropology of Space and Place: Locating Culture.* London: Blackwell.

Luhmann, Niklas. 1995 [1984]. *Social Systems.* Stanford: Stanford University Press.

Lukashok, Alvin K. and Kevin Lynch. 1956. 'Some childhood memories of the city'. *Journal of the American Institute of Planners* 22 (3): 142–152.

Lukes, Steven. 1985. *Emile Durkheim: His Life and Work. A Historical and Critical Study.* Palo Alto: Stanford University Press.

Lünen, Alexander von. 2013. 'Tracking in a new territory: re-imaging GIS for history'. In *History and GIS: Epistemologies, Considerations and Reflections,* edited by Alexander von Lünen and Charles Travis. Dordrecht: Springer, 211–239.

Lünen, Alexander von and Emmanuel Le Roy Ladurie. 2013. 'Immobile history: an interview with Emmanuel Le Roy Ladurie'. In *History and GIS: Epistemologies, Considerations and Reflections,* edited by Alexander von Lünen and Charles Travis. Dordrecht: Springer, 15–25.

Lünen, Alexander von and Wolfgang Moschek. 2011. 'Without limits: ancient history and GIS'. In *Geohumanities: Art, History, Text at the Edge of Place,* edited by Michael Dear, Jim Ketchum, Sarah Luria and Doug Richardson. London: Routledge, 241–250.

Lünen, Alexander von and Gunnar Olsson. 2013. '"Thou shalt make no graven maps!": an interview with Gunnar Olsson'. In *History and GIS: Epistemologies, Considerations and Reflections,* edited by Alexander von Lünen and Charles Travis. Dordrecht: Springer, 73–87.

Lünen, Alexander von and Charles Travis, eds. 2013. *History and GIS: Epistemologies, Considerations and Reflections.* Dordrecht: Springer.

Lynch, Kevin. 1960. *The Image of the City.* Cambridge, MA: MIT Press.

Lynch, Kevin, ed. 1977a. *Growing Up in Cities: Studies of the Spatial Environment of Adolescence in Cracow, Melbourne, Mexico City, Salta, Toluca and Warszawa.* Cambridge, MA and London: MIT Press.

Lynch, Kevin. 1977b. 'The spatial world of the child'. In *The Child in the City: Today and Tomorrow,* edited by William Michelson, Saul V. Levine and Ellen Michelson. Toronto, Buffalo and London: University of Toronto Press, 102–127.

Lynch, Kevin. 1981. *Good City Form.* Cambridge, MA: MIT Press.

Macdougall, E.M., ed. 1987. *Ancient Roman Villa Gardens, Dumbarton Oaks Colloquium on the History of Landscape Architecture* 10. Washington, DC: Dumbarton Oaks.

Mace, Alan. 2013. 'The future has come and gone: managing change in the ageing suburbs'. In *New Suburban Stories,* edited by Martin Dines and Timotheus Vermeulen. London and New York: Bloomsbury Academic, 85–96.

MacMullen, Ramsay. 1960. 'Inscriptions on armor and the supply of arms in the Roman Empire'. *American Journal of Archaeology* 64: 23–40.

Maliene, Vida, Vytautas Grigonis, Vytautas Palevicius and Sam Griffiths. 2011. 'Geographic information system: old principles with new capabilities'. *Urban Design International,* 16 (1): 1–6.

Malpas, Jeff. 2012. *Heidegger and the Thinking of Place: Explorations in the Topology of Being.* Cambridge, MA: MIT Press.

Manning, William. H. 1975. 'Roman military timber granaries in Britain'. *Saalburg Jahrbuch* 32: 105–129.

Marichal, Robert. 1992. *Les Ostraca de Bu-Njem, Suppléments de Libya Antiqua VII.* Tripoli: Grande Jamahira Arabe, Libyenne, Populaire et Socialiste Département des Antiquités.

Martinez, K., J.K. Hart and R. Ong. 2004. 'Environmental sensor networks'. *Computer* 37 (8): 50–56.

Massey, Doreen. 1998. 'The spatial construction of youth cultures'. In *Cool Places: Geographies of Youth Cultures,* edited by Tracey Skelton and Gill Valentine. London and New York: Routledge, 121–129.

Massey, Doreen. 2005. *For Space.* London: SAGE.

Matthias, Frank Richard. 1985. 'This journey of mine'. Edited by Gayle Hardie. State Library of Victoria MS 12532, box 3404/4.

Mattingly, David. 2006. *An Imperial Possession: Britain in the Roman Empire, 54 BC-AD 409.* London: Allen Lane.

Mau, A. 1899. *Pompeii: Its Life and Art.* Translated by F.W. Kelsey. New York: Macmillan.

Mauss, Marcel. 1979. *Seasonal Variations of the Eskimo: A Study in Social Morphology.* London: Routledge and Kegan Paul.

Mauss, Marcel. 2003. 'Concrete divisions of sociology'. In *The Nature of Sociology*, edited by W. Jeffrey and M. Gane. New York and Oxford: Durkheim Press, 55–74.

Mauss, Marcel. 2007. *Manual of Ethnography.* New York and Oxford: Durkheim Press.

McCarty, Willard. 2012. 'A telescope for the mind?' In *Debates in the Digital Humanities*, edited by Matthew K. Gold. Minneapolis: University of Minnesota Press, 113–123.

McCreedy, Alexander. 2005. 'Turnpike roads and the spatial culture of London, 1756–1830'. PhD Diss. University of London.

Meiggs, Russel. 1973. *Roman Ostia*, 2nd edn. Oxford and New York: Oxford University Press.

Merleau-Ponty, M. 1962 [1945]. *Phenomenology of Perception.* Translated by C. Smith. London: Routledge.

Merriman, Peter. 2007. *Driving Spaces: A Cultural-Historic Geography of England's M1 Motorway.* Oxford: Blackwell.

Milgram, Stanley. 1974. 'The experience of living in cities'. In *Crowding and Behavior*, edited by C.M. Loo. New York: Ardent Media, 41–54.

Miller, Daniel. 2010. *Stuff.* London: Polity.

Miller, Daniel. 2011. *Tales from Facebook.* London: Wiley.

Milnor, Kristina. 2014. *Graffiti and the Literary Landscape in Roman Pompeii.* Oxford: Oxford University Press.

Minton, A. 2009. *Ground Control: Fear and Happiness in the Twenty-First-Century City.* London: Penguin.

Mitchell, William. 1995. *City of Bits: Space, Place, and the Infobahn.* Cambridge, MA: MIT Press.

Mitchell, William. 1997. *City of Bits: Space, Place, and the Infobahn.* Fourth printing. Cambridge, MA and London: MIT Press.

Mongelluzzo, R. 2013. 'Maya palaces as experiences: ancient Maya royal architecture and its influence on sensory perception'. In *Making Senses of the Past: Towards a Sensory Archaeology*, edited by J. Day. Illinois: Southern Illinois University Press, 90–112.

Montgomery, Gary T. 1994. 'Slowed respiration training'. *Applied Psychophysiology and Biofeedback* 19 (3): 211–225.

Morozov, Evgeny. 2014. 'The rise of data and the death of politics'. *Observer*, 20 July.

Morrison, Arthur. 1896. *A Child of the Jago.* London: Methuen and Co.

Morwood, James. 2000. *The Pocket Oxford Latin Dictionary.* Oxford: Oxford University Press.

Motta, Wallis, Ava Fatah gen. Schieck, Holger Schnädelbach, Efstathia Kostopoulou, Moritz Behrens, Steve North and Lei Ye. 2013. 'Considering communities, diversity and the production of locality in the design of networked urban screens'. In *Human-Computer Interaction – INTERACT 2013*, edited by P. Kotzé, G. Marsden, G. Lindgaard, J. Wesson and M. Winckler. Berlin and Heidelberg: Springer.

Mouffe, Chantal. 2005. *On the Political.* London and New York: Routledge.

Mouffe, Chantal. 2013. 'Space, hegemony and radical critique'. In *Spatial Politics: Essays for Doreen Massey*, edited by David Featherstone and Joe Painter. Oxford: Wiley Blackwell, 21–31.

Moughtin, Cliff. 1991. 'The European city. Part I: paths and places'. *Town Planning Review* 62 (1): 51–77.

Mowlabocus, Sharif. 2010. *Gaydar Culture: Gay Men, Technology and Embodiment in the Digital Age.* England: Ashgate.

Mulhall, Stephen. 1996. *Heidegger and Being and Time.* London: Routledge.

Mumford, L. 1989. *The City in History: Its Origins, its Transformations, and its Prospects.* New York: Harcourt, Brace and World.

Mumford, K. and A. Power. 2003. *EastEnder: Family and Community in East London.* Bristol: Policy Press.

Munslow, A. 2014. 'On "presence" and conversing with the past: do historians communicate with the past?' *Rethinking History* 18 (4): 569–574.

Murdoch, Jonathan. 2006. *Post-Structuralist Geography: A Guide to Relational Space.* London: SAGE.

Nejad, Reza M. 2013. 'The discursive manifestation of past and present through the spatial organization of the *Ashura* procession'. *Space and Culture.* Accessed 7 April 2015. DOI: 10.1177/1206331213475747.

Némedi, Dénes. 1995. 'Collective consciousness, morphology, and collective representations: Durkheim's sociology of knowledge, 1894–1900'. *Sociological Perspectives* 38 (1): 41–56.

Netto, Vinicius M. 2008. 'Practice, space, and the duality of meaning'. *Society and Space – Environment and Planning D* 26: 359–379.

Netto, Vinicius M. 2015 forthcoming. '"What is space syntax not?" Reflections on space syntax as sociospatial theory'. *Urban Design International.*

Netto, Vinicius M. 2016 *The Social Fabric of Cities.* London: Routledge.

Neufert, Ernst and Peter Neufert. 2000. *Architect's Data.* 3rd edn. Oxford: Blackwell.

Nevola, F. 2010. 'Introduction: locating communities in the early modern Italian city'. *Urban History* 37 (3): 349–359.

Newman, O. 1972. *Defensible Space: Crime Prevention Through Urban Design.* New York: Macmillan.

Newsome, D. 2011. 'Introduction: making movement meaningful'. In *Rome, Ostia, Pompeii: Movement and Space,* edited by R. Laurence and D. Newsome. Oxford: Oxford University Press, 1–54.

Nicolet, Claude. 1991. *Space, Geography and Politics in the Early Roman Empire.* Ann Arbor: University of Michigan Press.

North, Steve, Holger Schnädelbach, Ava Fatah gen. Schieck, Wallis Motta, Lei Ye, Moritz Behrens and Efstathia Kostopoulou. 2013. 'Tension space analysis: exploring community requirements for networked urban screens'. In *Human-Computer Interaction–INTERACT 2013,* Part II, edited by P. Kotzé, G. Marsden, G. Lindgaard, J. Wesson and M. Winckler. Berlin and Heidelberg: Springer, 81–98.

O'Dell, Tom. 2009. 'My soul for a seat: commuting and the routines of mobility'. In *Time, Consumption and Everyday Life: Practice, Materiality and Culture,* edited by Elizabeth Shove, Frank Trentmann and Richard Wilk. Oxford: Berg, 85–98.

Official Programme of the Visit of Their Royal Highnesses the Prince & Princess of Wales August 16/17th, 1875. 1875. Local Pamphlet, 0425. Sheffield: Sheffield Local Studies Library.

Official Programme, Visit of her most gracious majesty Queen Victoria to Sheffield, May 21st, 1897. 1897. Sheffield: Sheffield Local Studies Library.

Olwig, Kenneth. 1996. 'Recovering the substantive meaning of landscape'. *Annals of the Association of American Geographers* 86: 630–653.

O'Mahony, Eoin. 2014. 'Problems with drawing lines : theo-geographies of the catholic parish in Ireland'. *Journal of the Irish Society for the Academic Study of Religions* 1 (1): 48–65.

O'Reilly, Carole. 2013. '"We have gone recreation mad": the consumption of leisure and popular entertainment in municipal public parks in early twentieth century Britain'. *International Journal of Regional and Local History* 8 (2): 112–128.

Orlandi, T. 1992. 'Ideas for a basic theory of humanities computing'. In *Towards an International Curriculum for History and Computing: A Workshop of the International Association for History and Computing, University of Glasgow, 15–17 May 1992,* edited by Donald Spaeth, Peter Denley, Virginia Davis and Richard Trainor. St. Katharinen: Scripta Mercaturae, 35–36.

Osborne, Jonathan F. 1996. 'Beyond constructivism'. *Science Education* 80 (1): 53–82.

O'Sullivan, Tim. 2011. *Walking in Roman Culture.* Cambridge: Cambridge University Press.

Oxford English Dictionary. [Online]. Accessed October 2014. www.oed.com/.

Pacione, M. 2005. *Urban Geography: A Global Perspective.* London: Routledge.

Paglen, Trevor. 2006. 'Recording carceral landscapes'. *Leonardo Music Journal* 16: 56–57.

Paliou, E., U. Lieberwirth and S. Polla, eds. 2014. *Spatial Analysis and Social Spaces: Interdisciplinary Approaches to the Interpretation of Prehistoric and Historic Built Environments.* Berlin: De Gruyter.

Palliser, D.M., ed. 2000. *The Cambridge Urban History of Britain,* vol. 1, *600–1540.* Cambridge: Cambridge University Press.

Palmer, Sue. 2006. *Toxic Childhood: How the Modern World is Damaging Our Children and What We Can Do About It.* London: Orion.

Parfitt, T. 2004. 'The ambiguity of participation: a qualified defence of participatory development'. *Third World Quarterly* 25 (3): 537–555.

Parkin, David. 1992. 'Ritual as spatial direction and bodily division'. In *Understanding Ritual,* edited by Daniel de Coppet. London and New York: Routledge, 11–25.

Parolin, Christina. 2010. *Radical Spaces: Venues of Popular Politics in London, 1790-c.1845.* Canberra: Australia National University Press.

Parr, A.E. 1969. 'Lessons of an urban childhood'. *The American Montessori Society Bulletin* 7 (4): no pagination.

Pascoe, Carla. 2011. *Spaces Imagined, Places Remembered: Childhood in 1950s Australia.* Newcastle upon Tyne: Cambridge Scholars Press.

Pavlovskaya, Marianna. 2006. 'Theorizing with GIS: a tool for critical geographies?' *Environment and Planning A* 38 (11): 2003–2020.

Pavlovskaya, Marianna. 2009. 'Non-quantitative GIS'. In *Qualitative GIS: A Mixed Method Approach,* edited by Meghan Cope and Sarah Elwood. London: SAGE, 13–37.

Pavolini, Carlo. 1980. 'Appunti sui "vasetti ovoidi e piriformi" di Ostia'. *Mélanges de l'Ecole française de Rome. Antiquité* 92: 993–1013.

Pavolini, Carlo. 1996. *La Vita Quotidiana a Ostia.* Bari: Economica Laterza.

Pearson, S. 2004. 'Houses, shops and storage: building evidence from two Kentish ports'. In *The Medieval Household in Christian Europe c.850-c.1550,* edited by C. Beattie, A. Maslakovic and S. Rees Jones. Turnhout: Brepols, 410–431.

Pearson, S. 2007. 'Rural and urban houses 1100–1500: "urban adaptation" reconsidered'. In *Town and Country in the Middle Ages,* edited by C. Dyer and K. Giles. Leeds: Maney, 43–64.

Pellegrini, Anthony D. 2009. *The Role of Play in Human Development.* Oxford and New York: Oxford University Press.

Peponis, John. 1985. 'The spatial culture of factories'. *Human Relations* 38 (4): 357–390.

Peponis, John. 2001. 'Interacting questions and descriptions: how do they look from here?' In Proceedings *Space Syntax 3rd international Symposium,* edited by John Peponis, Jean Wineman and Sonit Bafna. Atlanta: Gergia Tech, viii–xxvi.

Perkins, C. 2004. 'Cartography: cultures of mapping, power in practice'. *Progress in Human Geography* 28: 381–339.

Phang, Sara Elise. 2008. *Roman Military Service: Ideologies of Discipline in the Late Republic and Early Principate.* Cambridge: Cambridge University Press.

Phythian-Adams, C. 1979. *Desolation of a City: Coventry and the Urban Crisis of the Late Middle Ages.* Cambridge: Cambridge University Press.

Picard, Rosalind W. 1997. *Affective Computing.* Cambridge, MA: MIT Press.

Pickering, W.S.F. 1984. *Durkheim's Sociology of Religion: Themes and Theories.* London: Routledge and Kegan Paul.

Pickles, John. 2004. *A History of Spaces: Cartographic Reason, Mapping and the Geo-Coded World.* London and New York: Routledge.

Pink, Sarah. 2007. 'Walking with video'. *Visual Studies* 22 (3): 240–252.

Platts, H. 2011. '"Keeping up with the Joneses": competitive display within the Roman villa landscape'. In *Competition in the Ancient World,* edited by N. Fisher and H. Van Wees. Wales: Classical Press of Wales, 239–279.

Platts, H. 2017a. 'Experiencing sense, place and space in the Roman villa'. In *Between Words and Walls,* edited by J. Baird and A. Pudsey. Cambridge: Cambridge University Press (forthcoming).

Platts, H. 2017b. 'The Development and Role of the Roman Aristocratic *Domus*'. In *Blackwell Companion to the City of Rome,* edited by A. Claridge and C. Holleran. Oxford: Wiley-Blackwell (forthcoming).

Poehler, Eric E. 2006. 'The circulation of traffic in Pompeii's *Regio* VI'. *Journal of Roman Archaeology* 19: 53–74.

Porteous, J. D. 2006. 'Smellscape'. In *The Smell Culture Reader,* edited by J. Drobnik. Oxford: Berg, 89–106.

Povinelli, E. 2011. *Economies of Abandonment.* Durham, NC and London: Duke University Press.

Powell, W.R., ed. 1966. *A History of the County of Essex,* vol. 5. Accessed 1 December 2014. www.british-history.ac.uk/source.aspx?pubid=280.

Pred, Allan. 1977. 'The choreography of existence: comments on Hägerstrand's time-geography and its usefulness'. *Economic Geography* 53: 207–221.

Psarra, Sophia. 2013. 'Beyond analytical knowledge: the need for a combined theory of generation and explanation'. In *Proceedings of the Ninth International Space Syntax Symposium,* edited by Y.O. Kim, H.T. Park and K.W. Seo. Seoul: Sejong University, 065.1–22.

Pugh, Edwin. 1906. 'Some London street amusements'. In *Living London: Its Work and Its Play, Its Humour and Its Pathos, Its Sights and Its Scenes,* vol. 3, edited by George R. Sims. London: Cassell and Co.

Purcell, N. 1987. 'Town in country and country in town'. In *Ancient Roman Villa Gardens,* edited by E.B. Macdougall. Washington, DC: Dumbarton Oaks, 185–203.

Raban, Jonathan. 1974. *Soft City.* Glasgow: Collins.

Rabinow, Paul. 1995. *French Modern: Norms and Forms of the Social Environment.* Chicago: University of Chicago Press.

Ramadier, Thierry. 2004. 'Transdisciplinarity and its challenges: the case of urban studies'. *Futures* 36: 423–439.

Rappoport, A. 1990. *The Meaning of the Built Environment: A Nonverbal Communication Approach.* Tucson: University of Arizona Press.

Rappoport, A. 1993. 'Systems of activities and systems of settings'. In *Domestic Architecture and the Use of Space: An Interdisciplinary Cross-Cultural Study,* edited by S Kent. Cambridge: Cambridge University Press, 9–20.

Rasmussen, Kim. 2004. 'Places for children – children's places'. *Childhood* 11 (2): 155–173.

Rayment, Sean. 2012. 'I've been inside Camp Bastion – and it seemed like the safest place on earth'. *Telegraph,* 15 September: 12.

Reeh, Henrik, Jannie Rosenberg Bendsen, Henriette Steiner, Birgitte Bundesen Svarre, eds. 2012. *Spatial Culture: Studies in Urbanity and Aesthetics.* Copenhagen: Museum Tusculanums Press.

Rees Jones, S. 1987. *Property, Tenure and Rent: Some Aspects of the Topography and Economy of York,* 2 vols. Unpublished PhD thesis. York: University of York.

Rees Jones, S. 2008. 'Building domesticity in the city: English urban housing before the Black Death'. In *Domesticity: Home, Housing and Household in Medieval England,* edited by P.J.P. Goldberg and M. Kowaleski. Cambridge: Cambridge University Press, 66–91.

Reichenbach, Hans. 1958. *The Philosophy of Space and Time.* Translated by Maria Reichenbach and John Freund. New York: Dover. German original from 1927.

Ricciardi, Maria A. and Valnea Santa Maria Scrinari, eds. 1996. *La civilta dell' acqua in Ostia Antica,* vols I–II. Roma: Fratelli Palombi.

Ricketts, Dudley. c. 1981. 'My story'. Royal Historical Society of Victoria, MS 621, box 7/4, Part B.

Ricouer, P. 1970. 'The model of the text: meaningful action considered as a text'. *Social Research* 38: 529–562.

Rimmer, J. 2007. *Small Houses in Late Medieval York and Norwich.* Unpublished PhD thesis. University of York.

Rimmer, J. 2011. 'People and their buildings in the working-class neighbourhood of Hungate, York'. *International Journal of Historical Archaeology* 15: 617–628.

Robertson, E.C., J.D. Seiber, D.C. Fernandez and M.U. Zender. 2006. *Space and Spatial Analysis in Archaeology.* Calgary: University of Calgary Press.

Robinson, O.F. 1992. *Ancient Rome: City Planning and Administration.* London: Routledge.

Rodgers, Tara. 2006. 'Sound and the social organization of space'. *Leonardo Music Journal* 16: 49–50.

Rogers, Scott. 2013. 'Geography's digital turn?' Accessed 1 November 2014. www.publiclysited.com/geographys-digital-turn/.

Rosser, G. 1988. 'Communities of parish and guild in the late Middle Ages'. In *Parish, Church and People: Local Studies in Lay Religion 1350–1750,* edited by S.J. Wright. London: HarperCollins, 29–55.

Rothschild, N. 2006. 'Colonial and postcolonial New York: issues of size, scale, and structure'. In *Urbanism in the Preindustrial World: Cross-Cultural Approaches,* edited by G. Storey. Tuscaloosa: University of Alabama Press, 121–138.

Rothschild, N. 2008. *New York City Neighbourhoods.* New York: Percheron Press.

Round, Walker (pseud.). 1905. *Rambles around Enfield.* Enfield: Walker Round and Co.

Ryan, Mary. 1989. 'The American parade: representations of the nineteenth century social order'. In *The New Cultural History,* edited by Lynn Hunt. Berkeley: University of California Press, 131–152.

Ryan, Mary. 1990. *Women in Public: Between Banners and Ballots, 1825–1880.* Baltimore and London: Johns Hopkins University Press.

Sabin, Philip. 2000. 'The face of Roman battle'. *Journal of Roman Studies* 90: 1–17.

Sablayrolles, Robert. 1996. *Libertinus Miles. Les Cohortes de Vigiles.* Rome: L'École Française de Rome.

Sallust. 1931. *Jugurthine Wars.* Translated by J.C. Rolfe. London and New York: Loeb Classical Library 116.

Santaella, Lucia. 2010. 'Revisitando o corpo na era da mobilidade'. In *Comunicação e mobilidade: aspectos socioculturais das tecnologias móveis de comunicação no Brasil,* edited by André Lemos e Fabio Josgrilberg. Salvador: Editora da UFBA, 127–136.

Scarborough, Vernon and David R. Wilcox, eds. 1991. *The Mesoamerican Ballgame.* Tucson: University of Arizona Press.

Schatzki, Theodore. 2012. 'A primer on practices: theory and research'. In *Practice-Based Education: Perspectives and Strategies,* edited by Joy Higgs, Ronald Barnett, Stephen Billett, Maggie Hutchings and Franziska Trede. Rotterdam: Sense, 13–26.

Schiermer, Bjørn. 2011. 'Quasi-objects, cult objects and fashion objects: on two kinds of fetishism on display in modern culture'. *Theory, Culture & Society* 28 (1): 81–102.

Schmölder-Veit, Andrea. 2009. *Brunnen in den Städten des westlichen Römischen Reiches,* Palilia 19. Wiesbaden: Dr. Ludwig Reichert Verlag.

Schnädelbach, Holger. 2010. 'Adaptive architecture – a conceptual framework'. In *Proceedings of 'MediaCity',* edited by J. Geelhaar, F. Eckardt, B. Rudolf, S. Zierold and M. Markert. Weimar: Bauhaus-Universität, 523–555.

Schnädelbach, Holger. 2012. 'Hybrid spatial topologies'. *Journal of Space Syntax* 3 (2): 204–222.

Schnädelbach, Holger. 2013. 'Conceptual framework of adaptive architecture'. Accessed 19 February 2013. www.adaptivearchitecture.org.uk.

Schnädelbach, Holger, Kevin Glover and Alex Irune. 2010. 'ExoBuilding – breathing life into architecture'. In *Proceedings of NordiCHI.* Reykjavik: ACM Press, 442–451.

Schnädelbach, Holger, Ainojie Irune, David Kirk, Kevin Glover and Patrick Brundell. 2012. 'ExoBuilding: physiologically driven adaptive architecture'. *ACM Transactions in Computer Human Interaction* 19 (4): 1–22.

Schnädelbach, Holger, Alan Penn, Philip Steadman, Steve Benford, Boriana Koleva and Tom Rodden. 2006. 'Moving office: inhabiting a dynamic building'. In *Proceedings of Computer Supported Cooperative Work.* Banff: ACM Press, 313–322.

Schofield, J. 2011. *London 1100–1600.* Sheffield: Equinox.

Schofield, J. and A. Vince. 2003. *Medieval Towns.* London: Equinox.

Scholkmann, B. 2012. 'The anatomy of medieval towns'. In *The Archaeology of Medieval Europe,* vol. 2, edited by M. Carver and J. Klapste. Aahus: Aahus University Press, 379–407.

Schönberger, Hans. 1969. 'The Roman frontier in Germany: an archaeological survey'. *The Journal of Roman Studies* 59: 144–197.

Schroer, Markus. 2006. *Räume, Orte, Grenzen. Auf dem Weg zu einer Soziologe des Raumes.* Frankfurt/M.: Suhrkamp.

Schütz, Alfred and Thomas Luckmann, 1973 [1959]. *The Structures of the Life-World,* vol. 1. Evanston: Northwestern University Press.

Seamon, David. 1979. *A Geography of the Lifeworld: Movement, Rest, and Encounter.* London: Croom Helm.

Sennett, Richard. 1994. *Flesh and Stone: The Body and the City in Western Civilization.* New York: W.W. Norton and Company.

Serres, M. 2008 [1985]. *The Five Senses: A Philosophy of Mingled Bodies (I).* Translated by M. Stankey and P. Cowley. London: Bloomsbury.

Serres, M. with B. Latour. 1995 [1990]. *Conversations on Science, Culture and Time.* Translated by R. Lapidus. Michigan: University of Michigan Press.

Sewell, William. 2001. 'Space in contentious politics'. In *Silence and Voice in the Study of Contentious Politics,* edited by Ronald Aminzade *et al.*. Cambridge: Cambridge University Press, 51–88.

Shand, Pete. c. 2009–10. 'Children's voices'. Accessed 13 November 2014. www.peteshand.com/Children-s-Voices.

Shaw, Brent. 1983. 'Soldiers and society: the army in Numidia'. *Opus: Rivista internazionale per la storia economica e sociale dell'antichitá* 2: 133–159.

Shaw, Gareth and Allison Tipper. 1988. *British Directories: A Bibliography and Guide to Directories Published in England and Wales (1850–1950) and Scotland (1773–1950).* Leicester: Leicester University Press.

Shaw, Jon and Markus Hesse. 2010. 'Transport, geography and the "new" mobilities'. *Transactions of the Institute of British Geographers* 35 (3): 305–312.

Sheller, Mimi. 2004. 'Automotive emotions: feeling the car'. *Theory, Culture & Society* 21: 221–242.

Sheller, Mimi. 2011. 'Mobility'. *Sociopedia.isa,* DOI: 10.1177/205684601163.

Sheller, Mimi and John Urry. 2006. 'The new mobilities paradigm'. *Environment and Planning A* 38 (2): 207–226.

Shoemaker, Robert. 2004. *The London Mob: Violence and Disorder in Eighteenth-Century England.* London: Hambledon.

Shove, Elizabeth, Mika Pantzar and Matt Watson. 2012. *The Dynamics of Social Practice: Everyday Life and How It Changes.* London: SAGE.

Silverstone, Roger. 2005. 'The sociology of mediation and communication'. In *The Sage Handbook of Sociology,* edited by C. Calhoun, C. Rojek and B. Turner. London: SAGE, 188–207.

Simmel, Georg. 1976. 'The metropolis and mental life'. In *The Sociology of Georg Simmel,* edited by K. Wolff. New York: Free Press, 409–424.

Skeates, R. 2008. 'Making sense of the Maltese temple period: an archaeology of sensory experience and perception'. *Time and Mind* I (2): 207–238.

Skeates, R. 2010. *An Archaeology of the Senses: Prehistoric Malta.* Oxford: Oxford University Press.

Skinner, Quentin. 1969. 'Meaning and understanding in the history of ideas'. *History and Theory,* 8 (1): 3–53.

Slater, T.R. and J.P.P. Higgins. 2000. 'What is urban decline: desolation, decay and destruction, or an opportunity?' In *Towns in Decline,* edited by T.R. Slater. Aldershot: Ashgate, 1–24.

Sleight, Simon. 2009. 'Interstitial acts: urban space and the larrikin repertoire in late-Victorian Melbourne'. *Australian Historical Studies* 40 (2): 232–250.

Sleight, Simon. 2013. *Young People and the Shaping of Public Space in Melbourne, 1870–2014.* Aldershot and Burlington, VT: Ashgate.

Sleight, Simon. 2014. 'Simon Sleight on Thomas Kennington's *Homeless* (1890)'. SHCY web post. Accessed 13 November 2014. http://shcyhome.org/2013/06/guest-post-simon-sleight-on-thomas-kenningtons-homeless-2/.

Smith, J.T. 1997. *Roman Villas: A Study in Social Structure.* London: Routledge.

Smith, M. 2007. *Sensing the Past: Seeing, Hearing, Smelling, Tasting, and Touching in History.* Berkeley: University of California Press.

Snow, C.P. 1965. *The Two Cultures: And a Second Look.* Cambridge: Cambridge University Press.

Soja, Edward W. 1990. *Postmodern Geographies: The Reassertion of Space in Critical Theory.* London and New York: Verso.

Soja, Edward W. 1996. *Thirdspace: Journeys to Los Angeles and Other Real-and-Imagined Places.* Oxford: Blackwell.

Soja, Edward W. 2000. *Postmetropolis: Critical Studies of Cities and Regions.* Oxford: Blackwell

Solodkin, Ana, Petr Hlustik and Giovanni Buccino. 2007. 'The anatomy and physiology of the Motor Studio Gang Architects. (2009). Bengt Sjostrom Starlight Theatre / Studio Gang Architects'. Accessed 2 March 2015. www.archdaily.com/?p=28649.

Sombart, Nicolaus. 1992. 'Nachrichten aus Ascona. Auf dem Wege zu einer kulturwissenschaftlichen Hermeneutik'. In *Städtische Intellektuelle. Urbane Millieus im 20. Jahrhundert,* edited by Walter Prigge. Frankfurt/M.: Fischer, 107–117.

Speidel, Michael. 1984. 'Numerus or Ala Vocontiorum at Palmyra?' In *Roman Army Studies,* vol. 1. Amsterdam: J.C. Gieben, 167–172.

Speidel, Michael. 1996. *Die römischen Schreibtafeln von Vindonissa.* Brugg: Gesellschaft pro Vindonissa.

Speidel, Michael. 2006. *Emperor Hadrian's Speeches to the African Army: A new Text.* Mainz: Verlag des Römisch-Germanischen Zentralmuseums.

Spencer, Diana. 2011. 'Movement and the linguistic turn'. In *Rome, Ostia, Pompeii: Movement and Space,* edited by David Newsome and Ray Laurence. Oxford: Oxford University Press, 57–80.

Steadman, Phillip. 2014. *Building Types and Built Forms.* London: Matador.

Stevens, Quentin. 2007. *The Ludic City: Exploring the Potential of Public Spaces.* Abingdon: Routledge.

Stewart, Peter. 2003. *Statues in Roman Society.* Oxford: Oxford University Press.

Stiegler, Bernard. 1998. *Technics and Time: The Fault of Epimetheus.* Palo Alto: Stanford University Press.

Stobart, Jon, Andrew Hann and Victoria Morgan. 2007. *Spaces of Consumption: Leisure and Shopping in the English Town, c.1680–1830.* Abingdon: Routledge.

Stöger, Hanna. 2011a. *Rethinking Ostia: A Spatial Enquiry into the Urban Society of Rome's Imperial Port-Town.* Leiden: Leiden University Press.

Stöger, Hanna. 2011b. 'The spatial organization of the movement economy: the analysis of Ostia's *Schola*'. In *Rome, Ostia, Pompeii: Movement and Space,* edited by R. Laurence and D. Newsome. Oxford: Oxford University Press, 215–242.

Strathern, M. 1992. *Partial Connections.* Cambridge: Cambridge University Press.

Strenski, Ivan. 2006. *The New Durkheim.* New Brunswick: Rutgers University Press.

Stretton, Leonard. 1976. 'Judge Stretton's reminiscences'. *La Trobe Library Journal* 5 (17): 1–21.

Swain, Shurlee and Margo Hillel. 2010. *Child, Nation, Race and Empire: Child Rescue Discourse, England, Canada and Australia, 1850–1915.* Manchester: Manchester University Press.

Swanson, H. 1989. *Medieval Artisans.* Oxford: Blackwell.

Syal, Rajeev. 2014. 'UK troops hand over Camp Bastion to Afghan forces, ending 13-year campaign'. *Guardian,* 26 October: 4.

Synnott, A. 1991. 'Puzzling over the senses: from Plato to Marx'. In *Varieties of Sensory Experience,* edited by D. Howes. Toronto: University of Toronto Press, 61–76.

Tacchi, J. 2012. 'Digital engagement: voice and participation in development'. In *Digital Anthropology*, edited by H.A. Horst and D. Miller. London: Berg, 225–241.

Tacitus. 1951. *The Histories.* Translated by C.H. Moore. London: Heinemann. Loeb Classical Library 249.

Tacitus. 1971. *The Annals of Imperial Rome.* Translated by M. Grant. Harmondsworth: Penguin.

Tadmoor, N. 2010. *The Social Universe of the English Bible.* Cambridge: Cambridge University Press.

Tasker, George Edward. 1910. *Country Rambles by Field Path and Road in and around Ilford.* Ilford: South Essex Recorders.

Taylor, Lily Ross. 1952–4. 'The four urban tribes and the four regions of ancient Rome'. *Atti della Pontificia Accademia romana di archeologia: Rendiconti* 27: 225–238.

Thibaud, Jean. 2011. 'The sensory fabric of urban ambiances'. *Senses & Society* 6: 203–215.

Thompson, E.P. 1975. 'The crime of anonymity'. In *Albion's Fatal Tree: Crime and Society in Eighteenth-Century England,* edited by Douglas Hay *et al.*. London: Allen Lane, 255–344.

Thompson, E.P. 1991. *Customs in Common.* Harmondsworth: Penguin.

Thompson, Francis M.L., ed. 1982. *The Rise of Suburbia,* edited by F.M.L. Thompson. Leicester: Leicester University Press, 1–26.

Thompson, Francis M.L. 1988. *The Rise of Respectable Society.* London: Fontana Press.

Thrift, Nigel. 1983. 'On the determination of social action in space and time'. *Environment and Planning D* 1: 23–57.

Thrift, Nigel. 2003. 'Space: the fundamental stuff of human geography'. In *Key Concepts in Geography,* edited by Sarah L. Holloway, Stephen P. Rice and Gill Valentine. London: SAGE, 95–107.

Thrift, Nigel. 2006. 'Movement-space: the channging domain of thinking resulting from the development of new kinds of spatial awareness'. *Economy and Society* 33 (4): 582–604.

Thrift, Nigel. 2008. *Non-Representational Theory. Space. Politics. Affect.* London: Routledge.

Thrift, Nigel and J.D. Dewsbury. 2000. 'Dead geographies and how to make them live'. *Environment and Planning D: Society and Space* 18: 411–432.

Tilley, C. 1994. *A Phenomenology of Landscape.* Oxford: Berg.

Tilley, C. 2004. *The Materiality of Stone.* With the assistance of W. Bennett. Oxford: Berg.

Tilley, C. 2006. 'Objectification'. In *Handbook of Material Culture,* edited by C. Tilley, W. Keane, S. Küchler, M. Rowlands and P. Spyer. London: SAGE, 60–73.

Tilley, C. 2010. *Interpreting Landscape: Geologies, Topographies, Identities. Explorations in Landscape Phenomenology 3.* California: Left Coast Press.

Tomlin, Russel S., Linda Forrest, Ming Ming Pu and Myung Hee Kim. 1997. 'Discourse semantics'. In *Discourse as Structure and Process,* edited by Teun A. van Dijk. London: SAGE, 63–111.

Tonboe, Jens. 1993. *Rummets Sociologi.* København: Akademisk Forlag.

Toner, Jerry. 2009. *Popular Culture in Ancient Rome.* Cambridge: Cambridge University Press.

Tonkiss, Fran. 2005. *Space, The City and Social Theory: Social Relations and Urban Forms.* Cambridge: Polity Press.

Trench, W.S. 1863. *Annual Report of the Medical Officer of Health for Liverpool.* Liverpool: Port Health Authority.

Trifilò, Francesco. 2008. 'Power, architecture and community in the distribution of honorary statues in Roman public space'. In *TRAC 2007. Proceedings of the Seventeenth Annual Theoretical Roman Archaeology Conference,* edited by Corisande Fenwick, Meredith Wiggins and Dave Wythe. Oxford: Oxbow, 109–120.

Trifilò, Francesco. 2011. 'Movement, gaming, and the use of space in the forum'. In *Rome, Ostia, Pompeii. Movement and Space,* edited by Ray Laurence and David. J. Newsome. Oxford: Oxford University Press, 312–331.

Trifilò, Francesco. 2013. 'Text, space, and movement: a study of the Platea in epigraphy'. In *Written Space in the Latin West,* edited by Gareth Sears, Peter Keegan and Ray Laurence. London: Bloomsbury, 169–184.

Tschumi, Bernard. 1996. *Architecture and Disjunction.* Cambridge, MA: MIT Press.

Tuan, Yi-Fu. 1974. *Topophilia: A Study of Environmental Perception, Attitudes and Values.* New Jersey: Prentice Hall.

Turkle, Sherry. 2011. *Alone Together: Why We Expect More from Technology and Less from Each Other.* New York: Basic.

Turner, Alasdair. 2000–10. *Depthmap.* Computer application. London: University College London.

Twopeny, Richard. 1973 [1883]. *Town Life in Australia,* edited by Michael Cannon. Melbourne: Dominion.

UNICEF. 2014. 'Examples of CFC initiatives'. Accessed 25 November 2014. http://childfriendlycities.org/building-a-cfc/examples-of-cfc-initiatives/.

Urry, John. 2000. *Sociology beyond Society: Mobilities for the Twenty-First Century.* London: Routledge.

Urry, John. 2007. *Mobilities.* Oxford: Polity Press.

Vandeweghe, Evert. 2011. 'Staging urban history: festivities and the creation of historical townscapes in Belgium (1860–1958)'. *Environment, Space, Place* 3 (2): 122–160.

Vannini, Phillip. 2010. 'Mobile cultures: from the sociology of transport to the study of mobilities'. *Sociology Compass* 4 (2): 111–121.

Vannini, Phillip. 2012. *Ferry Tales.* New York: Routledge.

Varela, Francisco. 1979. *Principles of Biological Autonomy.* New York: North Holland.

Vaughan, Laura, ed. 2015 forthcoming. *Suburban Urbanities: Suburbs and the Life of the High Street.* London: UCL Press.

Vaughan, Laura., Muki Haklay, Sam Griffiths and Victor Buchli. 2010. *Adaptable Suburbs Case for Support.* Unpublished research document.

Vegetius. 1995. *Epitome of Military Science.* Translated by N.P. Milner. Liverpool: Liverpool University Press.

Vertovec, Steven. 2006. 'The emergence of super-diversity in Britain'. Compass Working Paper No.25. Oxford: University of Oxford.

Vertovec, Steven. 2010. 'Towards post-multiculturalism? Changing communities, conditions and contexts of diversity'. *International Social Science Journal* 61 (199): 83–95.

Vigne, L. 2009. 'The five senses in classical science and ethics'. In *The Sixth Sense Reader,* edited by D. Howes. Oxford: Berg, 107–118.

Virilio, Paul. 1991. *Lost Dimension.* New York: Semiotext(e).

Vout, Caroline. 2007. 'Sizing up Rome, or theorizing the overview'. In *The Sites of Rome: Time, Space, Memory,* edited by D.H.J. Larmor and D.J. Spencer. Oxford: Oxford University Press, 295–322.

Vout, Caroline. 2012. *The Hills of Rome.* Cambridge: Cambridge University Press.

Waelkens, Marc and Jeroen Poblome. 2011. *Sagalassos: Eine Römische Stadt in der Südwesttürkei.* Köln: Wienand.

Wagner, R. 1976. *The Invention of Culture.* Chicago: University of Chicago Press.

Waldock, Sarah. 1998. *A Re-Examination of the Evidence for Parade-Grounds at Auxiliary Forts in Roman Britain.* Durham: Durham University.

Wallace-Hadrill, Andrew. 1988. 'The social structure of the Roman house'. *PBSR* 56: 58–77.

Wallace-Hadrill, Andrew. 1994. *Housing and Society in Pompeii and Herculaneum.* Princeton: Princeton University Press.

Wallace-Hadrill, Andrew. 2003. 'The streets of Rome as a representation of imperial power'. In *The Representation and Perception of Roman Imperial Power,* edited by Lukas de Blois, Paul Erdkamp. Olivier Hekster, Gerda De Kleijn and Stephan Mols. Leiden: Brill, 189–206.

Ward, Colin. 1972. *Connexions: Work.* Harmondsworth: Penguin.

Ward, Colin. 1978. *The Child in the City.* New York: Pantheon.

Ward, Stephen V. 1998. *Selling Places: The Marketing and Promotion of Towns and Cities 1850–2000.* London: E. and F.N. Spon.

Warf, Barney and Santa Arias, eds. 2008. *The Spatial Turn: Interdisciplinary Perspectives.* Abingdon: Routledge.

Warnier, Jean-Pierre. 2001. 'A praxeological approach to subjectivation in a material world'. *Journal of Material Culture* 6 (1): 5–24.

Watson, A. and D. Keating. 1999. 'Architecture and sound: an acoustic analysis of megalithic monuments in prehistoric Britain'. *Antiquity* 73: 325–336.

Weaver, Warren. 1948. 'Science and complexity'. *American Scientist* 36 (4): 536–544.

Weiser, Mark. 1991. 'The computer for the twenty-first century'. *Scientific American* 265 (3): 94–104.

Welles, C. Bradford, Robert O. Fink and J. Frank Gilliam. 1959. *The Excavations at Dura-Europos: conducted by Yale University and the French Academy of Inscriptions and Letters / Final report V, Part 1, The Parchments and Papyri.* New Haven: Yale University Press.

Wener, Richard E. and Gary W. Evans. 2011. 'Comparing stress of car and train commuters'. *Transportation Research Part F: Traffic Psychology and Behaviour* 14 (2): 111–116.

Wessendorf, Susanne. 2013. 'Common place diversity and the "ethos of mixing": perceptions of difference in a London neighbourhood'. *Identities* 20 (4): 407–422.

Westin, Sara. 2014. *The Paradoxes of Planning: A Pscyho-Analytical Perspective.* Farnham: Ashgate.

Wheeler, Michael. 2005. *Reconstructing the Cognitive World.* Cambridge, MA: MIT Press.

Whittaker, Cuthbert Wilfred. 1911. *An Illustrated, Historical, Statistical & Topographical Account of the Urban District of Enfield.* London: G. Bell and Sons.

White, Jerry. 2007. *London in the Nineteenth Century: 'A Human Awful Wonder of God'.* London: Jonathan Cape.

White, William. 1841. *White's General Directory of the Town and Borough of Sheffield with Chesterfield, Rotherham and the Surrounding Villages and Hamlets.* Sheffield.

White, William. 1905. *White's Directory of Sheffield with Rotherham and the Parishes, Townships, Villages and Hamlets Within a Radius of Seven Miles From Sheffield.* London.

Whitzman, Carolyn, Megan Worthington and Dana Mizrachi. 2009. *Walking the Walk: Can Child Friendly Cities Promote Children's Independent Mobility?* Melbourne: Australian Centre for Governance and Management of Urban Transportation.

Whyte, William F. 1943. *Street Corner Society: The Social Structure of an Italian Slum.* Chicago: University of Chicago Press.

Whyte, William H. 1980. *The Social Life of Small Urban Space.* Washington, DC: Conservation Foundation.

Whyte, William. 2006. 'How do buildings mean? Some issues of interpretation in the history of architecture'. *History and Theory* 45: 153–177.

Wickstead, H. 2013. 'The goat boy of Mount Seething: heritage and the English suburbs'. In *New Suburban Stories,* edited by M. Dines and T. Vermeulen. London: Bloomsbury, 199–213.

Williams, J. 2013. 'Musical space and quiet space in medieval monastic Canterbury'. In *Making Senses of the Past: Toward a Sensory Archaeology,* edited by J. Day. South Illinois: Southern Illinois University Press, 196–221.

Wilmott, Tony. 1997. *Birdoswald – Excavation of a Roman Fort on Hadrian's Wall and its Successor Settlements: 1987–1992.* London: English Heritage.

Wilson, Helen. 2011. 'Passing propinquities in the multicultural city: the everyday encounters of buss passengering'. *Environment and Planning A* 43: 634–649.

Wirth, Louis. 1938. 'Urbanism as a way of life'. *American Journal of Sociology* 44: 1–24.

Wisser, Richard, ed. 1970. *Martin Heidegger im Gespräch.* Freiburg and München: Alber.

Witcher, Rob. 2005. 'The extended metropolis: *urbs, suburbium* and population'. *Journal of Roman Archaeology* 18: 120–138.

Withers, Charles. 2009. 'Place and the spatial turn in geography and history'. *Journal of the History of Ideas* 70: 637–675.

Wittgenstein, Ludwig. 1961. *Tractatus Logico-Philosophicus.* Translated by D.F. Pears and B.F. McGuiness. London: Routledge and Kegan Paul.

Wittgenstein, Ludwig. 2001 [1953]. *Philosophical Investigations.* London: Blackwell.

Wolf, E.R. 1990. 'Facing power – old insights, new wuestions'. *American Anthropologists* 92 (3): 586–596.

Wollscheid, Achim. 1999. 'Does the song remain the same?' In *Site of Sound: Of Architecture and the Ear*, edited by B. LaBelle and S. Roden. Los Angeles: Errant Bodies Press, 5–10.

Wood, Peter and Alan Latham. 2015. 'Inhabiting infrastructure: exploring the interactional spaces of urban cycling'. *Environment and Planning A* 47 (2): 300–319.

Woodruff, Allison, Sally Augustin and Brooke Foucault. 2007. 'Sabbath day home automation: "it's like mixing technology and religion"'. In *Proceedings of the SIGCHI Conference on Human Factors in Computing Systems*. San Jose: ACM Press, 527–536.

Woolgar, C.M. 2006. *The Senses in Late Medieval England*. London: Yale University Press.

Wrightson, K. 2007. 'The "decline of neighbourliness" revisited'. In *Local Identities in Late Medieval and Early Modern England*, edited by N.L. Jones and D. Woolf. Basingstoke: Palgrave Macmillan, 19–49.

Wylie, John. 2007. *Landscape*. Abingdon: Routledge.

Yaneva, Albena. 2009. 'Border crossings. Making the social hold: towards an actor-network theory of design'. *Design and Culture* 1 (3): 273–288.

Yaneva, Albena. 2010. 'The architectural as a type of connector'. *Perspecta* 42: 138–143.

Yaneva, Albena. 2012. *Mapping Controversies in Architecture*. Farnham: Ashgate.

Yiftachel, O. 1998. 'Planning and social control: exploring the dark side'. *Journal of Planning Literature* 12 (4): 395–406.

Young, Michael and Peter Willmott. 1962 [1957]. *Family and Kinship in East London*. London: Pelican.

Zanker, Paul. 1988. *The Power of Images in the Age of Augustus*. Ann Arbor: University of Michigan Press.

Zardini, Mirko, ed. 2005. *Sense of the City: An Alternate Approach to Urbanism*. Quebec: Lars Müller Publishers.

Zimmerman, Michael E. 1990. *Heidegger's Confrontation with Modernity: Technology, Politics, and Art*. Bloomington: Indiana University Press.

Index

For Product Safety Concerns and Information please contact our
EU representative GPSR@taylorandfrancis.com Taylor & Francis
Verlag GmbH, Kaufingerstraße 24, 80331 München, Germany